T0303868

GEOTECHNICAL AND GEOPHYSICAL SITE CHARACTERIZATION

BALKEMA – Proceedings and Monographs
in Engineering, Water and Earth Sciences

PROCEEDINGS OF THE THIRD INTERNATIONAL CONFERENCE ON SITE
CHARACTERIZATION ISC'3, TAIPEI, TAIWAN, 1–4 APRIL, 2008

Geotechnical and Geophysical Site Characterization

Edited by

An-Bin Huang
Department of Civil Engineering, National Chiao Tung University,
Hsin Chu, Taiwan

Paul W. Mayne
School of Civil & Environmental Engineering, Georgia Institute
of Technology, Atlanta, GA, USA

Taylor & Francis
Taylor & Francis Group

LONDON / LEIDEN / NEW YORK / PHILADELPHIA / SINGAPORE

Taylor & Francis is an imprint of the Taylor & Francis Group, an informa business

© 2008 Taylor & Francis Group, London, UK

ISBN 13: 978-0-415-46936-4 (Hbk)

Geotechnical and Geophysical Site Characterization – Huang & Mayne (eds)
© 2008 Taylor & Francis Group, London, ISBN 978-0-415-46936-4

Table of contents

Geotechnical and Geophysical Site Characterization – Huang & Mayne (eds)
© 2008 Taylor & Francis Group, London, ISBN 978-0-415-46936-4

Introduction

The Third International Conference on Site Characterization (ISC'3) was held in Taipei, Taiwan from April 1–4, 2008. ISC'3 is a continuation of a long list of successful conferences on various aspects of site characterization. The most recent events of the series can be traced back to ISC-98 (Atlanta, USA) and ISC'2 (Portal, Portugal). The idea of ISC was initiated by the technical committee on Ground Property Characterization by In-Situ Tests (TC-16) of the International Society of Soil Mechanics and Geotechnical Engineering (ISSMGE).

A rather diverse group of practitioners and researchers from all over the world have gathered in the conference to exchange their experiences in all aspects of geotechnical site characterization practice and research. Novel ideas in the use and developments of sampling as well as in situ testing tools such as SPT, CPT, PMT and DMT continued to be the core of the conference. Papers presented at ISC'3 revealed some of the remarkable developments in geophysical testing and imaging techniques that enable nondestructive profiling of underground conditions. A rise in the use of risk management and statistical analysis in geotechnical site characterization has also been noticed. Taiwan and some parts of Asia are prone to natural hazards such as earthquakes and typhoons. As a consequence, many papers submitted from this region deals with these concerns as they relate to site characterization.

The conference proceedings have been arranged under seven themes including: Case histories in field applications; Characterization of unusual/unsaturated geomaterials; Developments of new equipment & methods; Geophysical testing & imaging techniques; Interpretation/analysis of test data; Pavement geomechanics; and Sampling disturbance. A total of 207 papers that include 13 keynote/theme lectures were presented at the conference. In response to the call of saving the planet and reducing consumption of natural resources, the keynote/theme lecture papers are included in a single volume of printed proceedings. All papers are stored digitally in a CD.

The conference would not have been possible without the support of TC-16, members of the various organizing/advisory committees and the outstanding papers submitted by the authors. Financial support of ISC'3 provided by the National Science Council, Ministry of Economic Affairs of Taiwan, National Chiao Tung University and other organizations/companies are greatly appreciated.

An-Bin Huang   Paul W. Mayne
National Chiao Tung University   Georgia Institute of Technology

Keynote/Theme Lectures

Geo-environmental site characterization

R.G. Campanella
The University of British Columbia, Vancouver, British Columbia, Canada

ABSTRACT: The use of the piezocone to provide detailed stratigraphic information as well as the piezometric and hydraulic characteristics of the soil is discussed. A resistivity module of external electrode rings attached to the piezocone, which was developed at UBC (University of British Columbia) to log and assess groundwater quality, soil porosity and saturation, is also discussed together with basic theory and factors that affect in-situ electrical resistivity. Finally, the combination of the economical and rapid UBC-modified *BAT* groundwater penetration tool to provide 'specific-depth' groundwater samples for chemical and biological analysis and correlation with resistivity will also be explained. Field data from several case histories are presented to demonstrate the use of the resistivity piezocone in combination with groundwater sampling to provide screening data to locate permanent monitoring well systems or to develop remediation scenarios. The examples deal with acid mine drainage, creosote contamination at a pressure treatment plant, sea water intrusion along side a river outlet and seepage through a tailings dam.

1 INTRODUCTION

Geotechnical site characterization requires a full 3-D representation of stratigraphy (including variability), estimates of geotechnical parameters and hydrogeological conditions and properties. While traditional methods like drilling and undisturbed sampling can provide adequate stratigraphic details and estimates of geotechnical parameters, they can not provide useful estimates of hydrologic conditions like gradient and equilibrium pore water pressure. An installation of a network of piezometers would be required to determine pore water transport parameters. Drilling, undisturbed sampling and piezometer installations are all very costly procedures, often prohibitive, resulting in minimal drilling and sampling in most geotechnical projects. Rarely are pore pressure profiles correctly determined, even though an effective stress analysis requires it.

Environmental concerns add the need for also determining geochemical and transport conditions. The determination of the chemical properties of the pore water often requires sampling-wells which are very expensive to install and develop, dilute constituents moving in layers in stratified soils over the length of well screen and are difficult to know where to locate within the property of concern. Furthermore, environmental field studies require decontamination protocols, the use of drilling methods which do not use fluids, the removal of all cuttings to specially designated waste sites and the use of isolation and

safety protocols of all equipment and personnel at the site. All of this adds enormously to the cost of geo-environmental site investigations.

In-situ penetration test tools like the resistivity/seismic piezocone and *BAT* specific depth pore water-sampling systems provide particular advantages for a geo-environmental investigation. They displace the soil and do not create any soil cuttings (that must be removed), are relatively small tools causing minimal intrusion, do not require any fluids for penetration, are easily decontaminated, can incorporate continuous grouting to eliminate possible cross-contamination, can sample pore fluid and pore gases at specific depths, and can indicate chemical anomalies with continuous resistivity measurements. These penetration tools accurately and economically meet the requirements for environmental site characterization as set out by the US Environmental Protection Agency (US EPA, 1989), which includes measurement of stratigraphy, water level data, hydraulic conductivity, relative chemical distribution, and sources/receptors for potential and existing contaminants. Also, non-intrusive surface geophysical methods work effectively to guide and supplement data from any site investigation methodology, especially in-situ testing. An example is given in a subsequent case history.

This paper presents a brief review of selected penetration methods for environmental site characterization of soil deposits, and recent developments and experiences in the UBC In-situ testing group. These methods include the piezocone penetration test, the resistivity piezocone and the *BAT* groundwater system. Also presented are several case histories, which demonstrate the advantages of these methods.

2 IN-SITU TESTING

2.1 *Piezocone*

The piezometer cone penetration test (CPTU) involves the penetration of a 60° apex cone of typically 35.7 mm diameter (10 cm^2 area) as shown in Fig. 1. Pushing at a constant 2 cm/s ($\sim$1 m/min) is achieved by hydraulic force supplied typically by either a drill-rig or a specially outfitted cone-pushing vehicle. At UBC, all cone equipment was designed by the author and built in the Civil Engineering machine and electronics shops including the enclosed in-situ testing research truck which supplies the cone-pushing force (Campanella and Robertson, 1981). Davies and Campanella (1995) list typical pushing capabilities through clay and sand soils.

The UBC piezocone measures tip resistance (q_c), friction sleeve stress (f_s), and pore pressure response at up to three locations; on the cone tip face, immediately behind the cone tip and immediately behind the friction sleeve (referred to as U1, U2, and U3, respectively). The U3 location has a more sensitive pore pressure transducer to measure more accurate small dissipations and equilibrium pressures compared to U2 or U1. Most correlations and direct calculations assume measurement at the standard U2 location. Temperature (T) and inclination (I) are also measured simultaneously as the piezocone is advanced into the ground. All channels are continuously monitored and typically digitized at 25 or 50 mm intervals. Campanella and Robertson (1988) outline the piezocone's main advantages, limitations, and standard testing and recommended interpretation procedures. All UBC cones are equipped with either a seismometer or accelerometer to determine shear wave velocity profiles, which is routinely performed in most UBC piezocone soundings. Several studies combining seismic data with piezocone data were presented by Gillespie, 1990.

The piezocone test provides the following advantages for environmental studies:

(1) Minimum intrusion with no possibility of cross-contamination.
(2) Continuous grouting if necessary and easy decontamination of tools (Lutenegger and DeGrott, 1994).
(3) Rapid delineation of site stratigraphy to identify specific depth of coarse layers where water transport is most likely and where water sampling for chemical analysis is needed.
(4) Measurement of equilibrium pore pressure at full PPD to quantify vertical hydraulic gradients

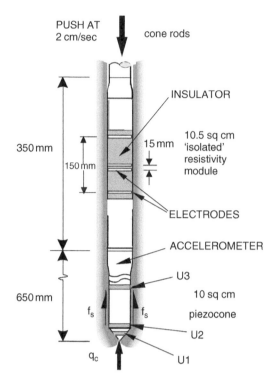

Figure 1. UBC Resistivity Piezocone (RCPTU).

(single sounding) and groundwater flow regimes (multiple soundings).
(5) Estimating hydraulic conductivity, K, from pore pressure dissipation data.
(6) Empirical and theoretical correlation of relevant piezocone measurements to soil parameters (ϕ, R_d, and G in sand S_u, OCR and G in clay).
(7) K-BAT for water sampling compatible with cone equipment.
(8) 'Add-on' modules measure resistivity, self-potential, gamma radiation, specific ions sensors, laser-induced fluorescence to detect hydrocarbons like oil and gasoline, etc.

2.2 *Resistivity Piezocone*

The resistivity piezocone (RCPTU) provides the ability to measure the electrical resistance to current flow in the ground on a continuous basis. This ability is extremely valuable due to the large effects that dissolved and free product constituents have on bulk soil resistivity (reciprocal of conductivity). The RCPTU consists of a resistivity module, which is added behind a standard piezocone (Fig. 1). Davies and Campanella (1995) give an overview summary of the RCPTU and its perceived application areas.

Measurements of bulk resistivity trends indicate whether some form(s) of dissolved or free product

4

constituent exist below or above background values. Background values are usually established from on-site testing. The areas where readings are very different (anomalies) from background values are then further evaluated with appropriate groundwater sampling at discrete depths for detailed chemical analyses. Of considerable practical value is the fact that the measured resistivity in *saturated* soil is almost totally governed by the pore fluid chemistry. Soil mineralogy, porosity, and particle size have a limited effect in most circumstances.

Hydraulic and electrical flow laws are similar and given by

- Hydraulic Gradient: $q = K \times h/L \times A$
- Electrical Gradient: $I = C \times V/L \times A$

or $(1/C) \times (L/A) = V/I = R$ (resistance) OHMS (Ω)
$1/C = \rho = $ RESISTIVITY with UNITS of OHM-meter ($\Omega - m$) and is a property of the medium.

Therefore $\rho = $ lab calibrated geometric constant for a given module $\times$ measured electrical resistance measured in the field.

Prior to 1990 bulk soil resistivity was only used as a means to determine the in-situ void ratio of saturated sandy soils in combination with pore fluid resistivity and laboratory calibrations (Delft Geotechnical Laboratory). The ring electrodes along the cone shaft allow continuous measurement of resistivity with depth where the electrodes are cleaned during advancement of the cone.

Figure 1 shows the current UBC resistivity module of electrode rings, which is attached behind the standard piezocone. Its excitation and response are electrically isolated from cone electronics, which gives very little current leakage and linear calibrations of resistivity from 0.01 ohm-m to very high values of 500 ohm-m for a given excitation current. The smallest electrode spacing (15 mm) is useful for detection of thin layers of contrasting bulk resistivity, whereas the largest electrode spacing is used for AC current excitation and measurement of average resistivity over a larger depth (150 mm). See Campanella and Weemees (1990) for the research and development of the resistivity module and Daniel et al. (2003) for determining engineering properties from resistivity measurements with the isolated module. A comparison of site data and laboratory equipotential data between isolated and non-isolated resistivity measurements will be demonstrated in a subsequent case history.

2.3 *BAT discrete-depth water sampling system*

A modification of the commercially available *BAT* System (named after the inventor, Bengt Arne Torstensson, 1984) is recommended for obtaining in-situ pore fluid samples. The original system consists of a sampling tip that is accessed through sterile

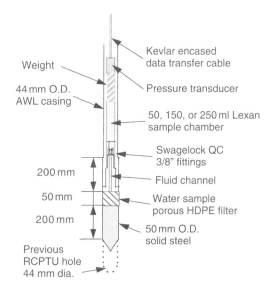

Figure 2.   UBC Modified *BAT* Groundwater Tool.

evacuated glass sample tubes and a double-ended hypodermic needle set-up pushed through septum seals. The tube sampler is lowered either by cable or electrical wire depending upon whether a pore fluid sample is taken with or without a pressure test being carried out. Figure 2 shows the modifications made at UBC (Wilson and Campanella, 1997) which include using a stainless steel or Lexan sampling carrier, a modified probe to push down a previous cone hole and replacing the hypodermic needle system with Swagelock quick-connect push-on valve fittings. This latter modification allows the direct measurement of permeability in sands and gives much more accurate and feasible sampling in higher TDS environments as experienced, for example, during water sampling in metallic mine tailings. The *BAT* is hydraulically pushed with the same equipment used for cone penetration testing. The surface of the HDPE filter is flush with the outside of the solid stainless steel probe and is effectively cleaned while being pushed through the soil. The 200 mm of steel above and below the filter seals off the filter from the previous open RCPTU hole.

The *BAT* probe is also able to take pore gas samples for collecting volatile contaminants. In California, where the water table may be very deep, oil and gasoline spills present a unique challenge. An environmental company using the CPTU developed a snifter cone (HFA Inc., 1994). At a point just behind the friction reducer behind the cone several holes collect air samples to the surface using a venturi where the air is quickly analyzed by a PID (Photo ionization detector) or similar detector sensitive to organic oil vapors. This snifter is used to rapidly identify 'hot spots' where remediation can be focused.

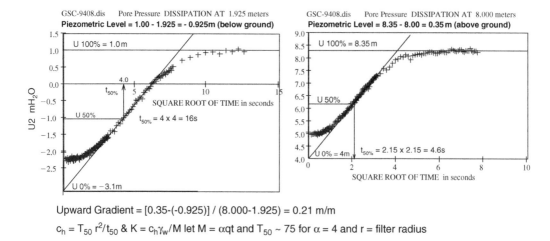

Upward Gradient = [0.35-(-0.925)] / (8.000-1.925) = 0.21 m/m

$c_h = T_{50} r^2/t_{50}$ & $K = c_h \gamma_w/M$ let $M = \alpha q t$ and $T_{50} \sim 75$ for $\alpha = 4$ and r = filter radius

Figure 3. High-speed pore pressure dissipation in sand to measure equilibrium pore pressure, gradient and hydraulic conductivity.

The US-EPA and other high conformance require-ment groups have adopted *BAT* technology as appro-priate and preferred for many environmental character-ization applications. The attraction of no drill cuttings and the repeatability of the data are cited as the key reasons for this preference. *BAT* technology has been scrutinized by many investigators and has met with widespread acceptance (e.g., Zemo et al., 1992).

After *BAT* water samples are retrieved to the ground surface, preliminary chemical tests should be con-ducted on-site and then the sample can be stored for further laboratory analyses. Field measurements should, at a minimum, include conductivity, tempera-ture, and pH. The first sample at a given depth is termed the 'purge' or discarded sample and any test data on this sample is only preliminary. Once enough sampling is carried out at a specific depth, the *BAT* probe is then pushed to the next depth and the procedure repeated.

2.4 *BAT hydraulic conductivity or permeability, (K), measuring system*

Recent studies at the University of British Columbia (UBC) (Wilson, 1996) have made use of the UBC-modified *BAT* to perform out-flow hydraulic conduc-tivity, K, tests, where sample-vial pressure changes with time are related to volume changes. The analyti-cal solution was verified in comparison testing where the *BAT* tip is made to function as an out-flow slug test. Not only were the results identical, but labora-tory tests in 5 m high water columns showed that the current limiting highest K of the measuring system with 50 mm long filter section and 3/8 inch valves was 0.0001 m/s (or a medium sand) as opposed to the original use of hypodermic needles which limited flow

to 0.000001 m/s. The limiting K was controlled by the porous HDPE filter material, which had a mean pore size of 125 μ. It was also found that high gradients cause turbulence and reduce K, thus controlled gra-dient tests are required. A recommended procedure is given in Wilson and Campanella, 1997.

An important finding in this study showed clearly that an in-flow K test in sandy soils often gave incor-rect and misleading K values which were from one to two orders of magnitude too low due to fines migrating through the sand and plugging the filter. This is usu-ally not a problem in clayey soils. All piezometer/slug testing to measure K in silty, sandy soils must be for outflow conditions under low gradients. Thus, water sampling (in-flow) should not be used to also give K of the soil. K measurement requires an out-flow test.

2.5 *Piezocone hydraulic conductivity and gradient measuring system*

The measurement of CPTU pore pressure dissipation in sandy deposits can also be used to determine the time for 50% dissipation, t_{50}, to estimate K. How-ever, high speed data logging is required in sandy soils where t_{50} can be 5 sec. or less. Figure 3 shows a typi-cal example in a sandy aquifer of rapid dissipation of excess pore pressure to equilibrium at two depths and the interpretation of results. The equation constant, T_{50}, needed to calculate, K, was directly calibrated using the out-flow K-BAT permeability determination at the same locations. This site-specific correlation is required for sandy soils.

Note that for the case in Figure 3, the difference in piezometric levels gave an average vertical gradient indicating upward flow. It should, however, be realized

Table 1. Summary of typical resistivity measurements of bulk soil mixtures and pore fluid (saturated mixtures only) (adapted from Davies and Campanella, 1995).

Material type	Bulk resistivity ρ_b, Ω-m	Fluid resistivity ρ_f, Ω-m
Deltaic sands with saltwater intrusion	2	0.5
Drinking water from sand	>50	>15
Typical landfill leachate	1–30	.5–10
Mine tailings (base metal) & oxidized sulphide leachate	0.01–20	.005–15
Mine tailings (base metal) no oxidized sulphide leachate	20–100	15–50
Arsenic contaminated sand and gravel	1–10	.5–4
Industry site: inorganic contaminants in sand	0.5–1.5	0.3–0.5
Industrial site: creosote contaminated silts and sands	200–1000	75–450
Industrial site: wood waste in clayey silts	300–600	80–200

Note: Conductivity (μS/cm) = 10,000 ÷ [Resistivity(Ohm-m)]

that the gradient would change each time the K value changes. Thus, in stratified soil the gradient, if one exists, would be the average over the measured distance. Any horizontal flow gradients can be evaluated by comparing the equilibrium pore pressure from two adjacent dissipations at the same depth using multiple soundings.

3 RESISTIVITY PIEZOCONE (RCPTU) FOR GEOENVIRONMENTAL SITE CHARACTERIZATION

As summarized by Davies and Campanella (1995), the resistivity piezocone can be used to evaluate the following environmental and geotechnical parameters: soil stratigraphy, soil density, undrained shear strength parameters, hydraulic conductivity, in-situ hydraulic gradients, and relative geochemical nature of pore water. The geochemical nature comes from evaluation of the continuous bulk resistivity signature from the resistivity piezocone compared with chemical analyses on samples obtained with the *BAT* sampling system.

Table 1 presents a small sampling of typical RCPTU bulk soil resistivity measurement values for soils beneath the water table and corresponding measurements of pore fluid resistivity. Note the wide range of values for different pore water chemical constituents.

While there are many other chemical sensors that can be put behind the cone (Lunne et al., 1997), they are always specific for given types of contaminants. The wide measurement range from 0.01 to 5000 ohm-m makes the resistivity a very useful parameter for screening sites for possible contaminants usually presented as anomalies. The practical advantages of the surface electrode rings are simplicity, robustness, direct coupling with soil, self-purging or cleaning of electrodes and continuous measurements.

3.1 Factors affecting bulk resistivity of soils

The measured bulk resistivity of soils is affected by:

(1) Pore fluid chemistry
(2) Degree of fluid saturation
(3) Porosity/Density of soil matrix
(4) Temperature
(5) Shape of pore space
(6) Clay content
(7) Mineralogy
(8) Dielectric properties may be important.

Archie (1942) proposed the following mixing relationship:

$$F = \rho_b / \rho_f = a(n)^{-m}(S_r)^{-s} \qquad (1)$$

where: F = formation factor, ρ = resistivity, b = bulk, f = fluid, n = porosity, S_r = degree of saturation, and a, m and s are constants for a given soil. The constants relate to above factors 5 through 8.

The relationship between temperature and electrical resistivity is a constant and all resistivity readings should be corrected to a given temperature like 25°C.

In a recent compaction study of factors affecting bulk soil resistivity, Daniel (1997), showed that it was possible to estimate porosity, n, and degree of saturation, S_r, from RCPTU tests. Lab calibration of two soils compacted over a wide range of densities and water contents yielded the following result:

$$F = \rho_b / \rho_f = 1.84\, n^{-0.5}\, S_r^{-1.45} \qquad (2)$$

(highly acid sulphide mineral tailings)

$$F = \rho_b / \rho_f = 0.60\, n^{-2.2}\, S_r^{-1.80} \qquad (3)$$

(quartz rock flour)

where: ρ_f = the measured resistivity of the water added for compaction, which had a range of from 1 to

7

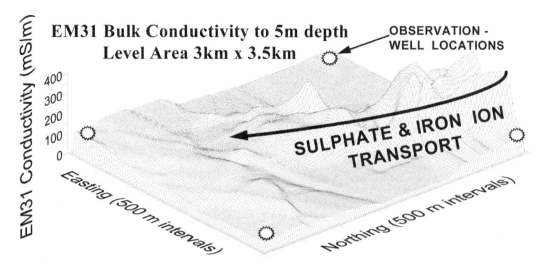

Figure 4. Surface plot of measured average bulk resistivity to a depth of 5 m in sulphide mine tailings showing plume. 10-year-old observation wells missed the plume completely. (after Davies, 1999).

11 ohm-m and ρ_b = measured bulk resistivity of each compacted sample.

It was found that the two very different soils fit Archie's proposed relationship very well, but the constants were very different. This difference is primarily due to very different soil mineralogy. It was recognized, however, that the actual pore fluid resistivity of the tailings might not be equal to the resistivity of the added water (1 ohm-m for the tailings) because of adsorbed soluble ions in the dry tailings.

For a given typical soil and fluid resistivity the change in measured bulk resistivity can be used to estimate degree of soil saturation (e.g., 70 Ω-m at 100% to 4400 Ω-m at 10% saturation) and evaluate density changes after vibro-densification (Campanella and Kokan, 1993, Daniel et al., 1999 and 2003).

The in-situ porosity of a saturated soil is easily determined (Eq. 1) from measurements of resistivity of in-situ bulk soil and pore fluid extracted in-situ from the soil in combination with Formation constants a and m determined from laboratory compaction tests for a particular soil. This, in fact, was the technique used by Delft Geotechnics (page 184, Lunne et al., 1997) to determine in-situ density of loose sea bottom sediments, which were essentially impossible to sample undisturbed.

4 THE USE OF RESISTIVITY PIEZOCONE SITE CHARACTERIZATION IN CONTAMINATED SOILS: CASE HISTORIES

The following examples of geo-environmental site characterization are taken from projects at the University of British Columbia, Civil Engineering Department from about 1990 to 2002. In all of the

case histories described the resistivity piezocones were equipped with accelerometers and down-hole shear wave velocity profiles were determined for each sounding. Presentation of that data is outside the scope of this paper and is therefore left out. However, it should be pointed out that the seismic data was necessary and very valuable in order to perform dynamic and seismic stability analyses including liquefaction.

4.1 *Mine tailings (base metal), oxidized sulphide leachate and Acid Mine Drainage (AMD)*

The site, which is relatively flat and consisting of tailings from a sulphide ore-body, had several geotechnical, hydrogeological and geochemical concerns. The old tailings are up to 100 years old and have oxidized leachate, which are highly acidic with pH values less than 1. At this low pH the metallic constituents are soluble and enter the groundwater to become mobile, thereby resulting in what is called Acid Mine Drainage (AMD). The resistivity piezocone and BAT sampling technology were selected for characterizing the site.

Because of the very large extent of the site, a portable surface geophysical tool called a ground conductivity meter (GEONICS™ EM31) was used to obtain a preliminary estimate of the locations of high ionic groundwaters and plumes. A single person walked the site with the meter collecting digital data every 2 m in a grid spacing. Figure 4 shows the effective conductivity to a depth of about 5 m in an area 3 km by 3.5 km, which was walked in one day. The higher the apparent conductivity, the higher the ion concentration in the groundwater and the lower the bulk resistivity. In this case the existing 10-year-old observation wells failed to identify the plume or its

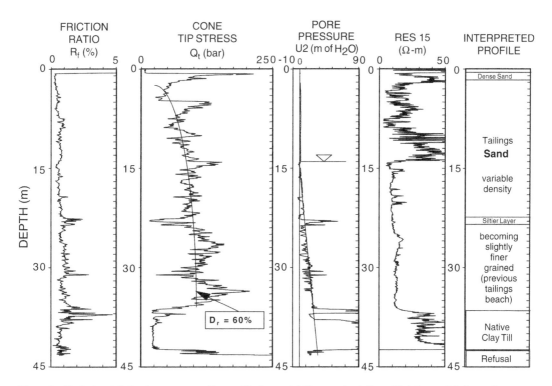

Figure 5. Typical resistivity piezocone sounding profile from sulphide ore mine tailings. Note low resistivity readings even above water table in the high oxidation area leading to acid generation. (after Davies, 1999).

direction of movement in an old buried stream channel, which is clearly identified by the EM31 survey.

Figure 5 shows a typical resistivity piezocone sounding from the site. From the continuous record of tip stress and penetration pore pressure, the strength and drainage characteristics of tailings could be accurately assessed. The tailings are reasonably free draining and dense. With the additional information of the friction ratio around 1, the tailings were shown to be largely contractant, sensitive fine sand-sized and possessing a high susceptibility to static or dynamic triggered liquefaction. This strength characterization work was used to optimize remedial works (berm placement) that were deemed necessary.

The resistivity profile shows the active nature of the tailings above the water table where oxidation is highest. Resistivity values approach 1 ohm-m in the upper zone, even in a partially saturated environment where saturations less than 100% causes resistivities to increase markedly, thus these very low values indicate the rapid onset of oxidation and subsequent acid generation. The saturated tailings below the water table from 14 m to 37 m depth have a fairly consistent resistivity value of about 10 ohm-m, have less sulphate concentration and have minimal on-going oxidation.

Geochemically, the UBC-BAT sampling program provided site-specific relationships between bulk resistivity piezocone values and chemical testing of porewater samples for the entire study site. The relationship between total dissolved solids (TDS) in pore water and bulk conductivity in saturated soil is linear. Specific ion correlations with RCPTU bulk resistivity values are most commonly site-specific in nature although sulphate anions and divalent iron have shown remarkable global correlation in our experience to date in mine tailings as shown in Figure 6 for sulphate.

With the aide of the EM31 data, the resistivity piezocone was used to delineate ionic rich plumes whose sampled characteristics included pH values as low as 1 ohm-m and TDS concentrations to 60,000 mg/l (ppm). The delineation from the resistivity piezocone allowed the future optimal spatial placement of regulatory required monitoring wells and the accurate depth location and length of discrete well screens. The site characterization also allowed the location of a cut-off catchment to collect the acid drainage. (See Boyd, 1996 and Davies, 1999 for in-depth site characterization of sulphide tailings and AMD.)

4.2   DNAPL and creosote contaminated saturated sediments

This 65-acre site, which is along side a river, has been treating timber with organic preservatives consisting

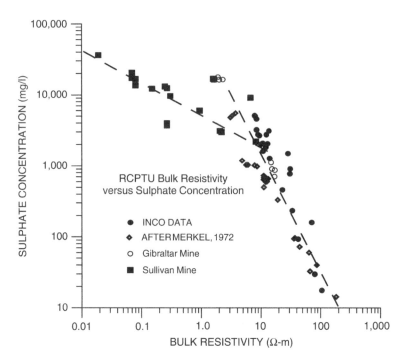

Figure 6. Sulphide ore tailings pore water chemistry of sulphate ion concentration. (after Davies, 1999).

mainly of mixtures of creosote tars, heavy and light petroleum products since 1930 and currently processes about 1.8 million cubic feet of timber per year. Creosote is electrically non-conducting, is not readily dissolved in water, has a density slightly greater than water and a viscosity 50 times that of water. Creosote is a DNAPL (dense non-aqueous phase liquid), which sinks in water and is not readily transported except for its lighter constituents and over long periods. It is toxic to fish and wildlife.

Figure 7 shows the results of two resistivity profiles from a similar location. RES 2 at 10 mm and 75 mm spacing was non-isolated and had a common ground with the cone. RES 3 at 10 mm and 150 mm spacing was isolated from CPTU electronics. All profiles show the thin clay layer around 10 m deep. The saturated sediments are mostly fine to medium sands with lenses of silty sands. The non-isolated RCPTU had values around 100 ohm-m and does not show any clear existence of the creosote. However, the isolated module gives spikes of resistivity values in excess of 250 ohm-m and clearly indicates locations where concentrated creosote had sunk through the sediments and was held on the silty lenses. Subsequent pore fluid sampling yielded pure product at the lens locations.

In an effort to understand the different responses of the resistivity modules, a laboratory test was conducted in a large PVC fish transport container, which measured about 4 ft by 6 ft and 4 ft deep. The tank was filled with 20 ohm-m water using ordinary salt and each resistivity module was in-turn supported horizontally at the center of the tank. The module was excited with the field electronics (1000 hz AC) and an AC voltage measuring tip probe was lowered to the mid-height of the module and moved horizontally to obtain the equipotential field. Figure 8 shows the equipotential fields for each module.

Figure 8(a) indicates a fairly high loss of current to the commonly grounded steel cone rod. Because of its inefficiency the linear calibration of output vs. resistivity as shown in Figure 9 only extends to about 75 ohm-m, where the excitation must be reduced to extend measurements to higher values of resistivity. Also, the smaller calibration cylinder interferes with the calibration as seen in Fig. 8(a).

Figure 8(b) indicates a very uniform set of equipotentials for the isolated module with essentially all of the current going between the outer excitation electrodes, totally ignoring the proximity of the steel cone rod. Also, the calibration as shown in Figure 9 is linear to a resistivity of about 500 ohm-m at the capacity of the voltmeter. This range will cover most investigations. It is also satisfying to see that the use of the convenient small cylinder does not affect the calibration.

While a non-isolated resistivity module works well in high ionic pore fluids, an isolated module is required to detect electrically non-conducting contaminants.

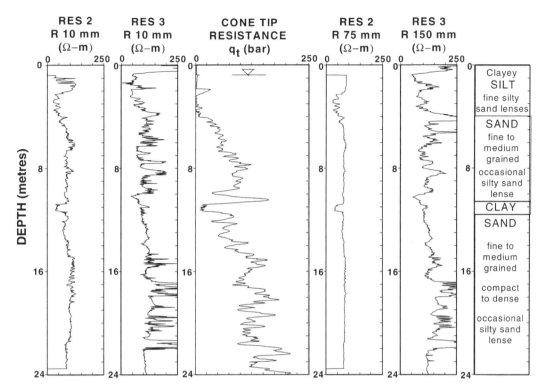

Figure 7.   Non-isolated (RES2) and Isolated (RES3) RCPTU at a creosote contaminated site (after Everard, 1995).

4.3   *Salt water intrusion*

Figure 10 shows the results of 5 RCPTU profiles along a 500 m line along the eastern boundary at the UBC primary field site at the BC Hydro Kidd2 station. This site is used by both the Hydrogeology program in Geological Sciences as well as the Geotechnical program in Civil Engineering for teaching as well as research.

The stratigraphy was determined by the piezocone data. Using the resistivity profile the outline of a salt water wedge is indicated by values of the order of 1 ohm-m. The site is adjacent to and near the mouth of the Fraser River where the salt water has intruded to the bottom of the river where a sand aquifer exists. The salt water being heavier than the fresh water has spread laterally into the fresh water aquifer creating a classic salt water wedge. Over the year the salt water wedge moves in and out of the aquifer depending on the outflow fresh water and its elevation. Also, there is often a small gradient moving river water into the site, as the river level is often higher than the land. Notice the dyke to the right of the figure.

4.4   *Embankment seepage*

Figure 11 shows the cross section of a tailings dam, which was built by the upstream method from cycloned

sand tailings. At this stage the embankment is about 175 ft high. Although there is no concern for acid mine drainage here, there is concern for the stability of the dam. A major lake is just below the downstream toe and the dam retains highly fluid mineralized tailings.

Because of the high seepage gradients through the embankment an RCPTU investigation was carried out to focus on in-situ pore pressures for stability analysis. In addition, this section had several open and closed piezometers with readings that were confusing to mine personnel who thought they were not working properly.

During penetration the piezocone pore pressures were allowed to come to full equilibrium at penetration pauses each meter when a rod is added. Gradient analyses (similar to Figure 3) were carried out for all piezocone soundings. Since the soil was highly stratified it was necessary to use average values, which yielded gradients from 0.7 to 0.4. The equipotential lines in Figure 11 were determined from all of the piezocone equilibrium pore pressure data supplemented by the in-situ piezometers. All of the fixed piezometers were shown to be working correctly when the gradient field was included in the analyses.

The resistivity profile proved to be very useful in determining the precise depth where saturation takes place. As the resistivity module passes into a saturated

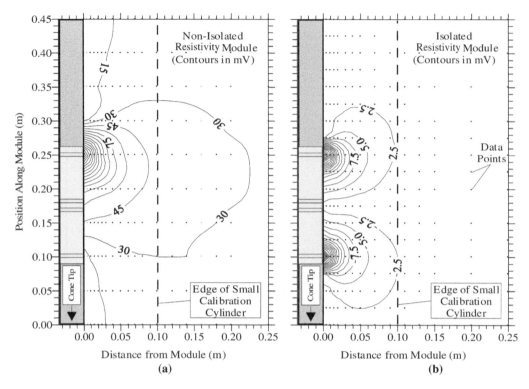

Figure 8. Electric potentials measured in salt-water tub around the UBC (a) non-isolated and (b) isolated resistivity modules. (after Daniel et al., 2003).

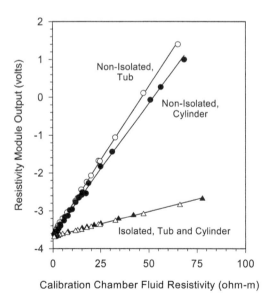

Figure 9. Typical tub and cylinder calibration results for isolated and non-isolated resistivity modules. (after Daniel et al., 2003).

soil the reading drops markedly and stays fairly constant. Resistivity was the only method that was able to identify the u = 0 water surface.

5 CONCLUSIONS

This paper has briefly summarized the main in-situ tools available for geo-environmental site characterization. The piezocone is used as a *screening* tool and is by far the most useful to determine stratigraphy, estimate strength and stability parameters, and to identify and estimate seepage parameters. The measurement of equilibrium water pressures and gradient field, and K estimates allows a transport model to be developed. The resistivity module, when added to the piezocone, is used to identify chemical anomalies (contaminant plume delineation), particularly in the piezocone identified coarse soil layers where contaminants are mobile and where water sampling is fastest. The original *BAT* concept developed by Torstensson (1984) has been modified by UBC to allow the direct measurement of hydraulic conductivity (permeability) in an outflow test for sands up to a value of 10^{-4} m/sec or 10^{-2} cm/sec. However, the *BAT* is used primarily to take water samples for analysis in coarse soils. When it is used to measure K, it is primarily

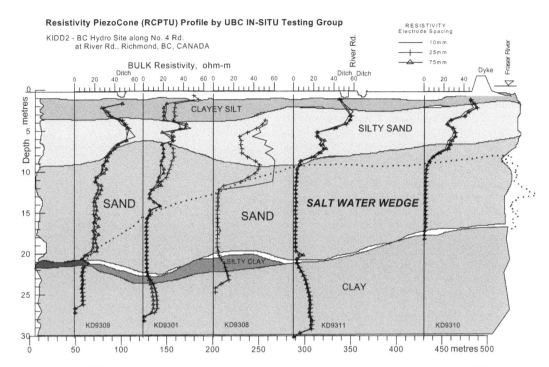

Figure 10. RCPTU profile showing estuary salt water intruding into sand aquifer (after Campanella et al., 1998).

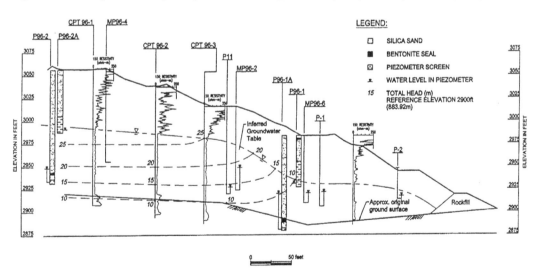

Figure 11. A cycloned sand tailings dam about 175 ft high and showing equipotential lines determined from piezocone data where resistivity was the only measurement to indicate full saturation and u = 0. (after Davies, 1999).

to validate correlations. The resistivity piezocone and *BAT* sampling technology are establishing themselves as the premier geo-environmental tools where ground conditions are appropriate.

Resistivity studies at UBC have also attempted to relate specific ion concentration to complex resistivity

at very high frequencies and induced polarization procedures; so far without success (Kristiansen, 1997).

These relatively inexpensive tools provide a simple In-situ testing methodology for site characterization of soils for both geotechnical and geo-environmental purposes. These penetration-type in-situ tests cause

13

the least disruption in the groundwater environment and the least risk of extended and cross contamination when compared with traditional and more costly borehole drilling methods and the development of water sampling wells.

Commercially available resistivity piezocone work is readily available in Canada and the USA for roughly $40 per meter or $12 per foot for 3 soundings of about 30 meters each, including seismic in two soundings and all holes grouted. This cost is for a typical deltaic deposit with no special problems like very hard or thick gravel layers. Many environmental characterization projects are carried out each year in materials well suited to the technology presented in this paper.

The addition of seismic to the resistivity piezocone makes a very powerful in-situ tool where one can rapidly profile 5 very repeatable, independent readings (q_t, f_s, u, V_s, ρ_b) in a single sounding (Mayne, 2006). However, it must be remembered that the cone penetrometer is an INDEX tool. Except for shear wave velocity and equilibrium pore pressure, which are directly measured, all other derived soil parameters are estimated from empirical correlations based on theoretical concepts. The best approach is to develop site-specific correlations. Unfortunately, this is often not possible. The use of published global correlations is often problematic and should be used with extreme caution, since they vary with geomorphology, mineralogy, drainage, stress history and undefined measurement errors to name a few; hence, their normal variation is very large.

The development of in-situ testing equipment, procedures and applications at UBC are well documented in theses and papers, which are listed and available for download at the author's home page at www.civil.ubc.ca. Also available as freeware is the final version of CPTINT, the UBC developed cone interpretation program. Alternately, the author may be contacted at rgcampanella@gmail.com.

ACKNOWLEDGMENTS

The author acknowledges the research support provided by the Science Council of British Columbia and the Natural Sciences and Engineering Research Council of Canada. Thanks also to our industry partners Klohn-Crippen, Placer Dome Inc., Cominco Ltd., BC Hydro and the Geological Survey Canada for providing site access, financial support and technical input. Of course I am grateful for the design, instrumentation and repair expertise of our electronic technician and machinist, Scott Jackson and Harald Schrempp, and Glen Jolly and Art Brookes, who preceded them. I am particularly indebted to my research students: Tim Boyd, Chris Daniels, Mike Davies, Jodi Everard, Don Gillespie, Henrick Kristiansen, Ilmar Weemees and Daryl Wilson, without whose efforts and dedication there would be little to report in the geo-environmental in-situ testing area.

Finally I would like to thank Professor Jim Mitchell for his guidance and encouragement during and after my studies at UC Berkeley. He gave me a keen interest and curiosity in fundamental soil behavior, which led to my career in experimental soil mechanics and in-situ testing.

REFERENCES

Archie, G.E. 1942. The electrical resistivity as an aid in determining some reservoir characteristics, Trans. Am. Inst. Min. Eng., Vol. 146, pp. 54–62.
Boyd, Tim, 1996, The use of the resistivity piezocone (RCPTU) for the geoenvironmental characterization of sulfide bearing tailings and native soils, MASc Thesis, Univ. of British Columbia, Civil Eng. Dept., October, 230 pgs.
Campanella, R.G. and Robertson, P.K. 1988. Current status of the piezocone test, ISOPT 1, Orlando, FL, Vol. 1, pp. 93–117, Balkema.
Campanella, R.G. and Weemees I. 1990. Development and use of an electrical resistivity cone for groundwater contamination studies, CGJ, 27(5): 557–567.
Campanella, R.G. and Kokan, M.J. 1993. A new approach to measuring dilatancy in saturated sands, Geotechnical Testing Journal, ASTM, 16(4), 485–495.
Campanella, R.G., Davies, M.P., Kristiansen, H. and Daniel, C., 1998. Site characterization of soil deposits using recent advances in piezocone technology. Proc. of the 1st Int. Conference on Site Characterization – ISC '98, Atlanta, Georgia, April 1998. pp. 995–1000.
Daniel, C.R. 1997. An investigation of the factors affecting bulk soil electrical resistivity, BASc Thesis, Univ. of British Columbia, Geological Engineering Program, Geological Sciences, 62 pgs.
Daniel, C.R., Giacheti, H.L., Campanella, R.G. and Howie, J.A. 1999. Resistivity piezocone: data interpretation and potential applications, Proc. XI Pan-American Conference on Geotechnical Engineering, Igusau Falls, Brazil.
Daniel, C.R., Campanella, R.G., Howie, J.A. and Giacheti, H.L., 2003, Specific depth cone resistivity measurements to determine soil engineering properties, Journal of Environmental and Engineering Geophysics, Vol. 8, No. 1, March.
Davies, M.P. 1999. Piezocone technology for the geoenvironmental characterization of mine tailings, 429 pgs, PhD Thesis, Univ. of British Columbia, Civil Eng. Dept.
Davies, M.P. and Campanella R.G. 1995. Piezocone technology: downhole geophysics for the geoenvironmental characterization of soil, Proc. SAGEEP 95, Orlando, Florida, April.
Everard, Jodi, 1995, Characterization of a hydrocarbon contaminated site using in-situ methods, MASc Thesis, Univ. of British Columbia, Civil Eng. Dept., 135 pgs.
Gillespie, Don, 1990, Evaluating shear velocity and pore pressure data from CPT, PhD Thesis, Univ. of British Columbia, Civil Eng. Dept., Sept., 201 pgs.
Henrik, Kristiansen, 1997, Induced polarization and complex resistivity effects in soils – Laboratory and In-Situ

14

Measurements, MASc Thesis, Univ. of British Columbia, Civil Eng. Dept., October, 159 pgs.

HFA, 1994, personal communication, Holquin, Fahan & Associates Inc., Environmental Consultants, Ventura, California.

Lunne, T., Robertson, P.K. and Powell, J.J.M. 1997. Cone penetration testing in geotechnical practice, Blackie Academic and Professional Press.

Lutenegger, A.J. and DeGroot, D.J. 1995. Techniques for sealing cone penetration holes, Canadian Geotechnical Journal, V32(5), October, 880–891.

Mayne, P.W. 2006, The 2006 James K. Mitchell Lecture: Undisturbed sand strength from seismic cone tests. Presented at the GeoShanghai Conference, China (6–8 June 2006): *Geomechanics and GeoEngineering* 1(4), Taylor & Francis Group, London: 239–257.

Torstensson, B.A. 1984. A new system for groundwater monitoring, Ground Water Monitoring Review, Fall 1984, pp. 131–138.

US EPA 1989. Seminar on site characterization for subsurface remediations, US EPA, Tech. Transfer, Report CERI – 89–224, September, ~350 pp.

Weemees, Ilmar, 1990, A resistivity cone penetrometer for groundwater studies, MASc Thesis, Univ. of British Columbia, Civil Eng. Dept., August, 86 pgs.

Wilson, Daryl, 1996, Analysis and modifications of KBAT hydraulic conductivity measurements, MEng Project Report for GEOL 598, Supervisor: Campanella, Univ. of British Columbia, Civil Eng. Dept., 62 pgs.

Wilson, D. and Campanella, R.G. 1997. A rapid in-situ hydraulic conductivity measurement in sands using a UBC modified BAT penetrometer, Proc. 50th Canadian Geotechnical Conference, Ottawa.

Zemo, D.A., Pierce Y.G. and Gallinatte J.D. 1992. Cone penetrometer testing and discrete-depth groundwater sampling techniques: a cost-effective method of site characterization in a multiple aquifer setting, Proc. 6th Outdoor Action Conference, National Groundwater Association, May, Las Vegas.

Geotechnical and Geophysical Site Characterization – Huang & Mayne (eds)
© 2008 Taylor & Francis Group, London, ISBN 978-0-415-46936-4

Geotechnical site characterization for Suvarnabhumi Airport

Chung-Tien Chin, Jung-Feng Chang, I-Chou Hu & Jie-Ru Chen
MAA Group Consulting Engineers, Taipei, Taiwan

ABSTRACT: To meet the growing demand for air travel in Southeast Asia, planning of construction of the Suvarnabhumi Airport, also called Second Bangkok International Airport (SBIA), at the Nong Ngu Hao site was initiated in the early 1970's. A series of investigations were carried out, including several phases of geotechnical engineering studies. As a result of the presence of soft clay layer (Bangkok Clay), most of geotechnical studies focused on the strength and deformation characteristics of the soils and used them for ground improvement design. This paper first introduces the subsoil and groundwater conditions at the SBIA site. Stress history, strength, compressibility, and time-dependent deformability of the soils are provided based on the results of a series of field and laboratory tests. A SHANSEP model is proposed for Bangkok Clay. The soils are found to exhibit slightly anisotropic deformation characteristics based on field PCPT dissipation and laboratory consolidation tests. Strength and deformation parameters characterized from the test results are applied to the design of ground improvement. The effectiveness of ground improvement using preloading with PVD is presented through the deformation and settlement of a series of test embankments and the change of soil properties as well.

1 INTRODUCTION

The Suvarnabhumi Airport, also called Second Bangkok International Airport (SBIA), is located in Racha Thewa of Bang Phli district, Samut Prakan Province of Thailand, approximately 30 km east of the capital Bangkok (Fig. 1). It occupies an area of approximately 3,100 ha, and is about 4 km wide in the east-west direction and 8 km long in the north-south direction. The area is formerly known as Nong Ngu Hao, translated as "Cobra Swamp", around which were primarily large number of fishponds, plants, and villages. The name "Suvarnabhumi" was chosen by His Majesty the King Bhumibol, which means "Golden Land". The "Golden Land" refers to the peninsular Indochina, a region including Thailand, Cambodia, Laos, and Burma.

The SBIA has been planned for more than 20 years to meet the demand for air travel in the region. During this period (1972–1992), a series of investigations were carried out to evaluate the feasibility of SBIA built at the Nong Ngu Hao site, including accessibility, ground surface hydrology, and subsoil conditions (e.g., Northrop/AIT, 1974; Engineers, 1984; STS/NGI, 1992; AIT, 1995). The Master Plan (Engineers, 1984) was developed for the site to ensure that it would become an integrated facility in the Greater Bangkok area, with road networks linking to the nearby expressways and arterial roads, connections to the railway system, and flood protection system

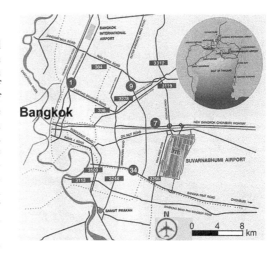

Figure 1. Location of SBIA site (after Suvarnabhumi Airport/New Bangkok Airport Guide, 2007).

and airport utilities provided for the associated airport support services.

Several major geotechnical engineering studies for the SBIA site have been carried out since the beginning of feasibility study in the early 1970's. Due to the presence of soft clay and the long-term ground subsidence history (Figure 2) at the site, most of geotechnical engineering studies focused on the evaluation of the subsoil and groundwater conditions and

17

the ground improvement techniques through a series of field and laboratory tests. In the stage of feasibility study, geotechnical engineering studies can be divided into the following four phases:

– Phase I: performance study of test sections by Northrop/AIT (1972–1974) (e.g., AIT, 1974)
– Phase II: master plan study, design, and construction phasing by NACO/MAA (1983–1984) (e.g., Engineers, 1984)
– Phase III: independent soil engineering study by STS/NGI (1992) (e.g., STS/NGI, 1992)
– Phase IV: full scale PVD (i.e., prefabricated vertical drains) test embankments by AIT (1993–1995) (e.g., AIT, 1995)

In all, there have been more than 150 boreholes, 250 field vane shear tests (VST), and 90 cone penetration tests (CPT) (including piezocone penetration tests, PCPT) performed at the site since the 1970's. Figure 3

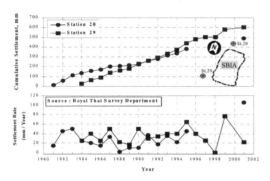

Figure 2. History of ground subsidence near SBIA site (after Moh & Lin, 2003).

illustrates the locations of subsoil investigation tests over the years.

At the site of SBIA, the low strength and high compressibility of the soft clay layer are the major concerns for the design and construction. The strength characteristics of the soils can be used to determine the stability of the embankments. The compressibility and hydraulic characteristics are, on the other hand, relevant to the magnitudes and rates of the settlement. This paper describes how the strengths and deformability of the soft clays are characterized. The subsoil and groundwater conditions are first introduced. The stress history, strengths, compressibility, and time-dependent deformability of the soils are then provided based on a series of field and laboratory test results. The strength and deformation parameters characterized from the test results are applied to the design of ground improvement. The effectiveness of ground improvement using preloading with PVD is evaluated by studying the deformation and settlement of a series of test embankments and the changes of soil properties as well.

2 SUBSOIL AND GROUNDWATER CONDITIONS

2.1 Local geology of Bangkok area

Bangkok is situated on the Chao Phraya plain, under which is a geological basin that contains alluvial, deltaic, and shallow marine sediments. These sediments, which are confined within a radius of 60 to 80 km from Bangkok, were formed during the Pleistocene and Holocene Period.

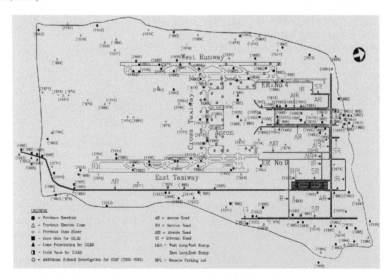

Figure 3. Locations of subsoil investigation tests at SBIA site (Moh & Lin, 2003).

18

The exact profile of bedrock is yet thoroughly understood. The level in the Bangkok area lies at depths ranging from 550 to 2000 m below ground surface. A fault zone trends north-south and follows the course of the Chao Phraya River, the major stream of Thailand (Rau & Nutalaya, 1981). Only a few boreholes have been drilled down to the bedrock, and a variety of rock types such as gneiss and quartzite are discovered (AIT, 1981).

2.2 *Subsoil conditions*

The subsoils underneath the SBIA site were primarily deposited by a sedimentary process. Figure 4 shows the variations of index properties (including total unit weight, water content, Atterberg limits, specific gravity, and grain size distribution) and representative undrained shear strength of the soils with depth. As illustrated by the figure, the subsoil profile within the top 20 m at the SBIA site can be divided into four soil layers. The profile, from the top to the bottom, includes: weathered crust, very soft to soft clay, medium stiff clay, and stiff clay. Brief introduction of the soil deposits are presented in the following paragraphs.

– The brownish grey weathered crust is the uppermost hard clay material formed by a cyclic wetting process together with natural cementation. The total thickness is about 1 to 2 m.
– The greenish grey soft clay material with total thickness of 8 to 10 m, as known as "Bangkok Clay", is found beneath the weathered crust. In general, this layer can be divided into two parts: the upper very soft clay layer and the lower soft clay later. The upper very soft clay layer lies within

depths of 1.5 to 8 m below ground surface, with over 100% of nature water content and as low as less than $1.25 \, t/m^2$ ($\approx 12.5 \, kP_a$) of undrained shear strength. The lower soft clay layer, on the other hand, lies between 8 to 11 m in depth, with average natural water content of about 85%, total unit weight of about $1.6 \, t/m^3$ ($\approx 16 \, kN/m^3$), and undrained strength of about $2.2 \, t/m^2$ ($\approx 22 \, kP_a$). This soft and highly compressible "Bangkok Clay" is the major concern for the construction of SBIA.
– The light grey medium stiff to stiff clay lies between 11 to 15 m below ground surface, with average natural water content of about 65%, total unit weight of about $1.7 \, t/m^3$ ($\approx 17 \, kN/m^3$), and undrained shear strength of about $6.0 \, t/m^2$ ($\approx 60 \, kP_a$).
– The light grey to light brown stiff clay is generally discovered at depths greater than 15 m. The average natural water content and total unit weight are about 40% and $1.8 \, t/m^3$ ($\approx 18 \, kN/m^3$), respectively. The undrained shear strength of the soil is commonly greater than $8.5 \, t/m^2$ ($\approx 85 \, kP_a$).

Underlying the stiff clay layer, a dense sand layer is discovered at about 21 to 26 m below ground surface (e.g., MAA, 2002; ADG, 1995). The natural water content is about 20%, and the SPT-N values are in general greater than 40.

2.3 *Groundwater and ground subsidence*

Based on the groundwater investigation at the SBIA site, the groundwater level is at a depth of about 0.5 m below ground surface. Figure 5 shows the distribution of pore water pressure within a depth of 35 m over the years. As illustrated by the figure, significant underpressure, as much as 20 m drop below the

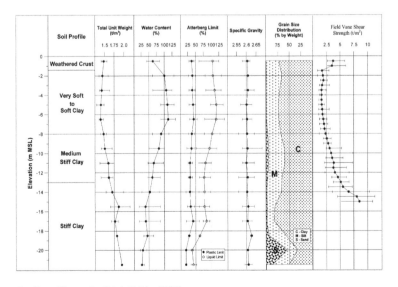

Figure 4. General soil profile on site (Moh & Lin, 2003).

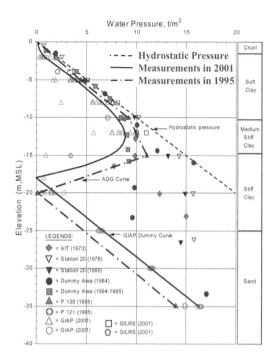

Figure 5. Pore water distribution with depth (after Moh & Lin, 2003).

corresponding hydrostatic pressure, is found to exist within depths between 10 and 20 m (i.e., in the medium stiff to stiff clay layers). Such a phenomenon as under-hydrostatic water pressure was first observed in 1973 and later verified by Engineers (1984). It is most probably due to the lowering of the piezometric head in the underlying sand layer as a result of deep well pumping. The increase of underpressure gives rise to an increase in effective stress, thus inducing consolidation settlement within the clay layers and causing ground subsidence.

For depths greater than 20 m to the maximum depth of 35 m, the water pressure varies linearly with depth. Figure 5 further depicts the most recent distribution obtained through the work of ADG (1995) and MAA (2002). The distribution of MAA (2002) is used as the basis of ground improvement design, as described in the latter section.

Ground subsidence in the Bangkok area has been monitored for more than 30 years. The study of Bangkok subsidence problem was first published in 1968 and was carried out continuously by Asian Institute of Technology (AIT), The National Environmental Board (NEB), Royal Thai Survey Department (RTSD), and the Department of Mineral Resources (DMR). According to Prinya et al. (1989), the maximum subsidence in Bangkok during the period of 1933 to 1987 was over 1.6 m and, thus, the elevation of ground surface at some areas was below the mean sea level

(MSL). Based on the study results for the land subsidence of 1978 to 1981 (AIT, 1981), the subsidence rate near the SBIA site was more than 0.1 m/year.

To reduce the subsidence rate, ground water pumping was restricted by the government in 1983. Low annual subsidence rate of about 20 to 50 mm/year was therefore observed around the SBIA site during the period of 1988 to 1990 (Figure 2). However, in the years of 1990 to 1994, the groundwater table was found to be lowered, and the subsidence rate was found to increase. Such an observation can be explained by the fact that the amount of groundwater pumping had increased since 1988. On the basis of survey data at the Station 20 and 29 around the SBIA site recorded by RTSD (Figure 2), a total of 0.5-m and 0.6-m subsidence has occurred at the Station 20 and 29, respectively, for the past 20 years.

3 STRESS HISTORY AND STRENGTH OF SOIL DEPOSITS

The stress history and the strengths of the soil deposits at the SBIA site are characterized through a series of field and laboratory test results. For the soil stress history, the preconsolidated stress (σ_p') and overconsolidation ratios (OCR) are obtained through the results of consolidation tests, such as oedometer (OED) and constant rate of strain (CRS) tests. The soil strengths were principally obtained from the results of field vane shear test (VST) and anisotropically-consolidated or isotropically-consolidated undrained compression triaxial tests (CAU and CIU, respectively).

3.1 Stress history

Figure 6 presents the distribution of σ_p' and OCR with depth derived on the basis of OED and CRS test results. As illustrated by Figure 6(a), the σ_p' values remain almost constant within a depth of 6 m, below which they increase with depth. The corresponding OCR values, as shown in Figure 6(b), exhibit a decreased trend over depths. The values decrease from 8 to about 1.5 as the depth increases from 1 to about 7 m. The values then remain almost the same, at about 1.5, for depths greater than 7 m. As the OCR distribution is compared to the soil profile (Figure 4), the top weathered crust is overconsolidated, with OCR values approximately varying from 5 to 8. The underlain very soft to soft clay layer is slightly overconsolidated, with OCR values approximately varying from 1 to 3. In addition, the medium stiff and stiff clay layers overlain by the very soft to soft clay layer are nearly normal-consolidated, with OCR values approximately varying from 1 to 1.5.

3.2 Undrained shear strengths

Figure 7 shows the distribution of the field vane s_u values obtained by the works of Engineers (1984),

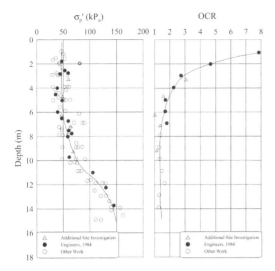

Figure 6. Stress history of soils on the site (after MAA/NECCO/PC, 1995).

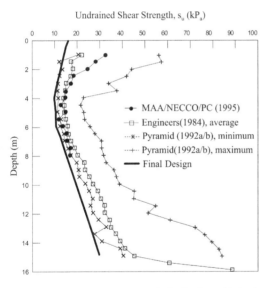

Figure 7. Undrained shear strength distribution with depth (after MAA/NECCO/PC, 1995).

Pyramid (1992), and MAA/NECCO/PC (1995). As illustrated in the figure, the results of the works all indicate that the s_u values first decrease with depth to a depth of 2 m, below which they remain almost constant and then rebound as the depth exceeds 6 m. The s_u values obtained by Engineers (1984) and MAA/NECCO/PC (1995) are close to the lower bound values obtained by Pyramid (1992a/b). The s_u values are in the range of 10 to 30 kP_a for the top 2 m, below which they are in the range of 10 to 20 kP_a for depths

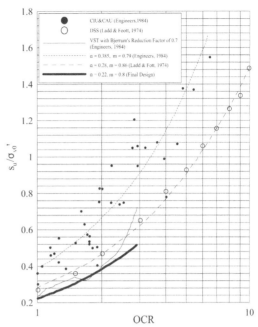

Figure 8. SHANSEP curves for Bangkok Clay (after MAA/NECCO/PC, 1995).

between 2 and 8 m. For depths greater than 8 m, the s_u values increase with depth. The figure further depicts the curve adopted in the final design which almost follows the trend of the lower bound s_u obtained by Pyramid (1992a/b).

3.3 SHANSEP model

The SHANSEP (Stress History and Normalized Soil Engineering Properties) model developed by Ladd and Foott (1974) was used to establish the relationship between shear stress and stress history, as stated in the following expression:

$$s_u / \sigma_{v0}' = \alpha \, (OCR)^m \qquad (1)$$

in which s_u and σ_{v0} are, respectively, undrained shear strength and effective overburden pressure; OCR is overconsolidation ratio; and α and m are constants.

Figure 8 shows the distribution of s_u/σ_{v0}' with respect to OCR for Bangkok Clay. From the figure, Ladd and Foott (1974) established the constants α and m, respectively, as 0.28 and 0.86 based on the DSS test results. On the other hand, Engineers (1984) established the constants α and m, respectively, as 0.39 and 0.79 based on the CIU and CAU test results. Based on the field VST results of Engineers (1984) and Pyramid (1992), MAA (2002) adopted α and m values as 0.22

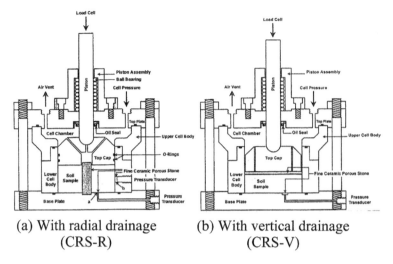

(a) With radial drainage
(CRS-R)

(b) With vertical drainage
(CRS-V)

Figure 9.   Constant rate of strain (CRS) consolidation test (after Seah & Koslanant, 2003).

and 0.8, respectively. It is noted that the proposed relationship more or less follows the design curve in Figure 7 and is subsequently used for the stage loading design of ground improvement.

4   COMPRESSIBILITY CHARACTERISTICS

The compressibility of the soft clay at the site is characterized by the compression index (C_c) and coefficient of secondary compression (C_α) through a series of OED and CRS tests. Considering anisotropy in the soils, CRS with radial drainage (CRS-R) is applied to derived C_c and C_α in the horizontal directions. The CRS-R test is first introduced in this section, followed by representative results of C_c and C_α characteristics.

4.1   CRS-R Testing

The C_c and C_α of clay layers in the horizontal direction for the SBIA site are derived from the CRS-R equipment developed by Seah & Juirnarongrit (2003), as shown in Figure 9(a). The assembly of CRS-R is similar to CRS used for vertical drainage (CRS-V), as shown in Figure 9(b), including a base plate, two parts of cell body (i.e., upper and lower cell body), top plate, top cap, and a loading piston. The difference between CRS-R and CRS-V exists in the sample preparation and the associated loading and measuring instruments.

In the CRS-R tests (Figure 9(a)), the soil specimen is hollowed out a hole in the middle. A cylindrical fine porous stone is then inserted in the hole to be used as the only drainage boundary of the soil sample during consolidation. After the specimen is prepared, the lower cell body is fixed to the base plate with a

small hole (1 mm in diameter) located close to the center (i.e., location a). This small hole is connected to a pressure transducer to monitor the pore pressure variation. Another pressure transducer is positioned at the outer boundary of the sample (i.e., position b). Furthermore, the top cap is fabricated with a hole in the center such that it can slide freely over the cylindrical porous stone during consolidation process. Detailed description of CRS-R is available in Seah & Koslanant (2003).

Once the equipment is assembled, a backpressure of 200 kP$_a$ is applied to the specimen for 24 hours for saturation. Subsequently, the sample is loaded with a gear-driven loading frame that forces the loading piston to move downwards at a constant speed. The vertical load, pore water pressure, and displacement are measured by a load cell, pressure transducers, and displacement transducers, respectively, and are recorded by a data acquisition system.

4.2   Characteristics of C_c and C_α

Figure 10 shows the representative compression curves derived from CRS-R, CRS-V, and OED tests. It is noted that all the soils are sampled in the vertical direction. As illustrated by the figure, the curves derived from the same depth are almost identical for these three consolidation methods. Such results indicate that the soil is less anisotropic in terms of compressibility; the C_c and C_α deduced from these three curves would be almost the same.

Figure 11 presents the C_c distribution with depth obtained by Engineers (1984) and MAA/NECOO /PC (1995). As illustrated by the figure, the C_c values of a soil specimen increase with applied pressure and reach a maximum at a pressure slightly higher than σ'_p. The

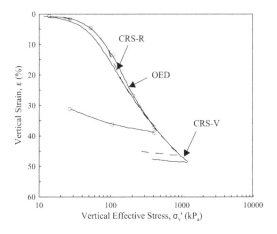

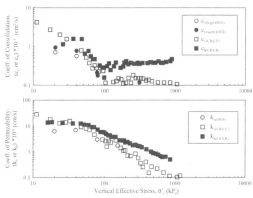

Figure 10. Compression curves from CRS-R, CRS-V, and OED tests (after Seah & Koslanant, 2003).

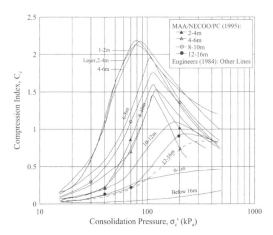

Figure 11. Compression index (C_c) of soils on the site (after MAA/NECCO/PC, 1995).

C_c values then decrease with further increase in pressure. The peak values of C_c for the top 6-m soils (i.e., in the weathered crust and very soft clay layer) vary between 2.1 and 2.2. As the depth increases, the peak C_c values decrease from about 1.9 for depths of 6 to 8 m (i.e., in the very soft clay layer) to about 0.8 for depths of 12 to 16 m (i.e., in the medium stiff to stiff clay layer).

According to the test results obtained by Engineers (1984) and MAA (2003), the C_α values vary at different loading stages. The values are also affected by the ratio of load-increment and load-increment duration in the OED tests. Further, Kulatilake (1978) and Ong (1983) point out that the C_α values for Bangkok Clay fall with a range between 0.011 and 0.021. Cox (1981) and Wongprasert (1990) suggest that the C_α/C_c values range from 0.04 to 0.05 for Bangkok Clay.

Figure 12. Coefficient of consolidation and permeability from CRS-R, CRS-V, and OED tests (after Seah & Koslanant, 2003).

5 TIME-DEPENDENT DEFORMABILITY

To characterize the time-dependent deformability (or to predict the rate of consolidation) of the soils at the site, it is essential to obtain reliable coefficients of consolidation in the vertical (c_v) and horizontal directions (c_h). The stress history of the soils, as described in Section 3, also plays an important role. In general, one needs to obtain four coefficients of consolidation, namely $c_{h(OC)}$, $c_{h(NC)}$, $c_{v(OC)}$, and $c_{h(NC)}$, where OC and NC in the parentheses represent overconsolidated and normally consolidated stress states, respectively. The c_v and c_h values were obtained from laboratory consolidation and field piezocone tests, as described in the following sections.

5.1 Consolidation Test Results

Figure 12 shows the distribution of c_v and c_h along with vertical effective stress (σ_v') obtained from the results of OED, CRS-V, and CRS-R tests. The soils were all sampled in the vertical direction, the same as those in the previous sections. It is noted that the c_v values of OED tests are derived based on the square root of time and logarithm of time fitting method proposed by Taylor and Casagrande, respectively. The c_v and c_h values of CRS-V and CRS-R, respectively, are on the other hand derived on the basis of the following equations:

$$c_v = \varepsilon \ H^2 / (2 \ u_b \ m_v) \qquad (2)$$

$$c_h = \beta \ v_p \ r_e^2 / (u_b \ H \ m_v) \qquad (3)$$

in which, ε = strain rate; H = height of sample (drainage path); u_b = excess pore pressure at impervious base; m_v = coefficient of volume compressibility; v_p = velocity of load piston; r_e = radius of sample;

and β = constant, as a function of r_w (radius of central drain) and r_e. Details of the parameters used for deriving c_h can be referred to Seah et al. (2004).

In addition, a back-analysis method proposed by Asaoka (1978) to estimate c_h was adopted for comparison. The back-analysis method was developed based on the measured settlement data of the embankment during construction.

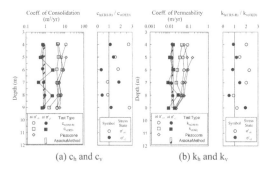

Figure 13. Distribution of coefficient of consolidation and permeability with depth (after Seah et al., 2004).

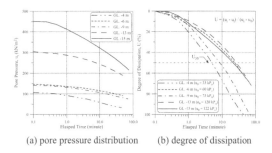

(a) pore pressure distribution   (b) degree of dissipation

Figure 14. PCPT dissipation test results (after IGN, 1995).

As illustrated by the figure, the value of $c_{h(OC)}$ is generally larger than that of $c_{h(NC)}$. The c_v and c_h values for OC state are almost the same for OED and CRS tests, whereas the c_h values are approximately 3 to 5 times larger than the corresponding c_v values for NC state. As the results are re-plotted to show the distribution along the depths (Figure 13), the c_h values are larger than the c_v values derived from the OED tests. As shown by the figure, the ratio of c_h to c_v varies from 0.3 to 2.8 along the depths. The coefficient of permeability in the vertical (k_v) and horizontal directions (k_h) derived from the c_v and c_h, respectively, (i.e., $k_{v/h} = c_{v/h} \gamma_w m_v$, where γ_w represents unit weight of water) is presented in Figures 12 and 13 for reference.

5.2 PCPT dissipation test results

To obtain data for the flow and consolidation characteristics of the cohesive soils at the site, dissipation tests were performed during piezocone penetration tests (PCPT) at various depths. Due to undrained condition at each cone penetration, the excess pore pressure is presumed the maximum at the time that cone penetrates. As pore water pressure is measured from the cone shaft right behind the cone tip, the coefficient of consolidation obtained from the piezocone test is in the horizontal direction (i.e., c_h) rather than in the vertical direction (i.e., c_v). Further, as the piezocone test release stress at the time of cone penetration the obtained c_h is for OC state (i.e., $c_{h(OC)}$). Representative dissipation results are shown in Figure 14, from which 60% to 90% of dissipation was generally achieved. It is of special noted that the u_0 used in Figure 14(b) for calculating degree of dissipation is the in-situ pore pressure, as illustrated in Figure 5.

To verify the $c_{h(OC)}$ values derived from the dissipation tests, they are compared to those derived from CRS-R tests, as shown in Figure 15. From the figure,

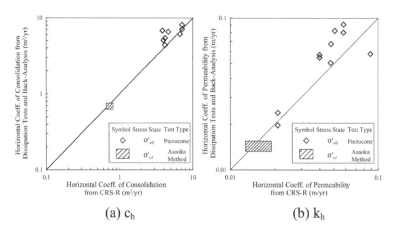

(a) c_h (b) k_h

Figure 15. Comparison between CRS-R and PCPT dissipation test results (after Seah et al., 2004).

Table 1. Summary of coefficient of Consolidation and permeability derived from various methods (after Seah et al., 2004).

(a) Coefficient of Consolidation (c_v/c_h)

Depth (m)	σ'_p (kPa)	σ'_{v0} (kPa)	At effective overburden pressure, σ'_{v0}				σ'_{vf} (kPa)	At final stress level, σ'_{vf}		
			c_h (m²/yr)			c_v (m²/yr)		c_h (m²/yr)		c_v (m²/yr)
			CRS-R	Piezoprobe test No.1	No.2	OED		CRS-R	Asaoka	OED
4	45	24	11.4	–	4.6	4.3	99	0.79	0.75 ± 0.05	0.7
5	53	28	6.6	11.3	6.1	4.6	103	0.73	0.75 ± 0.05	0.5
6	54	33	6.9	7.0	7.9	3.9	108	0.76	0.75 ± 0.05	2.4
7	59	38	3.8	5.0	6.7	4.4	113	0.82	0.75 ± 0.05	0.4
8	66	44	4.5	6.3	6.5	5.0	119	0.91	0.75 ± 0.05	2.2
9	85	47	4.1	4.3	5.3	1.5	122	0.79	0.75 ± 0.05	0.7

(b) Coefficient of Permeability (k_v/k_h)

Depth (m)	σ'_p (kPa)	σ'_{v0} (kPa)	At effective overburden pressure, σ'_{v0}				σ'_{vf} (kPa)	At final stress level, σ'_{vf}		
			$k_h \times 10^{-3}$ (m/yr)			$k_v \times 10^{-3}$ (m/yr)		$k_h \times 10^{-3}$ (m/yr)		$k_v \times 10^{-3}$ (m/yr)
			CRS-R	Piezoprobe test No.1	No.2	OED		CRS-R	Asaoka	OED
4	45	24	80	–	58	44	99	14	14 ± 1	16
5	53	28	50	154	83	35	103	13	14 ± 1	11
6	54	33	57	80	91	25	108	14	14 ± 1	17
7	59	38	47	51	67	24	113	13	14 ± 1	9
8	66	44	40	55	57	44	119	15	14 ± 1	28
9	85	47	21	20	24	17	122	19	14 ± 1	9

a favorable agreement is obtained. Further, Table 1 summarizes the c_h and the corresponding k_h values of CRS-R and piezocone tests and the c_v and the corresponding k_v values of OED tests with depth at in-situ and final stress levels. Figure 16 depicts the c_h (or c_v) and k_h (or k_v) variations with respect to σ'_v/σ'_p. In the figure, both c_h (or c_v) and k_h (or k_v) values decrease rapidly as σ'_v/σ'_p increases. There is a trend that the c_h (or c_v) and k_h (or k_v) values remain almost constant as σ'_v/σ'_p is greater than 1. Such a trend is also obtained by Aimdee (2002) using OED tests for Bangkok Clay.

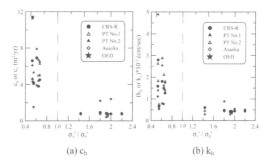

(a) c_h (b) k_h

Figure 16. Variation of c_h and k_h with σ'_v/σ'_p.

6 GROUND IMPROVEMENT

With the strengths and deformation characteristics of soft clay layers deduced from a series of field and laboratory test results, methods of ground improvement were evaluated as early as in the Phase II of feasibility study for the SBIA site. In the Phase II of feasibility study (Engineers, 1984), three test embankments with non-displacement sand drains installed underneath were constructed and ground improvement techniques such as surcharging, vacuum loading, and groundwater lowering were adopted to accelerate the rate of consolidation in the soft clay layers.

In the subsequent Phase III of the feasibility study, several ground improvement alternatives were studied, including preloading with the use of prefabricated vertical drains (PVD), deep soil improvement, the use of piles with free spanning plates at top, relief piles with caps and soil reinforcement, and light weight fills. In consideration of cost, schedule, and technical limitations, preloading with the use of PVD was recommended (STS/NGI, 1992). After a series of trials with different PVD arrangements underneath three test embankments in the Phase IV of the feasibility study, it was concluded that PVD is a favorable technique for ground improvement at the site (AIT, 1995).

Two projects, namely Airside Pavement Design (GIAP) and Landside Road System Design (GILRS), were carried out with the PVD systems designed by ADG (1995) and MAA/NECCO/PC (1995), respectively. Both of them verified the effectiveness of PVD used for ground improvement. Figure 17 shows the scope of working area for GIAP and GILRS, in which the former covers the west runway, apron,

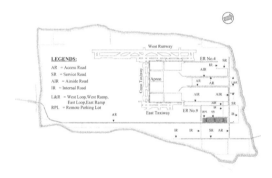

Figure 17. Scope of working area in GIAP and GILRS (TEC/MAA, 2002).

taxiways, part of east runway, and two emergency access roads and the latter includes 21 internal roads and remote parking lots. Basic information of these two projects is shown in Table 2. The scheme and monitoring system for the projects are briefed in the following sections. The effectiveness of PVD used for ground improvement is presented at the end.

6.1 Scheme

Ground improvement sequence for the projects GIAP and GILRS, as shown in Figure 18, primarily includes construction of sand blanket and drainage facilities, installation of monitoring system and PVD, and application of pre-loading. One major difference between these two projects was the introduction of filter fabric placed above and below the sand blankets for the GIAP. Furthermore, the test embankment in the GIAP was subjected to two stages of loading whereas three test embankments in the GILRS were separately subjected to one, two, and three stages of loading. Table 3 outlines ground improvement design for these two projects.

The major consideration for the GIAP is to achieve a minimum of 80% primary consolidation settlements of the underlying soft soils. The GIAP was further required to meet a maximum of 4% settlement ratio (ratio of last month settlement to accumulated settlement) for a surcharge of 75 kP_a and 2% for a surcharge of 85 kP_a. For the GILRS, the maximum monthly settlement of the road embankment prior to the laying of pavement was required not to exceed 5 mm.

In order to estimate the degree of primary consolidation, a graphical method developed by Asaoka

Table 2. Basic information of GIAP and GILRS (after Moh & Lin, 2003).

Project item	GIAP	GILRS
Total Construction Cost (Thai Baht)	8,419,205,000	1,767,488,000
Financial Source	Government Budget	OECF Loan (Japan)
Construction Period	01/11/97–30/04/02 (54 months)	01/12/00–19/04/03 (29 months)
Designer	Airside Design Group	Moh and Associates, Inc.
Construction Supervision	TMSUM (TEC/MAA/SIGEC/UIC/MTL)	TNM (TEC/MAA/NK)
Ground Improvement Area (m^2)	3,080,000 (west runway, apron, taxiways, part east runway, and emergency roads)	1,320,000 (21 internal roads and remote parking lot)
PVD (m)	33,580,000	10,889,600
Sand Blanket/Drainage Sand (m^3)	4,550,000	899,600
Preloading Material (m^3)	2,899,000 (crushed rock)	1,722,800 (sand)

(1978) was adopted to predict the ultimate primary consolidation settlement. The lengths of PVD in both projects were limited to 10 m to reduce the risk of hydraulic contact to the permeable layers of under-hydrostatic water pressure that might induce additional settlement as a result of suction.

6.2 Monitoring system

A thorough monitoring system was installed in and around the embankments, including surface settlement plates, deep settlement gauges, pneumatic and electric piezometers, inclinometers, and observation/pumping wells. Readings were taken weekly before and after each stage of fill loading and biweekly after the surcharge was removed. Typical cross-sections of instrumentation for GIAP and GILRS are as illustrated in Figures 19 and 20, respectively. The total amount of instruments installed for monitoring system is summarized in Table 4. The data obtained, such as vertical and lateral movements and pore pressure variation over loading stages, can be referred to ADG (1995) and MAA/NECCO/PCk (1995).

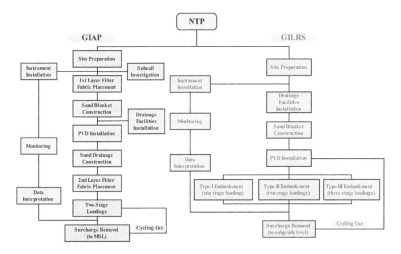

Figure 18.   Ground improvement sequence (Moh & Lin, 2003).

Table 3.   Ground improvement design for GIAP and GILRS (after Moh & Lin, 2003).

Project item	GIAP	GILRS		
		Type I	Type II	Type III
Design Criteria	Min. 80% primary consolidation reached	Consolidation settlement rate < 5 mm/month before pavement construction		
Sand Blanket	1.5 m	0.5 m	0.8 m	1.3 m
PVD	10 m deep with 1 m spacing in square pattern	10 m deep with 1 m spacing in triangular pattern		
Filter Fabric	Below and above sand blanket	None		
Preloading Material	Crushed rock	Sand		
Stage Loading	Two	One	Two	Three
Embankment Thickness	3.8 & 4.2 m	2.2 m	3.5 m	4.5 m
Berm	15 m wide & 1.7 m high with 1:4 side slope	No berm with 1:3 side slope		
Design Load	75 & 85 kPa	41.8 kPa	66.5 kPa	85.5 kPa
Removing Criteria	Min. 6 (or 11) months waiting period, min. 80% primary consolidation & 4% (or 2%) settlement ratio	Min. 6 months waiting period with max. 3 cm/month settlement rate		

6.3 Effectiveness of ground improvement

The effectiveness of ground improvement was assessed by comparing the settlement obtained in the PVD and non-PVD areas and the changes of soil properties. Figures 21 and 22 present representative results obtained from the project GIAP. The details are available in ADG (1995) and MAA/NECCO/PC (1995). As illustrated in Figure 21, the settlement rate for the PVD areas is about 3 times that for the non-PVD area after almost two-year loading (i.e., compared 1349 mm to 440 mm after 750-day loading). In Figure 22, the ground improvement led to significant change of soil properties primarily in the soft clay layer: the undrained shear strength increased by about 50%, water content decreased by about 30%, and total unit weight increased by about 7%.

7 SUMMARY AND CONCLUSIONS

To provide soil parameters required in the detailed design of ground improvement for the construction of SBIA, a series of field and laboratory tests were carried out to characterize the site conditions, including groundwater/pore water pressure and soil engineering properties. The results are itemized as follows.

- The soil profile within a depth of about 20 m includes: weathered crust (GL.0–GL.-2 m), very

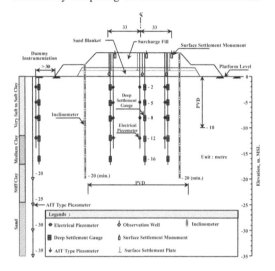

Figure 19.   Typical cross-section of instrumentation in GIAP (Moh & Lin, 2003).

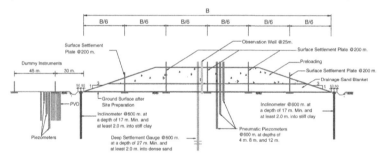

Figure 20.   Typical cross-section of instrumentation in GILRS (Moh & Lin, 2003).

Table 4.   Total amount of monitoring instruments (Moh & Lin, 2003).

Project item	GIAP Embankment	Dummy	GILRS Embankment	Dummy
Surface Settlement Plate	1,724	–	730	4
Surface Settlement Monument	553	–	–	–
Permanent Benchmark	–	2	–	–
Inclinometer	56	–	88	–
Deep Settlement Gauge	111*	11*	53**	–
Pneumatic Piezometer	–	–	159	6
Electric Piezometer	444	46	–	–
AIT-Type Piezometer	–	40	–	–
Observation Well	1,722	–	1,236***	–

*: One set includes 5 individual deep settlement gauges at 2, 5, 8, 12 & 16 m.
**: A deep settlement gauge includes sensor rings at every 2 m to min. depth of 27 m.
***: Also used as pumping wells

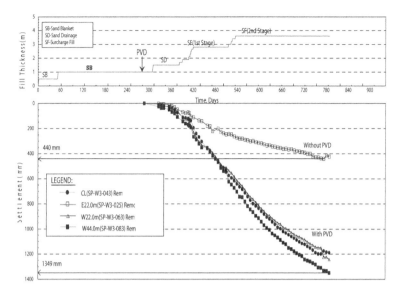

Figure 21. Comparison of settlements in the PVD and non-PVD areas of GIAP (Moh & Lin, 2003).

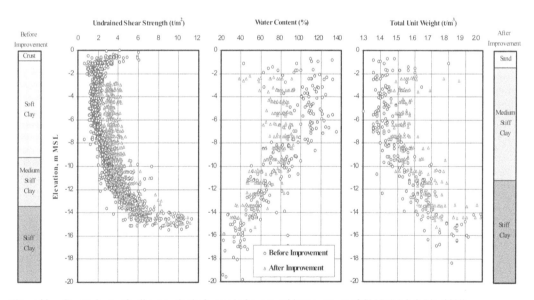

Figure 22. Comparisons of soil properties before and after ground improvement of GIAP (Moh & Lin, 2003).

soft to soft clay (GL.-2–GL.-11 m), medium stiff clay (GL.-11–GL.-15 m), and stiff clay (GL.-15–GL.-20 m), in which the very soft to soft clay layer is known as Bangkok Clay.
– The groundwater lies at a depth of about 0.5 m below ground surface, and pore water pressure is significantly lower than its hydrostatic counterpart for depths between 10 and 20 m.

– A SHANSEP model was provided to describe relationship between soil strength and stress history.
– Based on the results of OED, CRS-V, CRS-R, and PCPT tests, deformation characteristics were studied in both vertical and horizontal directions and slight anisotropy of the soil was verified.
– The use of PVD was proven to be successful as the settlement rate was increased by 3 times over the

29

2-year loading period and the soil properties were much improved.

ACKNOWLEDGEMENTS

The authors are grateful to the New Bangkok International Airport Co. Ltd., the Royal Thai Survey Department, and MAA Group Consulting Engineers for their kind support in providing data used in this paper. Sincere appreciation is due to Dr. Za-Chieh Moh, Dr. Richard N. Hwang, Dr. T.H. Seah, and Mr. Pen-Chi Lin of MAA Group Consulting Engineers and Dr. Der-Wen Chang of Tamkang University for their valuable advices on the preparation of this manuscript.

REFERENCES

Aimdee, W. 2002. *The Behavior and Mechanism of Time Dependency Settlement of Two Types of Bangkok Clay Having Different Basic Properties.* M. Eng Thesis. Chulalongkorn University. Bangkok: Thailand.

ADG. 1995. *SBIA Airside Pavements Design – Ground Improvement Design.* Final Design Report. Report Submitted to Airports Authority of Thailand. Airside Design Group. Bangkok: Thailand.

AIT. 1974. *Performance Study of Test Sections for New Bangkok Airport at Nong Ngu Hao.* Research Report to Northrop. Final Report. Asian Institute of Technology. Bangkok: Thailand.

AIT. 1981. *Investigation of Land Subsidence Caused by Deep Well Pumping in the Bangkok Area.* Asian Institute of Technology. Bangkok: Thailand.

AIT. 1995. *The Full Scale Field Test of Prefabricated Vertical Drain for the Second Bangkok International Airport.* Final Report. Volume I. Report Submitted to the Second Bangkok International Airport Co. Ltd. Asian Institute of Technology. Bangkok: Thailand.

Asaoka, A. 1978. Observation Procedure of Settlement Prediction. *Soils and Foundations.* 18(4): 87–101.

Cox, J.B. 1981. The Settlement of a 55 KM Long Highway on Soft Bangkok Clay. *Proceedings of The 10th International Conference on Soil Mechanics and Foundation Engineering.* 1: 101–104.

Engineers. 1984. *The Second Bangkok International Airport Master Plan Study. Design and Construction Phasing. Final Detailed Geotechnical Report. Volume II. Test Sections. Their Performance and Evaluation RP-GE6-143B.* Report Submitted to the Department of Aviation. Ministry of Communications. Engineers for the Second Bangkok International Airport. Bangkok: Thailand.

IGN. 1995. *Factural Report of Soil Investigation for Land Side Road System Second Bangkok International Airport.* IGN(Thailand) Limited.

Kulatilake, P.H.S.W. 1978. *Secondary Compression Study Related to Bangkok Subsidence.* M. Eng. Thesis. Asian Institute of Technology. Bangkok: Thailand.

MAA/NECCO/PC. 1995. *Consultancy Services for the Design and Preparation of Tender Documents for the Landside Road System for the Second Bangkok International Airport.* Final Design Report on Ground Improvement. Document No. LR2-FR-01/1/D. Volume I. Moh and Associates, Inc./MAA Consultants Co., Ltd/National Engineering Consultants Co., Ltd/Prasat Consultant Co., Ltd.

Moh, Z.C. & Lin, P.C. 2003. From Cobra Swamp to International Airport: Ground Improvement at Suvarnabhumi International Airport. *Ground Improvement.* 7(2): 87–102.

Ong, B. 1983. *Behavior of Secondary Settlement under Preconsolidation.* M. Eng. Thesis. Asian Institute of Technology. Bangkok: Thailand.

Pyramid. 1992a. *Report on Sub-Surface Investigation for the Design of Polder Dike (Vol. 1) for Second Bangkok International Airport.* Report Submitted to Airports Authority of Thailand. Pyramid Development International Co. Ltd. Bangkok: Thailand.

Pyramid. 1992b. *Progress Report on Sub-Surface Investigation for the Design of Polder Dike (Vol. 1) for Second Bangkok International Airport.* Report Submitted to Airports Authority of Thailand. Pyramid Development International Co. Ltd. Bangkok: Thailand.

Rau, J.L. & Nutalaya, P. 1981. Chloride Contamination in Aquifers of the Central Plain, Thailand. *Geotechnical Engineering.* 12: 123.

Seah, T.H. & Juirnarongrit, T. 2003. Constant Rate of Strain Consolidation with Radial Drainage. *Geotechnical Testing Journal.* ASTM 26(4): 432–443.

Seah, T.H. & Koslanant, S. 2003. Anisotropic Consolidation Behavior of Soft Bangkok Clay. *Geotechnical Testing Journal.* ASTM. 26(3): 266–276.

Seah, T.H., Tangthansup, B., & Wongsatian, P. 2004. Horizontal Coefficient of Consolidation of Soft Bangkok Clay. *Geotechnical Testing Journal.* ASTM. 27(5): 430–440.

STS/NGI. 1992. *Independent Soil Engineering Study for Second Bangkok International Airport Implementation Program.* Final Report. Report Submitted to the Airport Authority of Thailand. Ministry of Transportation and Communications. STS Engineering Consultants, Co. Ltd. and Norwegian Geotechnical Institute. Bangkok: Thailand.

Suvarnabhumi Airport/New Bangkok Airport Guide. 2007. *www.bangkokairportonline.com.*

TEC/MAA. 2002. *Construction Supervision Services for Ground Improvement for Airside Pavement for the Second Bangkok International Airport.* Geotechnical Evaluation Report. Report Submitted to the Second Bangkok International Airport Co. Ltd. Thai Engineering Consultants Co., Ltd/MAA Consultants Co., Ltd.

TRDG. 2004. *Second Bangkok International Airport Project: The Design of the Ground Improvement and Airfield Pavements for the 1st Midfield Satellite Aprons and 3rd Runway.* Final Ground Improvement Design Report. Doc. No. DG42-FR-01. Report Submitted to the New Bangkok International Airport Co. Ltd. Third Runway Design Group. Bangkok: Thailand.

Wongprasert, M. 1990. *Settlement Analysis of Steel Grids Mechanically Stabilized Earth Test Embankments and Embankments with Vertical Drains.* M. Eng. Thesis. Asian Institute of Technology. Bangkok: Thailand.

Geotechnical and Geophysical Site Characterization – Huang & Mayne (eds)
© 2008 Taylor & Francis Group, London, ISBN 978-0-415-46936-4

Forensic diagnosis for site-specific ground conditions in deep excavations of subway constructions

Kenji Ishihara
Chuo University, Tokyo, Japan

Wei F. Lee
Taiwan Construction Research Institute, Taiwan

ABSTRACT: In an attempt to address important but unheeded issues of soil conditions associated with deep subway construction, three cases of large-scale collapse that occurred in recent years are taken up for consideration. One of them is the failure of 31 m deep open excavation in Singapore which was underway for subway construction by utilizing the diaphragm wall-strut retaining structures. Amongst several reasons, unaccustomed presence of buried valley was pointed out in the forensic investigations. Thus, shortage of available information regarding deep-seated soil strata was found to be fatal for making conservative design and safe operation of the construction. Thus, scrutiny of ground conditions by closely spaced boring and logging and precise soil property characterization were a reminder to be seriously recognized as a lesson of supreme importance. The second case in this study is the large-scale cave-in at the site of subway construction in Kaohsiung, Taiwan, which occurred as a result of piping in a silty sand deposit at the bottom of the excavation. In view of the circumstances that, nowhere has the piping been observed more conspicuously than in the deposits in Kaohsiung, particular nature of the local silt was suspected to be a cause leading to the occurrence of the piping. The results of pinhole tests and specific surface measurements revealed peculiar characteristics of the local silt which is highly vulnerable to internal erosion subjected to seepage flow. Thus, as a result of forensic type study, the importance of exploring unknown nature of local soils was recognized as an important lesson to be paid attention particularly when construction projects are to be executed in regions of few experiences of deep excavations. The large-scale cave-in which occurred at the subway construction site in Shanghai is taken up as the third example of forensic consideration. The incident is reported to have occurred as a result of malfunctioning of equipments when the subway construction was going on at a depth of about 30 m. Although detailed information was not made available, it is considered likely that shortfall of investigations on the nature of locally prevalent sandy silts might have been an underlying reason behind the occurrence of unfortunate distress in Shanghai.

1 INTRODUCTION

The term Forensic Engineering is a relatively new word which has become known in the area of geotechnical engineering during the past 30 years. From the cases of this word being used, it is vaguely understood that its main aim is to unearth facts associated with occurrence of accidents or damages and to address them in a simple yet rational manner so that even non-professionals could understand and exercise a fair judgment. The outcome of the forensic investigation is often submitted to the court of law to make a decision or sentence by a judge. The needs for the forensic investigation in geotechnical engineering have arisen from frequent occurrences of trouble or dispute associated with shortcomings or defects in land development for residential use or accidents during construction works (Day, 1998). In the discipline of the geotechnical engineering, the state-of-the-art knowledge currently available is generally incorporated in the design and practice of construction. However, there are always shortcomings in many facets of soil investigations and misinterpretation of their outcome. Although small troubles are of occasional occurrence, construction works are executed in general without any vital mishap. However, incidents or collapse of fatal importance could occur sometimes inevitably in the foundation works involving deep excavations. Detailed investigations are carried out after the incident from the forensic standpoint. In the course of such investigation, various aspects of geotechnical importance hitherto unknown are

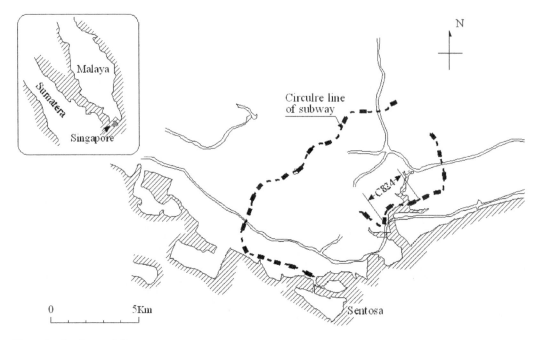

Figure 1. South part of Singapore.

unearthed regarding site-specific soil characteristics, soil-structure interaction, detailed process and control of construction.

The experiences of calamity as above leave important lessons to be profoundly remembered and to be reflected on future development of the state-of-the-art. However, it has been a general inclination that these precious experiences were neither exposed to eyes of professionals nor well-documented. In fact, to the authors' knowledge, there has been no single case of the big accidents associated with subway construction that was reported in the long list of literature in the discipline of geotechnical engineering.

In view of the circumstances as above, an attempt was made in this paper to collect as many information as possible from the investigation of big incidents that occurred in recent years and to arrange them in a precise manner focusing on unforeseen phenomena and circumstances existing in local soils and ground conditions.

2 COLLAPSE OF BRACED EXCAVATION IN SINGAPORE

2.1 Outline

At a site of braced excavation for a cut-and-cover tunnel construction in Singapore, a large scale collapse occurred on April 20, 2004 over the length of 100 m section of a carriage way called Nicoll Highway which is located in the southern coast of the Singapore Island

as shown in Figure 1. A closer map of the location is shown in Figure 2. The construction south of Nicoll Highway Station was underway involving deep braced excavation flanked by diaphragm walls. The depth of excavation at the time of the failure was 30.8 m. The diagram wall 0.8 m thick was braced by the steel struts with the help of walers and splays. The collapse took place when excavation was underway below the 9th step of struts already in place.

As a result, the soil mass moved towards the excavated space involving large slides on both sides accompanied by the subsidence in the surrounding area. The site is underlaid by thick deposits of soft marine clay to a depth of about 35 m. Following the incident, detailed studies have been made including borings, sounding, sampling, lab. testing and numerical analyses simulating conditions at the time of the failure. The outcome of the studies was reported by Tan (2006), Yong (2006) and Yong and Lee (2007). Although the probable causes of the incident were pointed out, the present paper will focus on one of the likely causes which is associated with poor understanding of soil conditions in the extremely soft soil area in Singapore.

2.2 Collapse of the strut-supported excavation

A section of temporary retaining wall structure adjacent to Nicoll Highway collapsed almost suddenly within a matter of a few minutes at 3:30 p.m. on April 20, 2004. An overview of the site of failure is shown in Figure 3. This is located at the seaward side of

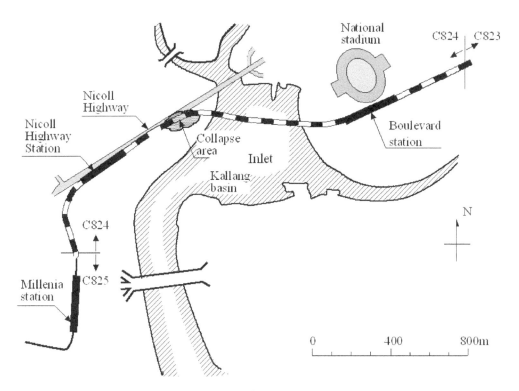

Figure 2.   Place of the collapse in the subway construction line.

Figure 3.   Oblique view of the site to the east after the collapse.

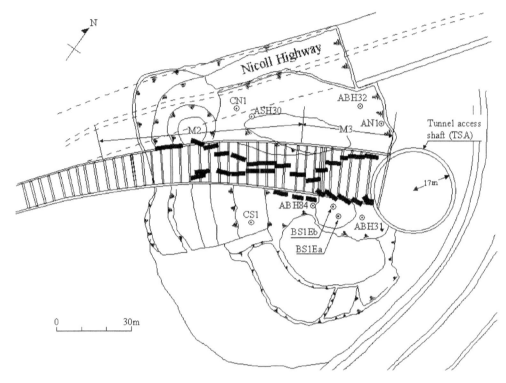

Figure 4. Plan view of the area of the failure and braced wall system.

the land reclaimed around 1970. The features of the braced excavation and collapse are roughly shown in the plane view of Figure 4 where it is noted that a large tunnel access shaft (TSA) 17 m in radius had been constructed to facilitate transport of structure elements and materials down to the floor of the excavation.

2.3 *Method of construction*

The section of the cut-and-cover tunnels were under construction by means of the bottom-up method with two lines of diaphragm walls supported by strut-waler-splay system.

The thickness of the diaphragm wall was 1.0 m at the location of cable crossing and at the end section next to the access shaft. Otherwise the thickness was 0.8 m. One element of the wall each 3 to 6 m wide was excavated in slurry by means of a basket-type digging machine in two or three bites. After lowering the reinforcement cages, concrete was poured in by Tremic pipes. Joints of neighboring elements of the wall was sealed against water penetration, but not rigidly connected as structures. The diaphragm walls, over the section Type M3, were for the curved stretch closest to the access shaft and constructed in May 2003. The walls were designed to be extended 3 m into the stiff base layer called Old Alluvium (OA-layer). A detailed plan view of the M3 section is shown in Figure 5 and a

typical side views of the strutting system for the cross section A-A′ is displayed in Figure 6.

The diaphragm walls were supported by the 10 levels of steel struts with discontinuous walers except at the top strut. To reinforce the strutting action, two soil layers at the depths of 28 to 29.5 m (sacrificial strut) and of 33.5 m to 36.5 m were stabilized as shown in Figure 6 by means of the jet grouting which had been installed from the ground surface before the excavation began. At the same time, with an aim of providing support for the railway tunnel structures to be placed later, four lines of bored piles each 1.4 to 1.8 m in diameter were installed from the ground surface at a spacing of 4 m or 6 m as shown in Figure 5. The depth of the pile embedment was 34.2 m on the north side counted downwards from the elevation of 74.2 m RL and, it was 29.6 m RL on the south side from the elevation of 69.6 m RL as shown in Figure 6, where RL (Reduced Level) means a local measure of elevation with a zero point taken at a depth 100 m from the mean sea level.

2.4 *Situations around the time of the collapse*

Excavation with strutting had gone without any hitch until the 9th strut was installed. It was envisioned at this stage that the entire lateral load near the bottom of the excavation was surely transferred to the 9th strut. Then, the excavation to the 10th level of strut was started,

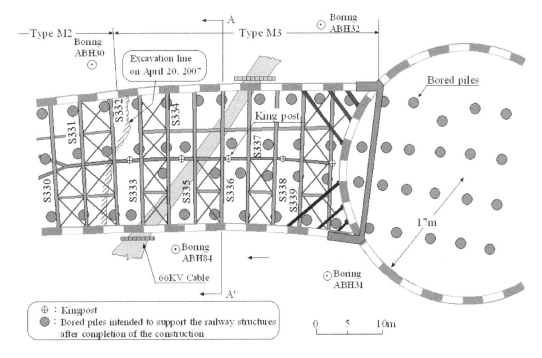

Figure 5. Plane view of the Type M3 section due west of the access shaft.

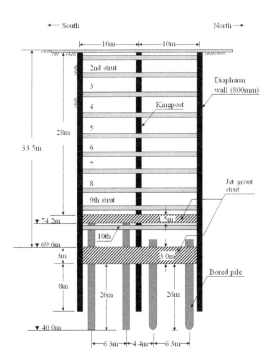

Figure 6. A typical side view of the strutted system for the A-A' cross section.

thereby taking out the jet-grouted portion 1.5 m thick. At this stage, the struts located in this stretch were S333 to S340 with a horizontal spacing of about 4 m. The width of the excavation in plan was about 20 m and the depth was 30.8 m. When the excavation had progressed to the location below the strut No. S332 as indicated in Figure 5, the collapse started to occur. That was at 3:33 p.m. on April 20, 2004. The pictures after the collapse, at 3:34 p.m. and at 3:41 p.m. are displayed in Figure 7. These photos looking over to the east were taken from the window of the apartment building just north of the Nicoll Highway.

In the morning of April 20, from 9:00 a.m. to 10:00 a.m. engineers on the spot is said to have heard sounds of metallic collision near the north side of the strut S338 and observed the inner flange of H-beam waler at the 9th strut having buckled and shifted about 5 to 6 cm downward. This shift aggravated to about 20 cm by 2:30 to 2:45 p.m. It was also witnessed around this time that the inner flange of the waler at the 9th strut S335 on the south side had also buckled. Between 2:00 p.m. and 3:00 p.m. the sounds of metallic collision were heard more frequently and eventually the collapse took place at 3:33 p.m.

2.5 Conceived causes of the failure

After the occurrence of the collapse, the Committee of Inquiry (COI) was established and probable causes to were investigated and pointed out. In May 2005 the

35

Figure 7. Photos showing the incident at 3:34 p.m. and 3:41 p.m.

COI report was released, thereby indicating the main reasons as described below.

(1) Under-design of the diaphragm wall using an inappropriate method of analysis.
(2) Under-design of strut-waler connection, particularly in the curved section due west of the SA shaft.
(3) Incorrect back analysis and problems with instrumentation and monitoring for conducting the observation-led construction control.

These would have been combined to induce the failure and detailed account is described for each of the above items in the papers by Tan (2006), Yong (2006), and Yong and Lee (2007). It is not the intention of this paper to debate on these items. The purpose herein would rather be to focus on features of soil conditions which have not been thoroughly discussed, but unearthed more recently through further investigations in details.

2.6 *Soil conditions at the site of the incident*

In the design stage before the incident, soil conditions were investigated by means of boring, sounding, and recovering undisturbed samples and testing in the laboratory. The locations of borings, standard penetration tests (SPT) and cone penetration tests conducted at the design stage in the vicinity of the failure zone are shown in Figure 8 with symbols ABH. Also designated by AC in Figure 8 are the places where the cone penetration tests (CPT) were carried out. Cone tests were also performed at the locations denoted by MC2026, 3006 and 3007. The results of the boring and SPT N-values at the site ABH30, 31, 32 and 84 are shown in Figure 9. It may be seen that the recent fills exist to a depth of 3 to 5 m underlaid by a thick deposit of soft marine clay, M, having a SPT N-value of zero. In the middle of this clay deposit, there exists a stiff silty clay layer of fluvial origin about 3 m thick at the depth of 18 to 20 m. This layer has a SPT N-value of 10 to 15 and denoted by F_2.

The deposits M and F_2 are called Kallang Formation. Under the Kallang Formation, there exists a stiff silty sand layer having variable SPT N-value from 10 to greater than 100.

This layer is an old alluvium and denoted by OA. Between the marine clay M and the OA layer, there exists occasionally the estuarine deposit consisting of organic soil and also the fluvial layer F_2.

As a result of physical and mechanical tests, the marine clay was found to possess the plasticity index of 30 to 50 and compression index of $C_c \cong 0.9$. The undrained shear strength was shown to be in the range of $S_u = 20 \sim 60$ kPa at depths of 40 m and the clay is still slightly under-consolidated state with OCR $= 0.8 \sim 1.0$. It was considered important to determine the elevation of the Old Alluvium (OA), as it was defined as the base of the bottom deposit into which the diaphragm wall was to be embedded. In the design, the toe of the diaphragm wall was set to be 3 m below the top of the OA layer.

In view of its importance, the contour lines indicating equal elevations of the top of the OA layer were also drawn in Figure 8, on the basis of the information available at the design stage. It is noted in Figure 8 that the distance between two adjacent boring logs was no shorter than about 20 m and consequently the buried valley near the southern wall close to the access shaft was not identified. This fact turned out to be vital for creating the worst situations to occur afterwards.

2.7 *Post-collapse site investigations for forensic diagnosis*

After the incident, no attempt was made to dig out the debris buried within the excavated space, because a new route for the subway construction was laid down at the southern side. Thus, the excavated space was buried instead by dumping soils and then in-situ boring and soundings were carried out. These included also the investigation of the post-collapse conditions of the

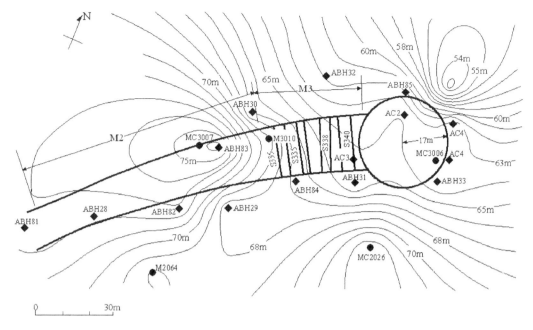

Figure 8. Contours of elevation at the top of the old alluvium (OA-layer) based on boring data at the design stage.

broken pieces of the diagram wall panels, construction machines buried within the excavated space.

(a) Borings

Outside the zone of the excavation several series of boring and soundings were performed to depths of about 40 to 50 m corresponding to an elevation of about 50 m RL (Reduced Level). Nearest the tunnel access shaft, a pair of boring, AN1 and AS1 was performed at sites north and south, respectively, of the excavation. Further westward, another pair of boring was conducted at the locations BN1 and BS1. When the drilling hit an obstacle before reaching a targeted depth, another hole was drilled. Likewise, still other pairs were conducted westwards as indicated by the heading letter C, D, E and F in Figure 10.

Some of the results of the borings conducted at points BS1Ea and BS1Eb about 7 m south of the collapsed wall is displayed in Figure 11(a) and (b) where it can be seen that remains of the grouted cement were encountered at an elevation of 73 m and 67 m RL. In another boring at CS1 shown in Figure 11(c), the soil stratification was found to be about the same as that before the failure event.

(b) Magnetic logging test

This test was intended to detect ferrous metal obstructions such as broken pieces of walers, struts or construction equipments buried in the debris within the sliding zone. The method consists of incremental lowering or rising of a magnetic probe through a borehole and to detect the magnetic reaction at varying depths. During the boring, the kind of soils was identified permitting depthwise pictures of soil profiles to be established. The magnetic logging was conducted at a spacing of about 2 m along the alignment 3 m behind the northern diaphragm wall. These are denoted by WN1 southward to WN34 as shown in Figure 12. Similarly, along the alignment 3 to 5 m behind the south diaphragm wall, the magnetic logging was performed. These are indicated by WS1 to WS33 in Figure 12. When the boring hits some obstructions, the drilling was made once again nearby. In the area between WS10 and WS14, it was not possible for the boring to reach the targeted depths because of obstructions encountered midway. Thus, several more borings were carried out until they reached the OA base deposit.

2.8 Damage features under the ground

After investigating the location of the exposed top of the diaphragm wall panels, it became possible to approximately figure out the scenario of the failure. Exact locations of the panel top in plan view are shown in Figure 12. Based on the features shown in this figure it may be mentioned that the collapse mechanism at panels M213, M212 and M306 were different from that elsewhere and almost certainly it involved the toe of the diaphragm wall panels kicking in, and flow of the soil underneath the diaphragm wall panels. This form of the failure will have allowed the maximum southward movement of the opposite north wall panels, namely M211 and M301.

37

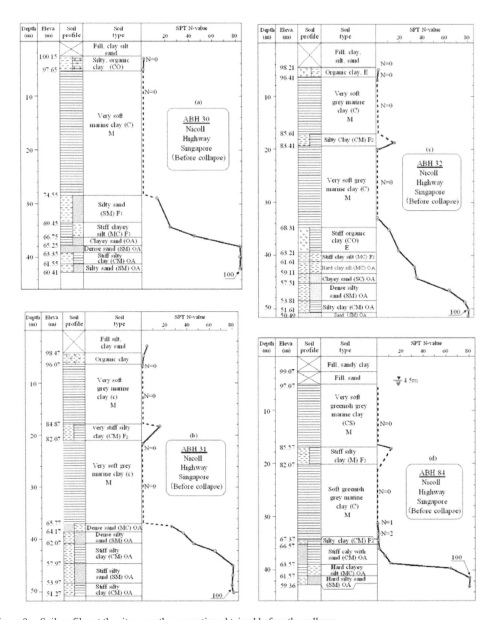

Figure 9. Soil profiles at the sites near the excavation obtained before the collapse.

As a consequence of inward collapse of the retaining diaphragm walls, it was speculated that a huge amount of soil mass moved into the excavated space. In fact, a subsidence of the ground surface as much as 10 m was observed on the south side in proximity to the wall and the settlement extended about 50 m southward, as seen in Figures 4 and 12. The area of the ground settlement extended also north towards the Nicoll Highway but to a lesser extent as can be seen in Figure 4. Thus, it was considered certain that the failure was initially triggered by the breakage of the diaphragm walls on the

southern side, thereby involving a landslide towards the excavated space. As a result of losing the horizontal support, the northern walls moved also towards the excavation, inducing sliding of soil mass towards the south but to a lesser extent.

This scenario of the event was substantiated by the borings conducted after the failure in which the elevation of the intermediate layer F_2 was found to be located almost 10 m lower than that found by the borings before the failure. Note that prior to the collapse, the elevation of the base of intermediate F_2 layer

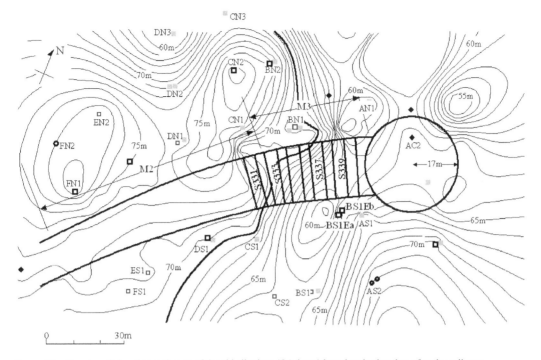

Figure 10. Counter of elevation at the top of the old alluvium (OA-layer) based on boring data after the collapse.

was remarkably consistent across the collapse area, and so provides a good marker for identifying the vertical displacement of the soil deposit. The location of the F_2 layers after the collapse is shown in the cross-section in Figure 11. It was then possible to depict the mode of soil movement involved in the slides on the south and north sides. The approximate locations of the sliding surfaces are displayed in Figure 13 for the cross section 6-6 of Figure 12.

As a result of the post-incident in-situ investigations as above, it became possible to figure out approximately the states of collapse to each of the cross sections indicated in Figure 14. Shown in Figure 15 are probable features of the collapsed waler-strut system buried within the ground which were speculated for the cross section 1, 5 and 7. It is to be noticed that in the cross section 1 and 5, the intermediate layer F_2 is inclined towards the excavation as verified by the boring data at ABH31 and 32 and also at AS1 and AN1 conducted after the failure. This is indicative of the settlement of the even deep-seated soil layer involved in the slide.

2.9 Existence of the buried valley

One of the key issues associated with generic cause of the collapse is the depth of embedment of the diaphragm wall toe into the stiff base layer OA which had been considered competent enough at the time the design was made. Thus, it is of crucial importance to identify exact configuration of the top of the OA layer, and its mechanical properties as well.

By compiling all the data made available after the collapse, the contour lines of the top of the OA layer were established as demonstrated in Figure 10 with a higher level of accuracy as compared to that shown in Figure 8 which was made up with the boring data at the time of the feasibility study and the design. The contour lines in Figure 10 unveil some important features as follows which had not been identified in the previous map in Figure 8.

(1) There exists a steeply depressed buried valley at the location on the south side a few meters south of the strut S336 and S337. The bottom of the buried valley has an elevation of about 59 m RL. The buried valley dips steeply towards the east from the south end of struts S331.

(2) The buried valley extends towards the north-east running on the west side of the tunnel access shaft (TSA).

(3) As typically observed in the soil profiles at BS1Ea and BS1Eb in Figure 11, the SPT N-value tends to increase at the elevation of 61 m RL upon hitting the base layer OA. The SPT N-values corresponding to the transition zone below the OA layer were collected from other borings and shown together in Figure 16 by choosing the nominally identified top of the OA layer as zero point. It may be seen that

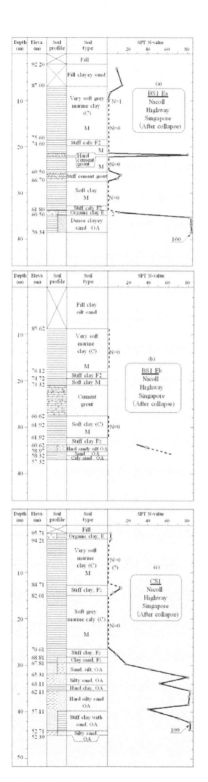

Figure 11. Soil profiles just south of the collapsed diaphragm wall panels.

the property of soils is not so competent with a SPT N-value less than about 30 to a depth of about 3 m from the top of the OA layer.

2.10 *Features of the toe-in of the diaphragm wall*

As a results of the detection of the buried valley, questions cropped up as to whether the toe of the diaphragm wall had been embedded sufficiently deep into the base layer of the Old Alluvium (OA), and also as to whether the stiffness of the infill materials near the top of the OA layer was strong enough to mobilize the resistance.

With regard to the latter suspect, closely-spaced soil investigations revealed that the upper part of the OA layer in the buried valley is comprised of sands, silts and clays sometimes containing organic materials with the SPT N-value not greater than 30 as demonstrated in Figure 16. This fact indicates that the infill materials were not necessarily as competent as they had been expected at the design stage.

To obtain a clear picture of the toe-in depth, the results of the borings at the time of the magnetic logging tests plus some others nearby were compiled and arranged side-by-side so that soil profiles along the alignment south of the diaphragm walls can be visualized. The picture arranged in this way is demonstrated in Figure 17 where the top of the OA layer considered at the design stage is indicated, together with the similar top line established after the collapse based on the detailed investigation by the magnetic logging. Note that the symbol such as M306 to M310 and M212 indicate the location of the diaphragm wall panel. It is also to be noted that for the section between M212 and M309, the estuarine deposit E and the upper part of the OA layer might have been displaced laterally being involved in the slide. At the same time, settlements of the order of 5–10 m are envisioned to have occurred in the upper part of the OA layer, as inferred from the settlements of the upperlying layer F_2, as shown in Figures 13 and 15, which was confirmed by the post collapse boring data. From the records of the diaphragm wall construction, the bottom line of the wall is known and it is shown together in Figure 17.

By comparing the elevations at the top of the OA layer with those of the bottom of the diaphragm wall, it can be seen from Figure 17 that in the eastern zone between WS1 and WS4A, the toe-in of the wall was more than 3 m. In the zone of collapse between WS5 and WS16, the toe of the diaphragm wall seems not to be embedded sufficiently deep. However, if allowance is made for the settlement of the OA layer of the order of 5~10 m, the depth of embedment into the OA layer may be considered reasonably deep almost in consistence with that considered in the design.

It is to be pointed out, however, that the quality of the soils at the upper part of the OA layer was less competent in stiffness or strength as compared to that assumed at the time of the design. This point

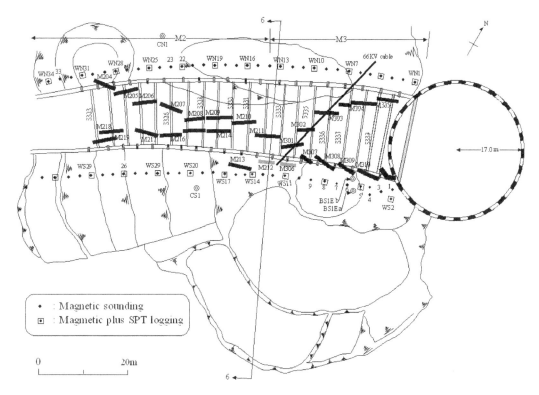

Figure 12.   Locations of the magnetic sounding and boring conducted after the collapse.

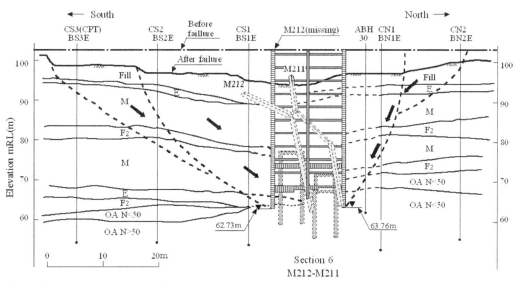

Note: All shapes of diaphragm walls underground
are unverified assumption

Figure 13.   Features of the ground movements towards the excavated space accompanies by the subsidence.

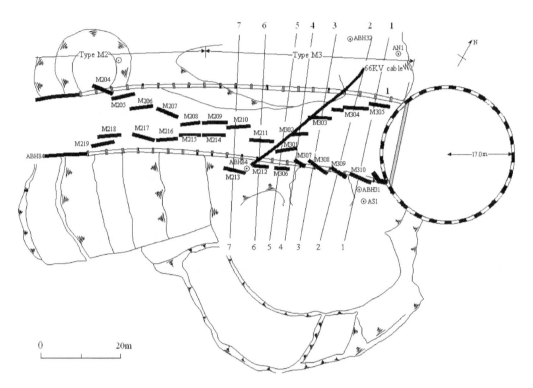

Figure 14.   Cross sections for which the damage features were depicted.

would have a more important bearing from the forensic aspect.

2.11   *Likely scenario leading to the collapse*

In the light of all pieces of available information as described above, the scenario leading to the collapse may be envisioned as follows.

(1) When the accident occurred at 3:33 p.m. on April 20, 2004, the front of westward excavation was located at the place of strut S332 as shown in Figure 18. At that time, the 1.5 m thick jet-grouted stiff soil (sacrificial strut) was being removed. The underlying soil layers of the OA having a SPT N-value less than about 30 at an elevation 61–63 m must have pushed the wall toe towards excavation. The second Jet-grouted stiff layer (JGP) at an elevation of 69.6 to 66.6 m was unable to resist against that earth pressure and thus the diaphragm wall panels M212 and M306 were forced to deform inwards. It is considered likely that the second JGP layer might have been possessed of horizontal stiffness lower than that expected before or it might have had brittle characteristics. It is to be noticed here that, because of the presence of the 66 KV electrical cable buried at shallow depth through the two panels denoted as GAP in Figure 18, it was not possible to install the bored piles from the ground surface

in advance underneath the elevation of 71 m. This is clearly indicated in the plan of Figure 5. Because of the absence of the bored piles, it was considered easier for the toe of the M212 to M306 to move towards the excavation.

(2) The inward movement of the wall panel toe must have caused slight uplift of the soil deposits within the excavation, which induced in turn a lift-up of the kingpost in the middle of the struts. In addition, excavation of the upper sacrificial jet-grouted layer is envisioned to have suddenly reduced friction between the jet-grouted lower layer and the diaphragm wall, and may have accelerated the soil heave and the uplift of kingposts. At the same time, sag is considered to have occurred in the diaphragm wall.

(3) This uplift, combined with the sag of the wall must have caused the local buckling or breakage of the flange of the H-beams at connection between the walers and struts, creating as much as 20 cm vertical offset as illustrated in Figure 19. As a result, the net length of the struts was shortened and diaphragm wall must have deformed further inwards.

(4) There were inclinometers installed behind the walls on the north and south sides. The lateral displacement measured on the south increased to 45 cm at the level of 10th strut on April 20. This appears to have been induced by the earth pressure

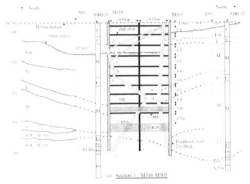

Note: All shapes of diaphragm walls underground are unverified assumption

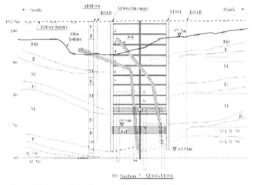

Note: All shapes of diaphragm walls underground are unverified assumption

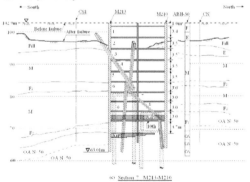

Note: All shapes of diaphragm walls underground are unverified assumption

Figure 15. Cross sections showing the mode of collapse in the zone of failure.

from behind which was mobilized by the inward movement of the wall.

(5) With the loss of functioning of the waler-strut system and inward movement of the wall panel toe, the integrity of the structural support as a whole was completely lost, leading eventually to the total collapse.

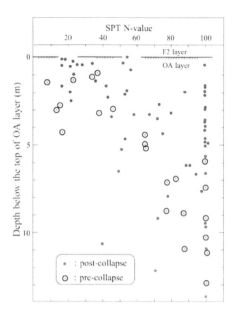

Figure 16. SPT N-values from the top of the OA layer.

(6) As a consequence of the collapse to the structural system, the diaphragm wall panels were twisted, displaced and rotated, accompanied by the slides of the surrounding ground as described in Figure 13.

The scenario of the collapse as described above is based on the point of view that the generic cause of the failure was the excessive inward movement of the toe of the diaphragm. Another scenario could be established alternatively on the basis of the assumption that the earth pressure acting on the diaphragm wall had been underestimated at the time of the design. A greater earth pressure than that considered in the design must have induced the large deflection of the diaphragm wall leading to the breakage of the water-strut system. This collapse scenario is addressed and thoroughly in the papers by Tang (2006), Yong et al. (2006) and Yong-Lee (2007). The authors, however, prefer to take the views descried in this paper.

2.12 *Lessons learned from the forensic diagnosis*

One of the most important lessons learned from the incident in Singapore seems to be the fact that, although the spatial distance of borings or sounding is reasonably short in view of the commonly adopted practice, there always are chances to miss irregularity in the deposits such as buried valleys. Therefore, special precaution should be paid to the soil conditions particularly in deep-seated deposits.

In the area of complicated geological setting, existence of buried valleys should be suspected at a the

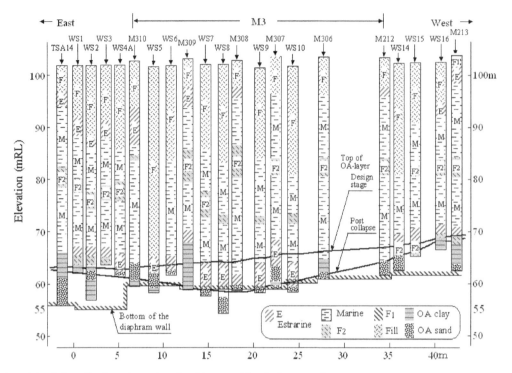

Figure 17. Soil profile along the south wall and bottom of the diaphragm wall.

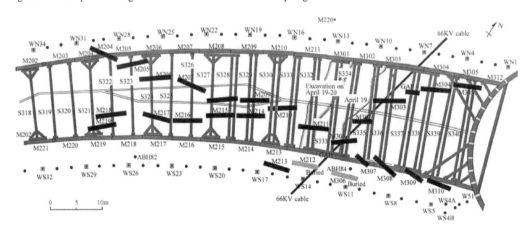

Figure 18. Detailed arrangements of the struts, diaphragm wall panel and locations of the magnetic logging.

stage when projects are put forward. In such a case soil investigations need to be carried out at spacing closer than that normally adopted. In addition, properties of soil materials within the buried valley should be investigated in details, if it is discovered. The outcome of these studies should be meticulously reflected on the grand planning as well as on detailed design of the soil-structure system considering each phase of construction processes.

3 COLLAPSE OF SUBWAY EXCAVATION IN KAOHSIUNG, TAIWAN

3.1 Outline of the incidence

Subway construction had been underway in east-west direction in the central part of Kaohsiung. Upon finishing the tunnel construction by the method of earth-balanced shield, the corridor connecting the two tunnels (up-line and down-line) was constructed by

44

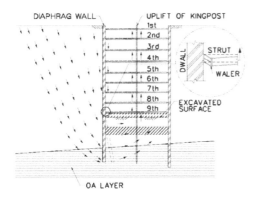

Figure 19.   Probable movements of soils near the toe of the diaphragm wall, and uplifting of the kingpost and sag of the wall leading to buckling and offset between strut and layer.

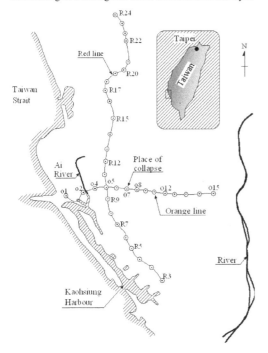

Figure 20.   The location of the collapse.

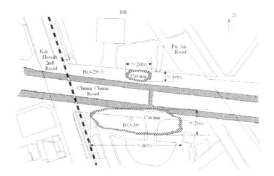

Figure 21.   Features of the cave-in in plan view.

Figure 22.   A bird-eye views over the cave-in in Kaohsiung.

undermining the already built-up tunnels, accompanied by an inflow of soil and water from the rips opened at the junction on the ceiling between the corridor and tunnel. This breakage culminated eventually in a large scale collapse of the tunnel structure involving formation of two cave-ins on the ground surface.

3.2   Description of the site conditions

The location of the collapse is shown in Figure 20 and the feature of the cave-in on the ground surface is displayed in Figure 21. A bird-eye view over the cave-in is displayed in Figure 22. The plan and side view of the tunnel are shown in Figure 23. It is to be noted that there was an underground roadway called Chung Cheng underpath just above the subway tunnel. The dual tunnels were constructed by the tunnel boring machine (TBM) which can advance by rotating a large steel disk equipped with cutting blade, while the cutting face is balanced by the mud pressure.

The soil profiles at the locations BO29 and BO30 shown in Figure 21 are indicated in Figure 24, where it can be seen that the deposits are comprised predominantly of silty sand with occasional layers of low-plasticity clay (CL) to a depth of 40 m. The blow

means of what is called the NATM method involving open excavation with the help of steel-framed support and injection. Then, a vertical shaft 3.3 m in diameter was excavated to provide a sump for water collection in the middle of the corridor in open dry conditions with the support of the H-shaped circular beams. When the shaft excavation reached a level 4.95 m from the floor of the corridor, a chunk of wet soil tumbled out from the southern wall of the shaft at the bottom.

The small collapse was followed by steadily increased outflow of mud water. The amount of water increased minute after minute. This breakage led to

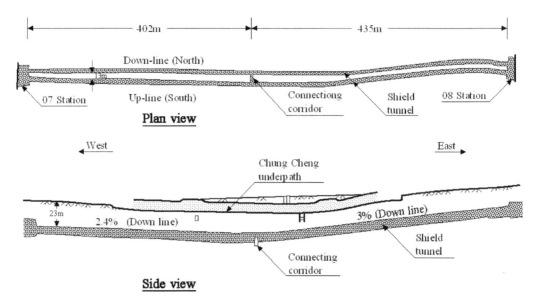

Figure 23.   Plan and side view of the tunnel.

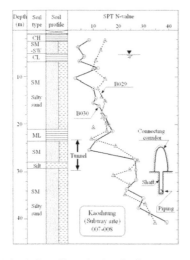

Figure 24.   Soil profile at the site of collapse.

count, N-value, in the standard penetration test is shown to increase with depth and to have a value of 20 to 30 at the depth where the sump for water collection was excavated.

3.3   Soil stabilization

The section of the ground where the corridor was to be installed had been stabilized by means of the jet grouting, where milky cement mortar is jetted out with compressed air horizontally into the ground from two nozzles attached 180 degree apart to the steel cylinder. The cylinder is lowered into a bored hole to a targeted depth and lifted stepwise while jetting the cement mortar with a pressure of 30 Mpa. During 2 to 4 rotations per minute, the cylinder was lifted 8.3 cm, thereby creating a solidified vertical column of soil-cement mixture with a diameter of 3.5 m.

The arrangement of the stabilized soil columns in plan view at an elevation of 85.83 m (zero point is taken at a depth 100 m below the median sea level) or at the depth of 31 m is shown in Figure 25. It may be seen that each column was installed so as to have mutual overlapping of 60 cm and to produce a huge stabilized massive zone 8.73 m by 25.487 m in plan which was considered strong and competent enough to permit excavation of the connecting corridor and the sump to be carried out in dry open conditions without any distress.

3.4   Piping failure of the tunnel

On December 4, 2005, the piping occurred at 3:30 p.m. at the very last stage of excavation of the sump. A block of silty sand was detached from the south wall at the bottom of the excavation which is located at a depth of 35 m from the ground surface. The circumstance at this time at the bottom of the sump excavation is illustrated in details in Figure 26.

A sequence of the events after triggering the piping which are likely to have occurred is illustrated in the cross sections in Figure 27. The blackish mud water continued to come out in the sump with increasing volume. Two men at work at the bottom strove in vein to clog the hole by dumping sand bags from the corridor. About one hour later, the sand bags were seen moving around in the sump as illustrated in Figure 27(b).

46

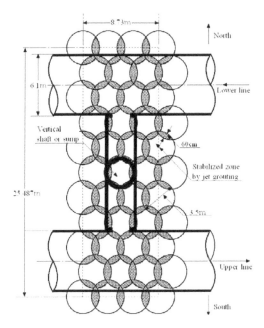

Figure 25. Plan view at the top of the stabilized zone (El. 85.83 m).

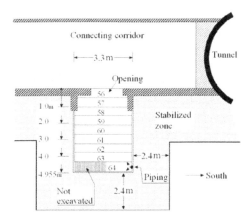

Figure 26. Details of the excavation for the sump and initiation of piping.

At this stage, the workmen closed the entrance to the sump by placing a steel lid and several bars for its support as illustrated in Figure 27(c).

It is believed that the mud water flowed into the sump from the bottom through a single hole probably with a diameter of about 30 cm. This assumption would hold true in the light of the fact that the zone stabilized by the jet grouting could not so easily broken down in the horizontal direction, and a weak zone might have existed in the form of vertical pipe.

About two hours later, engineers at spot heard squeaking sounds of breaks at segment joints of the tunnel on the south, accompanied by mud water cascading from the ripped joints of segments at the ceiling. It appears likely that because of the underscoring of the bottom portion, the stabilized zone subsided together with the tunnel body, resulting in the stepwise settlement in the longitudinal direction. The probable feature of the breakage at this stage is illustrated in Figure 27(c).

Around 10:20 p.m., the fall-off of the soil above the ceiling had spread upwards and, assisted by the slipping along the vertical wall of the motorway structure, soil mass fell down into the tunnel. This resulted in creation of a large cave-in on the ground surface on the south side. At this time, two mains 60 m and 30 m in diameter for water supply were broken releasing a large amount of water for a prolonged time. This breakage of buried pipes appears to have transferred the collapsed soil deposits into more flowable debris.

The feature of the collapse propagation as above would be understood more vividly by visualizing the sequence of events that might have occurred in the longitudinal direction. These features are described in Figure 28(a) through 28(d). Figure 28(c) and (b) shows the progress of the collapse from 3:30 p.m. to 5:30 p.m. on December 4, 2005. Figure 28(c) indicates the breakage of the tunnel, segment by segment, resulting in the fall-off the debris from the offsets of the segments in the ceiling on the east side. On the north side undermining of the tunnel body did occur as well as shown in Figure 27(d), resulting in the sinking of the tunnel together with the large block of the stabilized zone. This led to the breakage of the segment joints on the north side and allows the mud water cascading mainly through the opening at the junction between the portal and the adjacent segment. The length of the tunnel in which the debris packed in the entire cross section was as long as 130 m on the west side and 80 m on the east side in the up-line on the south, as shown in Figure 28(d). Thus, another cave-in occurred on the north side, as seen in Figure 21. The amount of soils and concrete dumped to fill the caves was 12,000 m^3 and the volume of water inflow from the 60 cm diameter water main lasting about 18 hours until 11:50 p.m. next day was approximately 2,000 m^3. Thus the total amount of debris was as much as 14,000 m^3.

3.5 *Investigations into causes of failure*

After the collapse, the Committee of Investigation was organized by the Kaohsiung Mass Transit Authority and its work commissioned to National Taiwan Construction Institute in Taipei. Even after comprehensive studies and discussions, it appeared difficult to single

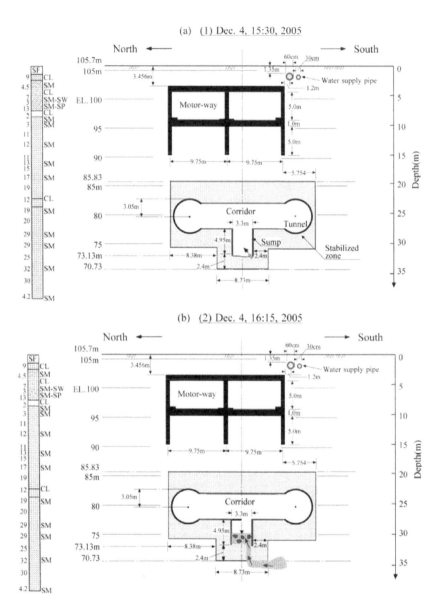

Figure 27. A sequence of likely scenario for progressing the piping failure.

out definitive causes with proven evidences. However there were several points agreed upon by the Committee which could be summed up as follows.

(1) It was obvious from the testimony of the two men at work that the phenomenon of piping was responsible for triggering the collapse.

(2) The reasons why the piping was initiated were conceived variously. One of them was existence of seepage-prone weak zones resulting from imperfect overlapping in the arrangement of the soil-cement

columns created by the jet-grouting method. In fact, because of the obstructions in the ground, it was difficult to install some of the jet-grouted columns exactly in plumb and the overlapping of adjacent columns at the elevation of the piping is estimated less than 60 cm.

(3) The hydraulic pressure at the depth of 34 m where the piping was initiated was estimated as being about 300 kPa and the length of conceivable shortest path for seepage was 2.4 m as counted from the bottom level of the stabilized

48

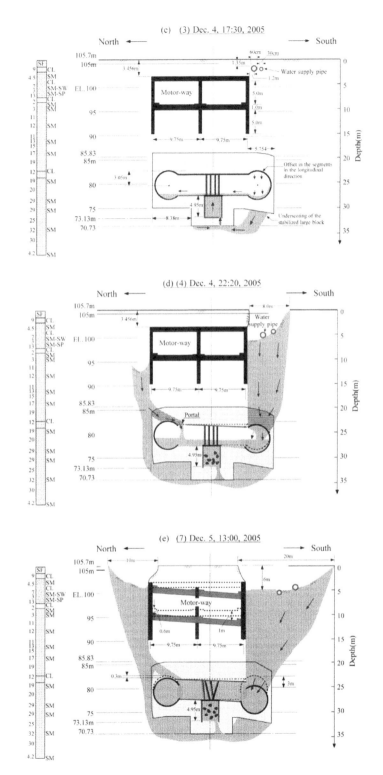

Figure 27. (*Continued*).

49

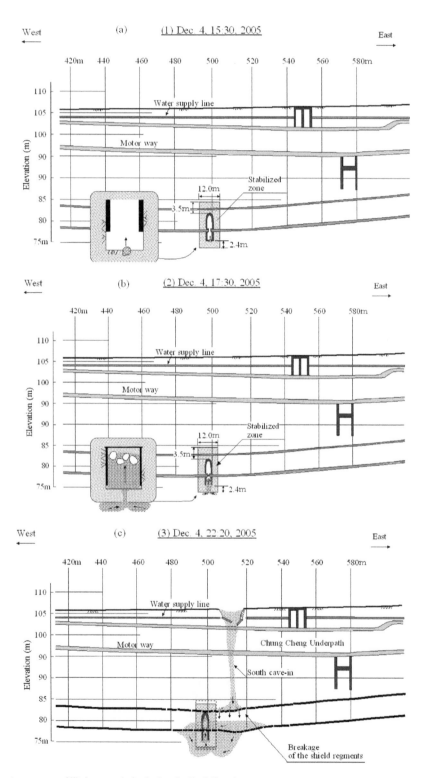

Figure 28.   A sequence of likely scenario in the longitudinal direction.

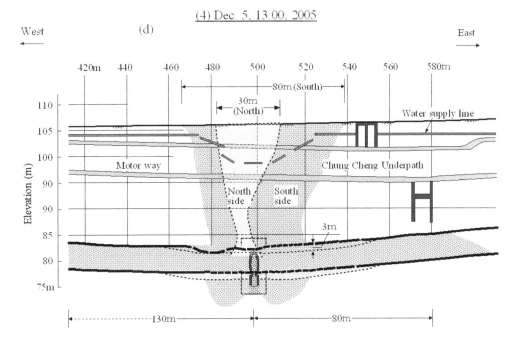

Figure 28. (*Continued*).

zone. Thus, the hydraulic gradient at the time of the piping is estimated as having been of the order of $i = 30\,m/2.4\,m = 12.5$ which was fairly high.

(4) Thus, it is expected that the critical hydraulic gradient for unstable seepage was about 12.5, although the reason for this to occur remained unidentified. It was also suspected rather strongly that the silty sand in the area of Kaohsiung might be possessed of a characteristic which is vastly prone to internal erosion as compared to soil deposits existing in other parts of the world. Thus, once the piping develops, the deposit is least self-healing and tends to become easily unstable.

3.6 *Proneness of local soils to internal erosion*

There have been several small collapses reported in other sections of subway construction in Kaohsiung. In fact, it was a concern among geotechnical engineers working there that the fines in sand deposits in Kaohsiung is highly non-plastic and easy to flow. Thus, one of the key issues for forensic diagnosis for piping of the local soil was considered to focus on the investigation into the extremely flowable nature of silts in Kaohsiung area which has not ever been addressed elsewhere.

As is well-known, the susceptibility of a given silty sand to internal erosion can be understood from two points of views, that is, generic reason and durability or resistively.

(1) If an aggregate of a soil is comprised of two major groups each having significantly different particle sizes, the grains with smaller size can move easily through pores of the matrix formed by larger particles. Thus, if such a soil with a gap grading is subjected to seepage flow, the smaller particles can be easily detached and washed away. This is the generic concept underlying the criterion for the design of filters in rockfill dams. However, it is normally applied to the type of soils in a range of particle sizes which are larger than those of silt sands encountered in Kaohsiung area.

(2) The condition as to whether or not a given soil actually suffers the piping collapse is expressed in terms of the critical hydraulic gradient. In fact, the experiments by Skempton and Brogan (1994) showed that the critical hydraulic gradient, ic, for the occurrence of piping in gap-graded sand-gravel mixture could be as low as 0.2–0.3 as against $i = 0.9$–1.0 for clean sands. This would hold true for soils with the range of particle size coarser than the silty sand in Kaohsiung. To the author's knowledge, there appears to be no study ever performed to clarify the vulnerability to piping or internal erosion for silty sands or sandy silts subjected to the seepage flow. Thus, this issue was taken up as a new problem area deserving further scrutiny

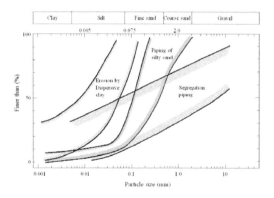

Figure 29.   Three groups of soils in terms of grain size range associated with erosion and seepage instability.

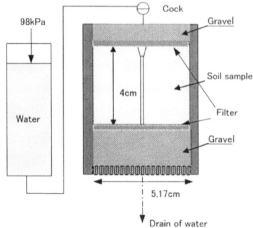

Figure 30.   Layout the pinhole test.

to identify causes of the collapse in deep-seated excavation.

3.7   *Laboratory tests on internal erosion*

With respect to the piping or erosion, there would be three types of problems to be distinguished depending upon the range of grain size of soil materials in question. These are shown in a graphical form in Figure 29. The segregation piping has been the target of extensive studies in the past in association with the filter criteria for the design of rock fill dams. There are many studies reported in this context such as those by Terzaghi (1939), Kenny-Lau (1985, 1980), Skempton-Brogan (1994) and Sherad (1979).

Erosion in clay-silt materials has been investigated by Sherad (1979) in response to the problem emerging from occasional failure of low-height earth dams which occurred upon water filling in Australia and U.S. The grain sizes associated with erosion or gully formation has been known to lie in the range indicated in Figure 29. It is known that dispersive nature of clayey soils is a dominant factor for inducing gulley or erosion tunnels in the clay fills. Tests for chemical analysis and pinhole tests were suggested as means to identify the dispersive nature of the fine-grained soils.

The outcome of the grain-size analyses for the silty sand at Kaohsiung has shown that the soil there belongs neither to the broadly graded gravelly sands for the segregation piping nor to the dispersive clay related to the erosion or gulley formation. In fact, the grain size in Kaohsiung soil lies midway in the range of silt and sand, as shown in Figure 29, which is coarser than the dispersive clay and finer than the soils related with segregation piping.

Thus, the collapse of the silty sand in the subway construction in Kaohsiung was considered to have posed a new challenge and addressed a novel problem area. The solution for this is not yet settled, but performing some tests was regarded as a useful attempt to

shed some light on this problem. In this context, a set of pinhole tests and measurements of specific surface test were performed as described below.

(a) Pinhole Tests

In order to examine the vulnerability of the silty sands at Kaohsiung to the internal erosion or piping, what might be called "Pinhole test" was carried out. The layout of the test system is displayed in Figure 30. The sample 4 cm thick and 5.17 cm in diameter was sandwiched by highly pervious gravel layers at the top and bottom with filter meshes placed between the sample and gravel. A vertical hole 3 mm in diameter was drilled as shown in Figure 30. Water was then circulated through the sample. In this type of test, water is supposed to flow mainly through the pinhole. If the silty portion is erodible, the water coming out is expected to be muddy, but otherwise the water transparent.

The silty sand from Kaohsiung tested had a grain size distribution as shown in Figure 31. For comparison sake, another material from a site in Chiba, Japan was secured and used for the same testing. The grading of this Japanese soil was almost the same as indicated in Figure 31. It is to be noticed that both silty sands had an average grain size of about $D_{50} = 0.075$ mm and fines content was about 50%. The physical characteristics of these two materials are shown in Table 1. The two samples were compacted so as to have the same wet density of 1.902 g/cm^3. A sample from Kaohsiung as prepared for the pinhole test is displayed in the middle of Figure 32 and gravel layers to be placed on top and at the bottom are shown on both sides of Figure 32. The procedures for testing were as follows.

(1) Water was circulated first slowly with a low pressure through the sample to ensure saturation.

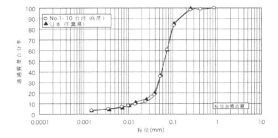

Figure 31. Grain size distribution curves of silty sands used for the pinhole tests.

Table 1. Grading of two materials used for the pinhole tests.

Material	Kaohsiung sand	Japanese sand from Chiba
Dry density ρ_d (g/cm³)	1.902	1.902
D_{60} (mm)	0.0651	0.0654
D_{50} (mm)	0.0746	0.0743
D_{10} (mm)	0.0168	0.0124
U_c	4.443	2.51
Specfic gravity Gs	2.733	2.708

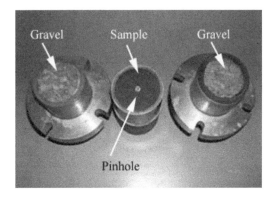

Figure 32. A samples from Kaohsiung for the pinhole test (the middle).

(2) Water was then circulated under a pressure of 98 kPa and colour of water coming out of the sample was observed. Pictures were taken every one-minute. Water percolation was continued for 5 minutes.

In the first series, the tests were conducted on disturbed samples, one from Kaohsiung and another from Japan. The result of the pinhole tests on a disturbed sample is presented in Figure 33. For the disturbed silt-containing sand sample from Kaohsiung, the pinhole is seen enlarging to a diameter of 5 mm from 3 mm as displayed in Figure 33(a). Correspondingly, the drained

water from the bottom was muddy for the first one-minute period of water percolation as seen in the water colour in the left-side cup of Figure 33(b). The result of the pinhole test on the Japanese silty sand is shown in Figure 34. In contrast to the Kaohsiung soil, the circulation of water through the Japanese soil did not exhibit any appreciable change in colour in the first one minute of percolation as seen in Figure 34(b). There was no change in the pinhole diameter for the Japanese soil before and after water permeation except for a local collapse at the top. Thus, it may be conclusively mentioned that the silty sand from Kaohsiung is potentially susceptible to internal erosion as compared to the soil from Chiba, Japan, although they had almost the same grading. This is believed to stem from the highly non-plastic nature of the Kaohsiung silt contained in the sand.

In the subsequent series, similar pinhole tests were conducted on undisturbed samples. The undisturbed sample from the depth of 14 m in Kaohsiung also demonstrated dirty blown colour in the first one-minute of percolation accompanied by the enlargement of the pinhole from 3 to 4 mm, as displayed in Figure 35.

Two fine sand samples from a deposit at depth 22 m in Kaohsiung were tested likewise with the result that it is also vulnerable to internal erosion.

From the series of the pinhole tests as described above, it may be concluded that, no matter whether the soil is disturbed or undisturbed, the silt ingredient contained in the Kaohsiung sand is not only non-plastic but also highly erodible being susceptible to piping as compared to other silts with the same gradation such as those in Japan.

(b) Specific surface test

To obtain another kind of index property for identifying the highly erodible nature of silty sand at Kaohsiung, what is called specific surface tests were conducted. In the Brunauer-Emmett-Teller (BET) method, the surface area is measured per unit weight of silt-size particle and expressed in terms of square meter divided by the weight in gram. If the specific surface is large, the particle is more of angular shape, and thus envisaged to become more difficult to move through the pores of the skeleton formed by coarser grains. In contrast, the particle of round shape with smaller value of the specific surface would be easier to move in the pores and therefore more erodible. Two tests were performed for the same batch of No. 1–10 silty sand from Kaohsiung having the grading shown in Figure 31. For comparison sake, the tests were conducted also for the Chiba sand in Japan of which the grading is shown in Figure 31. The results of the tests are shown in Table 2, where it may be seen that the specific surface area for Kaohsiung soil is about half

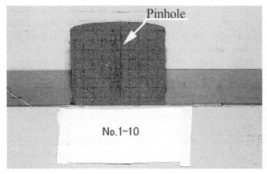

(a) Pinhole enlarged from 3mm to 5mm, visible erosion

(b) In the first one minute percolation, water was mud as seen in the left cup

Figure 33.　Silt-contained sample from Kaohsiung.

(a) Diameter of the pinhole remained unchanged except local collapse at the top

(b) In the first one minute percolation, water was seen almost transparent as observed in the left cup

Figure 34.　The results of the pinhole test on the Japanese silty sand.

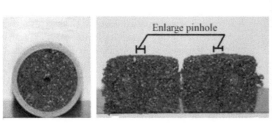

(a) Diameter of the pinhole enlarged from 3mm to 4mm

(b) In the first one minute percolation, water was seen becoming muddy in the left bowl

Figure 35.　Pinhole tests on undisturbed samples from Kaohsiung.

of the value for Chiba soil from Japan. In unison with the results of the pinhole tests as mentioned above, this observation of Kaohsiung soil shows that it is considered more susceptible to internal erosion due to seepage flow as compared to other soils.

3.8 Lessons learned from the incident in Kaohsiung

In the practice of subway construction in Kaohsiung, it has been known that the silty sand in that area is of peculiar nature being highly susceptible to erosion

Table 2. Results of specific surface tests (m²/g).

	Sample 1	Sample 2	Average
Kaohsiung silty sand, No. 1–10	3.2669	2.9540	3.1100
Silty sand from Chiba, Japan	6.2228	6.1469	6.1849
Silt from Shanghai	4.4584	4.1095	4.2840

due to seepage. Although the silt portion is known to be non-plastic, there has been no way further on to scrutinize the nature of the silt. In an effort to grope for some gauge, what is called pinhole test and specific surface test were conducted for the Kaohsiung silt and also for silt from Japan. The results of these tests have shown that the silt from Kaohsiung is more erodible and has a smaller value of specific surface area than the silt in Japan. This fact is indicative of the tendency of the Kaohsiung silt to be more liable to erosion as compared to other silts. The piping failure in the subway construction in Kaohsiung as descried above is considered as a consequence of such a peculiar nature of the local soil which had not explored until now. This incident is to be regarded as a typical example addressing an issue of new challenge in the area of geotechnical engineering.

4 INCIDENT IN SUBWAY CONSTRUCTION IN SHANGHAI

4.1 Outline of the incidence

A large-scale collapse took place on July 1, 2003 in the south part of Shanghai at a site where a new No.4 line of subway was under construction, leading to a large cave-in and subsidence of the ground. The location of the city near the mouth of Yangzu River is shown in Figure 36. As indicated in a more detailed map in Figure 37, the incident occurred at the location on the west bank of the Huangpu River. As shown in Figure 38(a), the curved sections across the river had been constructed by means of the mud-type shield tunneling machine. The river channel is 340 m wide and 17 m deep at this location as indicated in Figure 38(b). As a result of the cave-in, the city area on the west bank, about 70 m wide and 150 m long, suffered the ground subsidence of the order of 1–2 m, as indicated in Figure 38. The settlement extended into the riverbed, thereby involving the sinking of the river dike accompanied by inundation of the river water. Figure 39(a) shows sinking of two buildings in the forward and tilting of 7–10 storey buildings in the back ground. A high-rising building 30 storey high is also seen tilting in the back ground of Figure 39(b). Figure 39(c) shows the flooding in the city by inundation of water from the river.

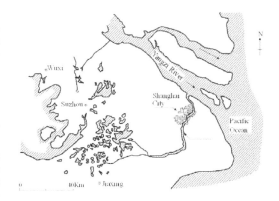

Figure 36. Location of Shanghai.

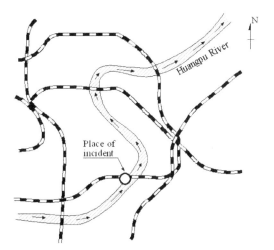

Figure 37. Subway lines in Shanghai and location of the incident.

4.2 Features of the incident

Somewhat detailed cross section along the tunnel line is displayed in Figure 40 where it can be seen that the vertical shaft for ventilation purpose had been installed above the corridor. There were two lines of diaphragm walls near the ventilation shaft which had been installed about 50 m away to protect the shaft from probable movement of the surrounding ground. There was the connecting corridor and water collecting sump at the depth of about 32 m. It is reported, although unofficially, that the excavation was going on for the horizontal corridor through the artificially frozen soil deposits. Because of malfunctioning of machines or cutoff of electric power supply, it is said that the frozen soils lost its strength and breakage took place near the junction between the tunnel and the corridor. As a consequence, a huge amount of water-saturated silty soils flowed down into the tunnel. The collapse propagated

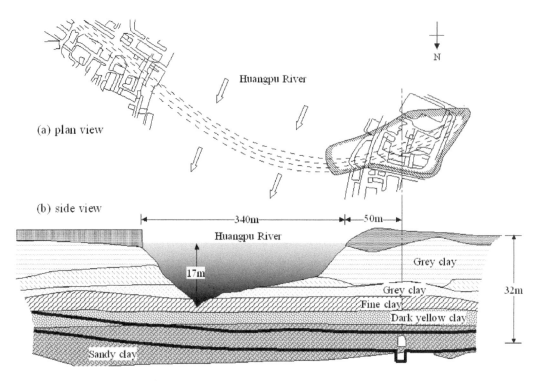

Figure 38.   Plan and side views of the tunnel across the Huangpu River.

up to the ground surface, accompanied by a large cave-in. The area of the subsidence in the city part near the west bank of the Huangpu River is shown in Figure 41. As a result of the ground subsidence, many buildings and infra-facilities for shopping and offices suffered titling, differential settlements. A high-rise building as tall as 30 storeys is said to have experienced a slight tilting.

4.3   Site conditions

The city of Shanghai is located in the deltaic flood plain near the mouth of the Yangzi River. The whole area is covered by a thick fluvial deposit of alluvial or Holocene origin as deep as about 40 m predominantly comprised of silt, clay and occasional sand. This alluvium is underlaid by the Pleistocene or diluvial deposit to a depth of about 70 m. The whole soft soil deposit consists of multi-layers of dark grey silty clay and yellowish fine sand. The plasticity index of the silty clay is known to be about 15 and its cohesion is 15 to 25 kPa.

4.4   Occurrence of collapse

It is purported that the construction of the dual tunnels and the installation of the ventilation shaft had been finished and the excavation was underway at the depth of 30 m to construct the corridor connecting the up- and down-lines of the tunnels. The silty sand deposits around the corridor tunnel had been frozen, in advance, by circulating coolant by a pipe system to permit the cutting to be made in open and dry conditions. When the excavation proceeded to the place of junction between the tunnel and the corridor, it is reported that malfunctioning or electricity cut-off occurred, resulting in thawing of the frozen soils. It seems likely that the saturated soils fell down and a kind of piping might have taken place where a small hole became gradually enlarged, leading to the large-scale fall-off and flow-in of the saturated soils.

There are no data available for the authors concerning measured values of SPT or CPT indicative of stiffness of silty sands at the location of the incident. It is, however, generally reported that the sandy silts in Shanghai are of low plasticity stiffness with the SPT N-value of 5–15 at shallow depth.

For the reference sake, the specific surface test was conducted on disturbed samples of silt from Shanghai, together with other soils as described in the foregoing section 3.7. The results of the test are listed in Table 2 where it can be seen that the specific surface in the BET-test takes a value of about 4.3 which lies between the two silts, that is, one from Japan and another from Kaohsiung. It may be mentioned that the silt from Shanghai could be somewhat more prone to internal erosion than the silt from Chiba, Japan.

(a) Sinking and tilting of buildings in the area of the ground subsidence

(b) Tilting of nearby building

(c) Flooding due to breach of the river dike

Figure 39.   Damage by ground settlement due to underground collapse in underground excavation, Shanghai.

Judging overall from the information as above, it may be mentioned that the silty sands or sandy silts prevailing over the Shanghai area are more or less susceptible to a piping type failure if they are subjected to the seepage field, and this was one of the factors accounting for in the enlarged scale of the catastrophe which occurred at the subway construction in Shanghai.

5   CONCLUSIVE REMARKS

The three examples of collapse as described above have some features in common, that is, (1) they occurred in the area of soft soil deposits of alluvial origin and (2) they are all associated with deep excavations at depths 30 to 35 m.

The case of the collapse in Singapore appears to come from the presence of an unforeseen buried valley which is generally difficult to detect in advance at the spacing interval of boring currently in use.

This incident gives us a lesson for needs for thorough investigations of site-specific soil conditions, if found necessary, in view of complexity of local geology and its history of formation.

The collapse in Kaohsiung was induced by unpredictable occurrence of piping in the sandy silt deposit subjected to a seepage field with a high hydraulic gradient. Although difficult to foresee from many past experiences in other sites, the local sandy silt in Kaohsiung was found to possess a specific characteristics in that it is more prone to the piping due to internal erosion than other silts with similar grading. Thus, the incident in Kaohsiung left a message that a local soil could possess specific properties which are difficult to identify from existing knowledge generally incorporated in the design and practice in deep excavations in deposits of low-lying area. It is necessary to investigate characteristics of local soils if they are found to exhibit peculiar behavior judging from the common sense in soil mechanics.

There is no exact information available on soil characteristics in Shanghai, but it seems likely that the low-plasticity nature of the local soil was prone to failure such as piping at deep depths under high hydraulic

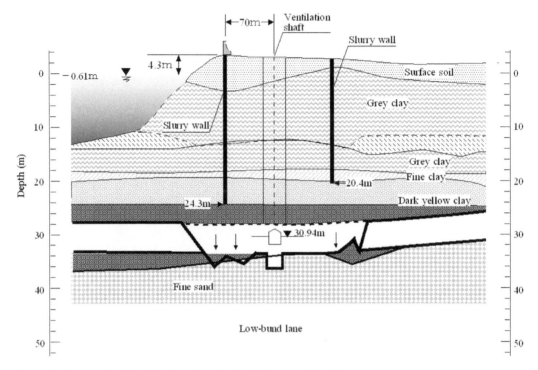

Figure 40.   Cross section along the subway line showing the feature of the collapse.

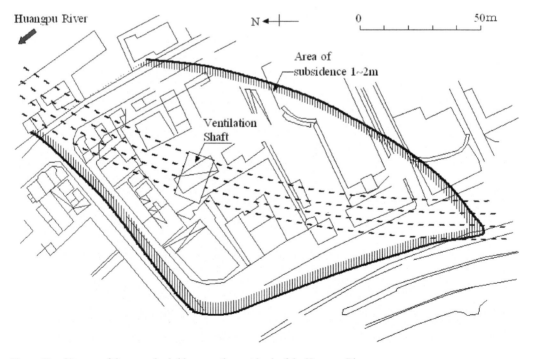

Figure 41.   City area of the ground subsidence on the west bank of the Huangpu River.

pressure and flowable nature of the local soil might have perhaps been responsible particularly for enlarging the scale of the incident. Once again, it must be emphasized that nature of local soils be investigated in details and its outcome be incorporated meticulously in the design and operation of construction works.

ACKNOWLEDGEMENTS

In providing the draft of this paper, cooperation was obtained from many officers, engineers and researchers who have been earnestly involved in forensic investigations after the incidents. The precise information regarding the features of the collapse in Kaohsiung was offered by the courtesy of the Kaohsiung Rapid Transit Construction (KRTC). The permission to include the results of the investigations was also given by KRTC. With respect to the information on the failure in Shanghai, a set of diagrams and pictures was given by Professor Feng Zhang of Nagoya Institute of Technology. Professor E.L. Peng of Tongji University, Geotechnical Engineering Depart, provided additional data and figures. Most useful discussions and comments were given by these two professors.

The authors wish to extend the most sincere thanks to the organizations and professionals as above for their overall cooperation and kind advice in preparing this paper.

The authors also wish to express a mind of gratitude and appreciation for the Organizing Committee of the Conference for their choosing the subject of forensics-oriented investigation as one of the new initiatives of focus in the realm of site characterization.

REFERENCES

Chen, C.H. and Lee, W. (2007), "Forensic Investigation of a Disastrous Failure at the Arrival End of a Station of the Kaohsiung Mass Rapid Transit System", Proceeding of the International Conference on Forensic Engineering, Mumbai, India

Day, R.W. (1998), Forensic Geotechnical and Foundation Engineering, McGraw Hill, pp. 1–10.

Geotechnical Consulting Group (GCG) (2005), "Post COI Report on Future Investigations into the Collapse of the Nicoll Highway".

Hwang, N.H., Ishihara, K. and Lee, W. (2007), "Forensic studies for failure at an underground station", Proceedings of the Thirteenth Asian Regional Conference on Soil Mechanics and Geotechnical Engineering, Calcutta, India.

Kenny, T.C. and Lau, D. (1986), "Internal Stability of Granular Filters," Canadian Geotechnical Journal, Vol. 22, No. 2, pp. 215–225.

Sherard, J.L., Dunnigan, L.P., Decker, R.S. and Steele, E.F. (1576), "Pinhole Test for Identifying Dispersive Soils," Journal of Geotechnical Engineering. GT1, ASCE, pp. 69–86.

Sherard, J.L. (1979), "Sinkholes in Dams of Coarse, Broadly Graded Soils," Transaction of the 13th International Congress on Large Dams, New Delhi, Vol. 2, pp. 25–34.

Skempton, A.W. and Brogan, J.M. (1994), "Experiments on Piping in Sandy Gravels," Geotechnique, Vol. 44, No. 3, pp. 449–460.

Tan, H. (2006), Lessons Learnt from Nicoll Highway Collapse, Haiti Conference Indonesia Geotechnical Society, Jakarta.

TCRI (2006). "Applications of Resistively Image Profiling Technique for Evaluating Integrity of Diaphragm Walls and Effectiveness of Ground Treatment", Taiwan Construction Research Institute. (in Chinese)

Terzaghi, K. (1939), Soil Mechanics; A New Chapter in Engineering Science," Journal of Institution of Civil Engineering, 12, pp. 106–141.

Yong, K.Y., Teh, H.S. and Wong, K.S. (2006), "System Failure of Temporary Earth Retaining Structure Leading to Collapse of Nicoll Highway," International Conference on Deep Excavations", Singapore.

Yong, K.Y. and S.L. Lee (2007), Collapse of Nicoll Highway-A Global Failure at the Curved Section of Cut-and-Cover Tunnel Construction," Proc. Southeast Asian Geotechnical Conference, Kuala Lumpur, Malaysia, pp. 23–36.

Crosshole electric tomography imaging in anisotropic environments

Jung-Ho Kim
Korean Institute of Geoscience and Mineral Resources, Daejeon, South Korea

ABSTRACT: Even though the anisotropy inverse problem is much more difficult to be solved than the isotropy one, we can extract more information on the subsurface distribution of material properties in principle since several times more is the number of components defining a material property in anisotropic medium. This paper discusses the crosshole tomography of radar and electric resistivity in anisotropic environment. In order to understand anisotropy phenomena of two different methods, the anisotropy effects observed in traveltime and apparent resistivity data are firstly discussed. The inversion algorithms of two anisotropy imaging techniques are introduced based on the similar objective functions consisting of squared error, anisotropy penalty functional, etc. After discussing these basic topics of the anisotropy effects and inversion algorithms, the real application case histories of crosshole radar are introduced, which were performed in the different geological conditions and with the different purposes of investigation. Finally, a case history of crosshole resistivity tomography is given, performed for the ground safety analysis of a high-storied building constructed over an abandoned mine. Through these discussions, it is demonstrated how the anisotropic characteristics themselves are the important information for understanding the internal status of basement rock.

1 INTRODUCTION

Ohm's law is the most important equation in electric prospecting methods, relating current density to electric field intensity with a constant, electric conductivity. Since both the current density and the electric field intensity are vectors, the electric conductivity must be a tensor in general; the material in nature is anisotropic (Stratton 1941). So is the electric permittivity relating electric field intensity and electric displacement. Even when the constituent minerals are perfectly isotropic, a rock or soil may be anisotropic due to so-called macro-anisotropy (Parasnis, 1979). Thus, the materials encountered are mostly anisotropic on view of geophysical measurements, although the degree of anisotropy (Keller & Frischknecht 1966) varies significantly, material by material.

If the subsurface material in a surveyed area reveals anisotropy and if we do not properly incorporate the anisotropy in the interpretation stage, the resultant subsurface image will be greatly distorted. On the other hand, if a proper approach for considering the anisotropy is applied to processing and interpreting the geophysical data, we can get the subsurface image accurately reflecting the underground structure. Furthermore, we can extract more information than when the subsurface material is isotropic: subsurface distribution of anisotropy characteristics. Suppose that the electric anisotropy of rock is caused by some preferential orientation of constituent materials. When anisotropic rock is weathered, altered, or fractured, the preferential orientation will be diminished, so that the degree of anisotropy will become less. Accordingly, we have another clue to interpret whether basement rock is weathered, altered, or fractured. In the case of a cavity, the filling material, such as air, clay, water, etc., is perfectly isotropic. In the tomographic image of material property, such as resistivity or velocity, cavities may be imaged as anomalies of lower or higher value of material property, depending on the contrast between the basement rock and the filling material of cavity. On the other hand, in the image of the anisotropy ratio or coefficient, cavities will always be imaged as anomalous zones of very low anisotropy degree. Accordingly, we will have another clue to locate cavities in anisotropic environment.

Solving the anisotropy problem is inherently more difficult than solving the isotropy mainly because the number of variables in an anisotropy problem is several times more than that in an isotropy problem; the number of the independent components of a conductivity or permittivity tensor is six, and becomes three even considering only its principal components. Consequently, the anisotropy inverse problem needs much huger amounts of computation space and time and is more non-unique and unstable than the isotropy one. This implies that a stable and elegant algorithm of anisotropy inversion is prerequisite in order to get additional useful information from the anisotropy of subsurface material.

In this paper, I discuss the geophysical imaging techniques using electrical properties, such as conductivity and permittivity, in anisotropic environments, particularly focusing on the crosshole tomography methods of radar and resistivity. Firstly, the anisotropic phenomena which can be observed in the field measurements are discussed, since the examination of anisotropic effects will be the first step of data interpretation. After that, the anisotropy inversion algorithms of the two tomography methods are introduced, which are the essential tools to process the anisotropy data. As for the actual applications of the anisotropy radar tomography, I introduce three case histories which were performed in the different geological conditions and with the different purposes of investigation. Finally, a case history of crosshole dc resistivity tomography (ERT) is given, which was performed for the ground safety analysis of a high-storied building constructed over an abandoned mine. Through these case histories, it is demonstrated how the anisotropiy characteristics themselves are the important information for understanding the internal status of basement rock.

Since the midst of the 1990's, Geoelectric Imaging Laboratory of Korea Institute of Geoscience and Mineral Resources (KIGAM) has continuously developed and updated the anisotropy inversion algorithms of crosshole radar and ERT data. The main materials covered in this paper are owed to their study results and publications.

2 ANISOTROPIC EFFECTS OBESEVED IN FIELD DATA

The first step of interpreting the field data acquired in anisotropic environment will be to examine the anisotropic effects observed in the field data. If the data were measured along a profile line on surface, it would be very difficult to recognize anisotropic effects in the data. On the other hands, the anisotropic effects in crosshole tomography data are more easily recognizable than in surface measurements, since the crosshole tomography method itself assumes wide variety of source-receiver configurations and measurement directions. Comparing the two crosshole imaging methods discussed in this paper, the anisotropic effects in the radar data are easier to be identified than those of dc resistivity, since the propagation of radar wave can be approximated by ray theory.

2.1 Crosshole radar tomography

Figure 1 shows the typical examples of travel time curves obtained in isotropic and anisotropic environments. One average velocity is enough to approximate

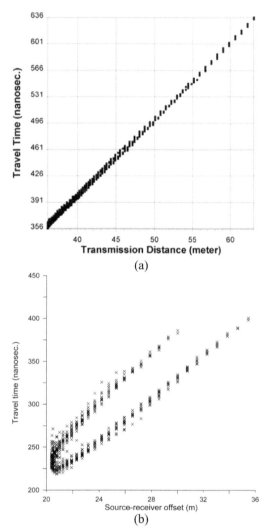

Figure 1. Examples of travel time curves acquired in isotropic (a) and anisotropic media (b).

the material properties of isotropic medium, while at least two values are needed for describing the characteristics of anisotropic medium as shown in Figure 1b. The strong dependence of velocity on the propagation angle is the main reason why the travel time data in Figure 1b are divided into two straight lines.

In Figure 2, which draws the same data of Figure 1b in different way, we can more clearly demonstrate the dependence of velocity on transmission angle. The horizontal and vertical axes represent ray angle from the source to the receiver, and residual velocity, respectively. A transmission angle of zero means that the ray propagates horizontally, i.e., the source and receiver are at the same level, while negative and positive values

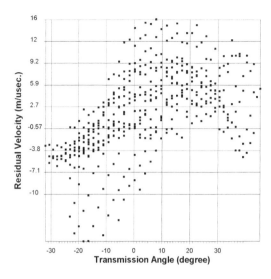

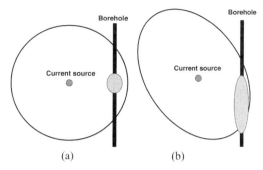

Figure 3. Schematic presentation of potential distribution in isotropic (a) and anisotropic (b) media. The gray colored zones along boreholes are the intervals where the negative apparent resistivities can be recorded when potential difference is measured along boreholes.

Figure 2. Plot of residual velocity versus ray angle for the radar tomography data shown in Figure 1b. This plot clearly shows velocity dependence on transmission angle of radar wave.

represent upward and downward propagations, respectively. Residual velocity is calculated by subtracting the average velocity on the tomography section from the apparent velocity obtained by dividing the distance from the source to the receiver by the first arrival time. As shown in Figure 2, the velocity of a downward propagating ray is higher than that of an upward ray: anisotropy.

2.2 Crosshole resistivity tomography

The response of radar tomography is approximated by ray theory and the anisotropy effect in travel time data is explained by transmission angle so that we can easily identify and understand the effect of anisotropy in field data. The anisotropy effect in electric resistivity data, however, is different from that of ray tomography, since the response due to a direct current source cannot be explained merely by the current path from the source to a measurement location.

In the homogeneous isotropic whole-space, the equipotential surface due to a current source must be a sphere (Figure 3a), thus the potential can be simply described by the distance from the source. On the other hand, since the equipotential surface in the anisotropic case shows a form of ellipsoid (Figure 3b), the potential cannot be described by the distance only. The potential value can increase even when the distance between the source and receiver increases. Therefore, even in the case of homogeneous earth, some parts of crosshole data can have negative values of apparent resistivity due to anisotropy if we adopt an electrode array measuring potential difference such as

pole-dipole array. In isotropic space, negative apparent resistivity in crosshole measurements is not uncommon if the subsurface medium is very inhomogeneous. However, we can easily understand that it is much more likely to encounter negative apparent resistivity in the anisotropic medium than in the isotopic one. Figure 4b shows that the negative apparent resistivity is widespread in the crosshole measurements, which implies that very strong anisotropy prevails in the surveyed area.

3 ALGORTHM OF ANISOTROPIC INVERSION

3.1 Traveltime radar tomography

The extensive studies on crosshole ray tomography inversion algorithm in anisotropic media have been performed in seismic exploration. Despite the relatively recent appearance of GPR (Ground Penetrating Radar) technique, anisotropy phenomena observed in the GPR data have been continuously reported in many studies (Tillard 1994, Lesmes et al. 2002, Seol et al. 2002, Kim et al. 2004a) and anisotropy tomography inversion algorithm has been studied by several authors. Vasco et al. (1997) presented an approach to solve the radar tomography inverse problem assuming weak anisotropy and successfully applied their method to image fractures in basaltic rock at the Idaho National Engineering Laboratory. Jung & Kim (1999) proposed a tomography inversion method assuming a simple elliptic anisotropy medium and used their algorithm to investigate the physical rock property changes due to the tunnel excavation activities. Kim et al. (2006a) extended the algorithm of Jung & Kim (1999) and showed several successful case histories of anisotropy radar tomography. A different approach was introduced by Senechal (2000) for reconstructing

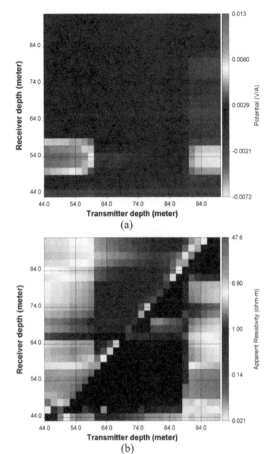

(a)

(b)

Figure 4. These cross-plots are the crosshole data measured between two boreholes by pole-dipole array. Potential difference (a), and apparent resistivity (b). Receiver depth means the depth of the plus potential electrode of potential dipole, while transmitter depth the one of the current source pole. Negative apparent resistivity is represented in black on the cross-plot of apparent resistivity (b).

a reasonable tomography image via correcting the artifacts due to medium anisotropy.

In this section, an inversion algorithm assuming a heterogeneous elliptic anisotropy (Kim et al. 2006a) is discussed since the anisotropy model is simplest and easy to be understood. Figure 5, an alternative display of the data shown in Figure 2, shows that the travel time data can be approximately represented by an ellipse, and the investigated basement rock can be described as a heterogeneous elliptic anisotropic medium in which the symmetry axis is not parallel to the surface.

On view of these phenomena appearing in the field data, inversion programs to process the data sets which show anisotropic features in the radar velocity data have been encoded and continuously updated

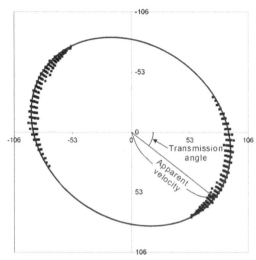

Figure 5. Polar presentation of apparent velocity distribution of travel time data of Figure 2, which can be approximated with an ellipse. The radial distance of each data point is proportional to the apparent velocity, and the polar angle corresponds to the ray angle.

by Geoelectric Imaging Laboratory of KIGAM. The algorithm is based on the fact that the subsurface medium can be approximated by a heterogeneous elliptic anisotropic medium, which consists of many elliptic anisotropic cells defined by the three material properties; maximum and minimum velocities, and symmetry axis (Figure 6). Thus, the final inverted results can be visualized by three kinds of tomograms: tomograms of maximum and minimum velocities, and the direction of maximum velocity. In theoretical and strict sense, the elliptic anisotropy model cannot be applied in general, but in practical sense, it is enough to be used in the interpretation of the field data showing anisotropic effects.

The smoothness constraint least-squares inversion (Sasaki 1989, Jung & Kim 1999) was the basic frame to fit the measure data to the data theoretically calculated. Additional features have been added in the inverse model regularization of the original algorithm. The sensitivity of maximum or minimum velocity may be greatly different from that of the direction of symmetry axis. Moreover, the resolution of each inversion parameter depends not only on the spatial coordinate but also on the way of data acquisition, i.e., the ray coverage. For example, an inversion parameter at the uppermost part of the tomography section will be poorly resolvable, while the one at the center of the tomography section will be highly resolvable. To incorporate these in inversion process, Active Constraint Balancing technique (Yi et al. 2003) was adopted, which uses spatially varying Lagrangian multipliers, calculated automatically using the resolving kernel of each

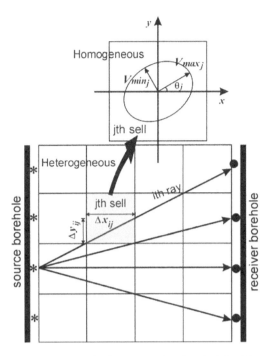

Figure 6. Schematic diagram showing the principle of the anisotropy tomography algorithm developed by Jung & Kim (1999).

inversion parameter. Another regularization to minimize the anisotropy ratio was also adopted to stabilize inversion. To reduce the possible static time shifts associated with source and receiver antenna positions, the source and receiver static terms were also included in the inversion parameters.

The above concept can be expressed as the following objective function to be minimized in an inversion process:

$$\Phi = \mathbf{e}^T \mathbf{e} + \Psi + \Gamma + \mathrm{H} , \qquad (1)$$

where $\mathbf{e}$ is the error vector between observed and synthetic data, and Ψ is the functional for the smoothness constraints adopting ACB technique. Γ and H are those for the anisotropy constraints and the problem of the source and receive static, respectively.

The functional for the smoothness constraints can be expressed as the amount of roughness. The Laplacian of the parameter perturbation vector was selected as the measure of the model roughness. Thus Ψ for the smoothness constraint in spatial domain is expressed as

$$\Psi = (\partial^2 \Delta \mathbf{v}_{max})^T \mathbf{\Lambda}_1 (\partial^2 \Delta \mathbf{v}_{max})$$
$$+ (\partial^2 \Delta \mathbf{v}_{min})^T \mathbf{\Lambda}_2 (\partial^2 \Delta \mathbf{v}_{min}) , \qquad (2)$$
$$+ (\partial^2 \Delta \boldsymbol{\theta})^T \mathbf{\Lambda}_3 (\partial^2 \Delta \boldsymbol{\theta})$$

where $\mathbf{v}_{max}$, $\mathbf{v}_{min}$, and $\boldsymbol{\theta}$ are the model vectors of maximum velocity, minimum velocity, and symmetry axis angle, respectively. The functional for the regularization to minimize the anisotropy can be expressed as

$$\Gamma = \lambda_4 (\Delta \mathbf{v}_{max} - \Delta \mathbf{v}_{min})^T (\Delta \mathbf{v}_{max} - \Delta \mathbf{v}_{min}) . \qquad (3)$$

Finally, the regularization for the minimization of source and receiver static is defined as

$$\mathrm{H} = \lambda_5 (\Delta \mathbf{t}_s)^T (\Delta \mathbf{t}_s) + \lambda_6 (\Delta \mathbf{t}_r)^T (\Delta \mathbf{t}_r) , \qquad (4)$$

where $\mathbf{t}_s$ and $\mathbf{t}_r$ are the source and receiver static. Note that the source and receiver static are included in the forward modeling scheme thus automatically calculated by the inversion process.

Six weighting coefficients λ_1 through λ_6 in the above three functions are the Lagrangian multipliers to balance various kinds of regularization terms. The first three of them (Eq. 2) are the coefficients to control the smoothness of material property distribution in space. Note that these three are expressed as diagonal matrices to introduce the Lagrangian multiplier as a spatially-dependent variable, i.e., $\mathbf{\Lambda} = diag(\lambda_i)$. The spatial distribution of Lagrangian multipliers can be automatically determined via the parameter resolution matrix and spread function analysis (Yi et al. 2003).

Since the angle of symmetry axis is perfectly irresolvable in the isotropic medium, the inversion process may be unstable if the isotropic medium is adopted as an initial model. In order to avoid this possibility, the program first calculates three anisotropy constants assuming homogeneous elliptic anisotropic medium before an actual inversion is attempted to obtain the three model vectors of heterogeneous medium. This homogeneous anisotropic model is used as the initial model of the actual inversion.

The crosshole travel time data showing very strong anisotropy effects were tested by the inversion code based on the algorithm described above. For the comparison, the data were also inverted by an isotropic inversion code, after the anisotropic effects were removed on the travel time data, based on the assumption of uniform anisotropy throughout the whole section. The results are illustrated in Figure 7.

As shown in Figure 7a, the results of the simple isotropic inversion show only the dependence of velocities on the ray angles, and cannot reflect the subsurface structure at all. An approach to correct the anisotropy effects can provide some improvement, but as in Figure 7b, we can recognize that the anisotropy effects still remain in the tomograms and distort parts of images. Figure 7c, the tomogram of maximum velocity from the anisotropic inversion, can be regarded as the best result, as inferred from the reasonable continuity of adjacent tomograms and the

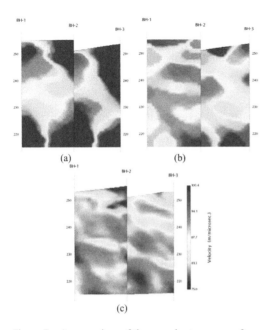

(a)   (b)

(c)

Figure 7. A comparison of the georadar tomograms from different tomography inversion approaches. (a) Inversion based on isotropic velocity parameterization. (b) Isotropic inversion after removing the anisotropic effects based on the assumption of a uniform anisotropy medium throughout the whole section. (c) Anisotropic inversion based on a heterogeneous elliptic anisotropy model. The vertical axis annotates the elevation along the vertical borehole axis in meter.

higher resolving power. In Figure 7c, anomalous zones of low velocity develop with gentle slopes, interpreted as fracture zones in which cavities are more likely to develop than elsewhere. Actually, BH-3 detected a cavity filled with clay at a depth of about 30 m. Here the bottommost anomalous zone of low velocity intersects the borehole BH-3 on the tomogram.

3.2 Crosshole resistivity tomography

The theory on the crosshole ERT has been developed since the late 1980s (Daily & Owen 1991, Sasaki 1992, Shima 1992). Since the bulk resistivity of subsurface material is greatly dependent on the porosity and the resistivity of pore fluid, ERT has been being successfully and actively used for solving environmental problems, particularly for subsurface monitoring (Daily & Ramirez 1995, Daily & Ramirez 2000, Slater et al. 2000, LeBrecque et al. 2004). At the same time, geotechnical engineering has also become one of the important application fields of the ERT method (Shima 1992, Mallol et al. 1999, Kim et al. 2004b). Extensive studies on this technique brought about the study on the ERT technology in more complex surroundings; crosshole

ERT in anisotropic environments. Although numerous studies have been performed for a long time on the electrical anisotropy of subsurface material, research results on the inversion algorithms of the crosshole anisotropic ERT have been published in recent years. Labrecque & Casale (2002) developed a three-dimensional inversion algorithm and showed that more reasonable subsurface images could be constructed through anisotropic inversion. Labrecque et al. (2004) used their anisotropic inversion algorithm to process monitoring data obtained by a specially designed autonomous system. Rigorous study on the anisotropic least-squares inversion algorithm can be found in Herwanger et al. (2004a), particularly on the effects of various kinds of regularizations. Herwanger et al. (2004b) compared and jointly interpreted the crosshole anisotropic images of seismic and electrical tomography surveys. Kim et al. (2006b) developed an anisotropic inversion algorithm to solve the problem of the very strong anisotropy owing to the very thin graphite layer developed along the foliation planes of the basement rock.

Generally, the conductivity in anisotropic medium is defined as a full symmetric tensor having six independent parameters.

$$\boldsymbol{\sigma} = \begin{bmatrix} \sigma_{xx} & \sigma_{xy} & \sigma_{xz} \\ \sigma_{yx} & \sigma_{yy} & \sigma_{yz} \\ \sigma_{zx} & \sigma_{zy} & \sigma_{zz} \end{bmatrix} \tag{5}$$

Using the conductivity tensor instead of a scalar one as in the isotropic case, we can obtain the response in the anisotropic medium. In the inverse modeling, the subsurface is divided into many blocks and elements of conductivity tensor for each block are parameterized as inverse variables to be calculated. Accordingly, the number of inverse variables is six times more than the conventional isotropic inversion scheme. Since crosshole tomography inversion requires huge amounts of computation space and time, it was assumed that the principal axes of the tensor coincide with the spatial Cartesian coordinate system and ignored the off-diagonal terms of the conductivity tensor. Actually, principal axes of the tensor can be obtained by the rotation of coordinates when they differ from Cartesian coordinate system. So, the inverse parameters could be reduced to three diagonal elements only, namely $\sigma_{xx} = \sigma_1$, $\sigma_{yy} = \sigma_2$, and $\sigma_{zz} = \sigma_3$. Thus the inverse problem becomes finding the model vector $\mathbf{P} = \{\sigma_1, \sigma_2, \sigma_3\}$, where σ_i is the model vector consisting of each diagonal element of conductivity tensor. If we set the number of inversion blocks as n, then the number of inversion parameters becomes $3n$.

We can imagine that the sensitivity of a component of a resistivity tensor may be significantly different from others. Moreover, it may vary depending not only on the spatial coordinate but also on the way of data acquisition. For example, an inversion parameter at the lowermost part of the tomography section will be poorly resolvable, while the one adjacent to an electrode location will be highly resolvable. In order to incorporate these into the inversion process as in the case of radar travel time tomography data inversion, Active Constraint Balancing (ACB) technique (Yi et al. 2003) was adopted. In addition to this, another regularization to minimize the anisotropy (Labrecque & Casale 2002, Herwanger et al. 2004a, Kim et al. 2006b) was also adopted to stabilize inversion process.

The above concept can be expressed as the following objective function to be minimized in an inversion process.

$$\Phi = \mathbf{e}^T \mathbf{e} + \Psi + \Gamma, \tag{6}$$

where $\mathbf{e}$ is the error vector between observed and synthetic data, and Ψ is the functional for the smoothness constraints while Γ for the anisotropy constraints. Recalling the objective function of radar traveltime inversion previously discussed (Eq. 1), the two objective functions of two different inverse problems are identical except that the inverse problem of traveltime tomography includes the term related with source and receiver static while that of resistivity not. Adopting the Laplacian of the parameter perturbation vector as the measure of the model roughness and also adopt ACB technique, the functional for the smoothness constraints of dc resistivity inversion is identical to Eq. 2 of the inverse problem of traveltime tomography when replacing the parameter vector of travel time tomography, $\{v_{max}, v_{min}, \theta\}$ with that of ERT, $\{\sigma_1, \sigma_2, \sigma_3\}$.

The last term of the objective function corresponds to the regularization to minimize the anisotropy degree and is defined by the following function

$$\Gamma = \beta(\mathbf{M}\Delta\mathbf{P})^T (\mathbf{M}\Delta\mathbf{P}), \tag{7}$$

where $\Delta\mathbf{P}$ is the parameter perturbation vector, i.e., $\Delta\mathbf{P} = \{\Delta\sigma_1, \Delta\sigma_2, \Delta\sigma_3\}$, and β is a Lagrangian multiplier for controlling the amount of anisotropy constraint. $\mathbf{M}$ is $3n \times 3n$ matrix and an operator for measuring the amount of the anisotropy of the model perturbation vector. Only the diagonal and four subdiagonal elements of this matrix have values such as

$$M_{ij} = \begin{cases} 2 & (if \ i = j) \\ -1 & (if \ j = i \pm n, \ j = i \pm 2n) \end{cases}. \tag{8}$$

Γ defined by the Eq. (7) can be regarded as the l_2 norm of the second order difference among three diagonal elements of conductivity tensor, which is nearly same as the anisotropy penalty functional of Labrecque & Casale (2002) and Herwanger et al. (2004a).

Figure 8 shows an example of the tomographic image derived from the anisotropic inversion. From the figures, we can recognize that there is a close correlation between the electric logs patterns and the resistivity variation adjacent to borehole BH-2. Moreover, the result of the short normal resistivity log is closer to the inverted ERT image than the long normal resistivity log. Figure 8c, the isotropic inversion result, on the other hand, shows quite different and unreasonable features such that very high resistivity is dominant through out the whole image and only the parts adjacent to two boreholes show relatively lower value of resistivity. When comparing these two images by the two different inversion approaches, we can also understand that the result from anisotropic inversion shows the much more reasonable subsurface structure.

4 CASE HISTORIES OF RADAR TOMOGRAPHY

In this section I introduce three case histories on the application of anisotropic radar tomography in Korea (Kim et al. 2006a). The first case history was conducted for the construction of infrastructure and the main objective was to locate cavities in limestone area. The other two were performed in gneiss and granite areas. All these three case histories showed that anisotropic characteristics provided useful and important information to understand the subsurface structures.

4.1 Limestone cavities and anisotropy ratio

Several cavities were found by excavation work for a planned highway bridge in a limestone area in Korea. Consequently the construction work was temporarily stopped for more than one year. A serious concern on the safety of the planned bridge led us to conduct an integrated geophysical investigation including dc resistivity, GPR, borehole radar reflection, and radar tomography surveys. The purpose of the investigation was to delineate possible cavities and weak zones and to provide basic information for designing a reinforcement method of the bridge foundation.

As borehole deviation could distort the resultant tomogram seriously (Peterson 2001) and can produce false anomalies of anisotropic effects, the measurement of borehole deviations and correction is very important. Before discussing the results of the case histories, I briefly explain a simple approach which was used to correct deviation effects in the following case histories. A crosshole tomogram is the two-dimensional image on the section containing two boreholes but the borehole deviation cannot be confined in the two-dimensional plane. The deviation effects, therefore, must be corrected three-dimensionally. To do this, the actual propagating

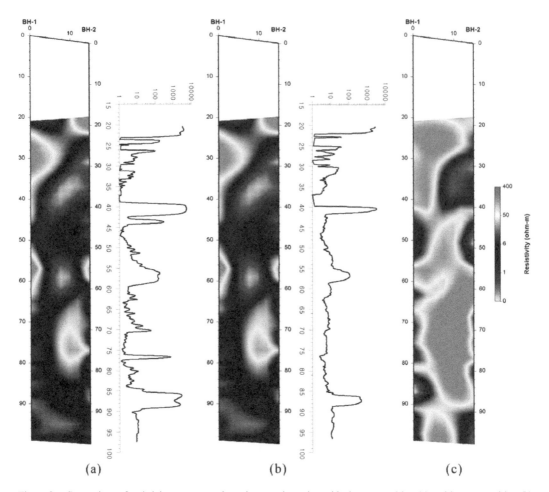

Figure 8. Comparison of resistivity tomograms by anisotropy inversion with short normal log (a) and long normal log (b) obtained at borehole BH-2. A tomogram obtained by the inversion assuming isotropic medium (c).

distances were calculated using the borehole deviation information, and imaginary propagating distances were computed assuming that the boreholes were on a perfect two-dimensional section (Figure 9). The traveltimes were corrected on the base of the difference between the actual and imaginary propagating distances. This approach of correcting borehole deviation effects was applied to the data of all case histories of radar tomography described in this paper. In case histories 1 and 3, the tomograms corrected for the deviation effects were similar to the uncorrected ones, while the tomography images of the case history 2 were greatly changed and became much more reasonable after correcting the deviation effects because the distances between adjacent boreholes were only about 1.4 m.

Since the bridge piers are the most important part in the safety of bridge, eight piers were selected and borehole radar surveys were applied to investigate the

basement rock under the piers. From the total of eight piers, the tomography travel time data of six piers showed anisotropy effects. Thin-bedded limestone is distributed under the planed bridge construction site and calcareous shale is partly intercalated. It was concluded that the anisotropy resulted from thin and parallel bedding planes, if the electromagnetic (EM) wave propagates faster in the direction of bedding plane.

On a velocity tomogram, a cavity may be imaged as a high or low velocity anomaly, depending on the filling material. All the cavities detected during the drilling for the borehole radar investigation were filled with clay, sometimes mixed with water, i.e., low velocity material. On the other hand, when anisotropic rock is weathered, altered, or fractured, the alignment in the direction of bedding planes or of constituent minerals will be diminished, so that the degree of anisotropy will become less. In the case of a cavity, because

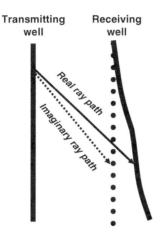

Transmitting well   **Receiving well**

Real ray path

Imaginary ray path

Figure 9.   A schematic diagram explaining the approach to correct the borehole deviation effects, which was used in the data processing of the case histories introduced in this paper.

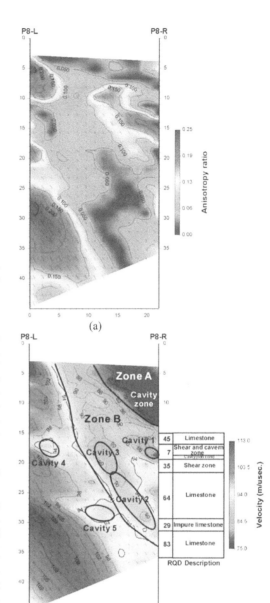

the filling material, such as air, clay, water, etc., is isotropic, anisotropy effect will be weak where the cavity develops. Therefore, the main focus was on identifying isolated anomalies showing lower velocity and lower degree of anisotropy simultaneously in the velocity tomogram. To do this the anisotropy ratio was defined as the difference between maximum and minimum velocities normalized by the maximum velocity, and made tomograms of the anisotropy ratio as shown in Figure 10a.

$$\alpha = (v_{max} - v_{min})/v_{max} \qquad (9)$$

This concept is supported by the observation that the degree of anisotropy in Figure 10a is very small in the uppermost part of the tomogram, corresponding to weathered rock or soil, and in all the cavity zones hit by drilling. In Zone B represented by green color in Figure 10b, we can recognize some regions (Cavity 2 and 3) which show slightly low velocities but very low anisotropy ratio, and they were interpreted as cavities or cavity zones. Cavity 4 and 5 in Figure 10b were interpreted to be cavities based on the appearance on the tomograms, i.e., low velocity and weak anisotropy.

Figure 10b shows that the weak zone of lower velocity extends deeper under the right part of the surveyed pier than the left part. Moreover, the frequency of occurrence of cavities in the right part is higher than in the left part. These tendencies could also be recognized in the tomograms acquired at the adjacent piers (Figure 11). If this feature were not considered in the construction of bridge, it would be likely for the bridge to suffer excessive subsidence and differential settlement.

Figure 10.   Tomograms of anisotropy ratio (a), and maximum velocity (b). The oblique short line in a tomogram points in the direction of maximum velocity and its length represents the anisotropy ratio. Note that the RQD (Rock Quality Designation) attached to the borehole axis of P8-R is matched fairly well with the velocity distribution.

4.2   *Monitoring physical rock property changes due to tunnel excavation*

The velocity changes of EM waves in gneiss were monitored through borehole radar tomography to

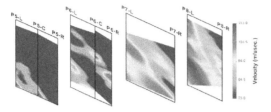

Figure 11. Comparative display of the maximum velocity tomograms beneath Piers 5, 6, 7, and 8.

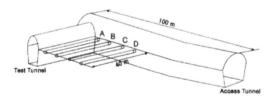

Figure 12. Schematic diagram of horizontal boreholes drilled parallel to the test tunnel.

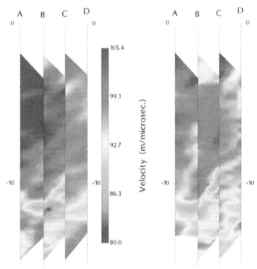

Figure 13. Comprehensive tomograms of maximum velocity between the boreholes A and D. (a) before excavation (b) after excavation. The oblique short line in a tomogram points in the direction of maximum velocity and its length represents the anisotropy ratio.

investigate rock behavior due to the excavation of a test tunnel in an underground laboratory in Korea. As shown in Figure 12, four horizontal boreholes spaced 1.4 m apart were drilled parallel to the planned test tunnel before excavation. Using these four boreholes, crosshole radar tomography surveys with a central frequency of about 500 MHz were performed at the pre- and post-excavation phases of the test tunnel. HP 8753 network analyzer was used for acquiring data in frequency domain. The measurement interval for transmitting and receiving EM wave was 0.3 m and eleven traces were obtained per each transmitter location.

The bedrock of the survey area is mainly composed of banded biotite gneiss which consists of biotite and amphibole in dark layers alternating with quartz and feldspar in light layers. In addition to the banded gneiss, there are small amounts of hydrothermally altered biotite schist with well developed schistocity. The anisotropy detected by radar tomography might be the result of these alternating occurrences of different minerals and schistocity.

Figure 13 shows comprehensive images of the survey area composed of three separate maximum velocity tomograms connecting the four boreholes. Significant differences in the pattern of velocity distribution were not observed before and after the excavation of the test tunnel. The direction of anisotropy, however, changed. Additionally, the velocity of EM waves decreased slightly and the low velocity zones appeared to expand after excavation. Moreover, the anisotropy ratio also changed on the whole. In particular, it decreased remarkably in the tomogram of the boreholes A and B, which were the closest boreholes to the test tunnel. These reductions of velocity and

anisotropy ratio might be attributed to micro-cracks in the rock that were created due to excavations and then saturated with groundwater.

4.3 Anisotropy and fine fissures in granite

To delineate inhomogeneities including fractures and to estimate the freshness of rock, borehole radar (reflection and tomography methods) and GPR surveys were conducted at a granite quarry. The borehole reflection survey using the directional antenna was also conducted to get the spatial orientations of reflectors (Kim et al. 1998).

Crosshole radar data were acquired for seven sections. Among the seven sections, three sections showed strong anisotropy, which is uncommon in granite quarries in Korea. By comparing the tomography data and acoustic televiewer images it was concluded that the anisotropy effect in this area is closely related to fine fissures aligned in the same direction. As shown in the televiewer 3-D logs attached to Figure 14, fine fissures are aligned in the same direction in BH-1 over considerable depth intervals. On the other hand, there are little fissures in BH-3, and moreover, they are not aligned in the same direction. Comparing the televiewer logs and the velocity tomogram, we can easily find that the anisotropy ratio becomes higher where the occurrence rate of fine fissures is fairly high and the EM wave velocity lower. The directions of the alignment of fine fissures are nearly the same as that of the symmetry axis, i.e., the direction showing maximum

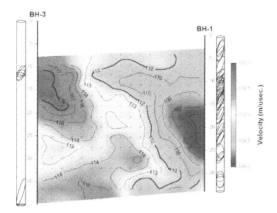

Figure 14. Tomogram of maximum velocity acquired at a granite quarry. Attached to the borehole axes are the televiewer 3-D logs, interpretation results of the televiewer images. The oblique short line in a tomogram points in the direction of maximum velocity and its length represents the anisotropy ratio.

velocity, which implies that the EM wave propagates faster along the direction of fissure alignment.

5 CASE HISTORY OF CROSSHOLE RESISTIVITY TOMOGRAPHY

For the case history of anisotropic crosshole ERT, discussed is the one performed for the ground safety assessment of a high-storied building. Due to the very thin graphite layers developed along the foliation planes of the basement rock under the apartment, very strong electrical anisotropy was observed in the measured data. As in the first case history of crosshole radar tomography, the locations of cavities were delineated more precisely owing to the anisotropic characteristics. In this section, rather than discussing the results of the anisotropy ERT survey only, the whole procedure conducted for the ground safety analysis is discussed.

5.1 Data acquisition and processing

Eight boreholes were drilled around the apartment and tomography data comprising of 20 sections were acquired over the site. The pole-pole array is most frequently adopted for crosshole ERT because it has a great advantage of high signal-to-noise (S/N) ratio, and is suitable for low resistivity environment. On the other hand, it has a disadvantage of low resolving power compared to pole-dipole or dipole-dipole arrays (Sasaki 1992). In order to overcome this inherent problem of pole-pole array, pole-dipole array data as well as pole-pole data were acquired over the entire 20 tomography sections. All the data were simultaneously

inverted using the anisotropic inversion code discussed in the previous section. The data acquired were subjected to careful editing before the inversion process was carried out; this is necessary in order to remove bad data points so as to enhance the S/N ratio of the input data of the inversion program.

By the simple resistance measurement of the core samples in the field site, it was known that the basement rock has very strong anisotropy, which is attributed to the very thin graphite layers developed along the foliation planes of the basement rock, Precambrian biotite-gneiss. In spite of this, the isotropic inversion code was applied but could not provide geologically meaningful subsurface images (see Figure 8). The error level of the final inversion results, moreover, did not converge to the reasonable level. As discussed in the section of anisotropic effects observed in field data, the measured data were carefully examined on view of anisotropy (see Figure 4). All these efforts were to confirm whether the anisotropy prevails in the surveyed area or not.

5.2 Survey results: establishing subsurface model

The most important objective of this investigation was to accurately locate any existing cavities beneath the building and to determine their sizes where possible. Two mining tunnels at 75 and 100 meter depths were encountered by the boreholes drilled for the site investigations purposes. According to the old mining record available, three mining tunnels had been excavated at this site, two at 75 meter depth, and the remaining one at 100 meters. Thus, the interpretation was orientated to delineate the presence of the remaining mining tunnel at 75 meter depth, and to locate mine-cavities, if they exist.

It was presumed that an empty cavity will be imaged as a high resistivity anomaly in an ERT image, while a cavity filled with water will be imaged as low or high resistivity anomaly depending on the resistivity of the basement rock. Because all the tunnels detected by drillings are filled with water and the resistivity of groundwater from well logging is higher than that of basement in most parts, a cavity may be imaged as a higher resistivity anomaly.

As in the cases of anisotropic radar tomography, anisotropy characteristics of electric conductivity can also provide valuable information in identifying cavities. Thus we can use the same philosophy in the interpretation of the anisotropic inversion results of ERT data; the filling material of the cavities, such as water or air, is usually perfectly isotropic, whereas the surrounding material in this area, which is biotite-gneiss, reveals very high anisotropic characteristics. Therefore, we mainly focused on identifying and isolating anomalies showing higher resistivity and lower anisotropy degree simultaneously. In order to interpret

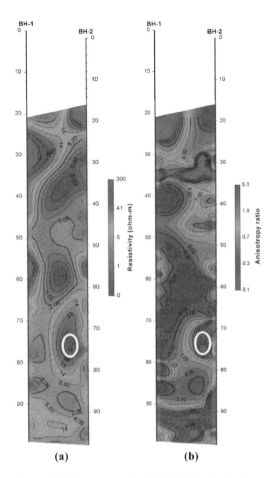

Figure 15. Tomograms of resistivity (a) and anisotropy degree (b). An ellipse on each tomogram depicts the delineated old mining tunnel.

the data in such way, effective resistivity σ_{eff} and anisotropic ratio α were defined:

$$\sigma_{eff} = \sqrt[3]{\sigma_1 \cdot \sigma_2 \cdot \sigma_3},$$

$$\alpha = \begin{cases} a-1, & if \ a > 1 \\ 1/a-1, & if \ a < 1 \end{cases}, \text{ where } a = \sigma_1 / \sigma_{eff}. \quad (10)$$

Figure 15 shows the image of the cavity identified using this criterion. In Figure 15, we can recognize that there are several places showing high resistivity or low anisotropy degree, but only the part designated by an ellipse shows high resistivity and low anisotropy degree simultaneously. From this, we can confirm that this part must be the image of a cavity. This delineated cavity corresponds to the mining tunnel not detected by the 8 boreholes, and the location matches quite well with the old mining records.

In the similar manner, all the 20 tomography sections were carefully examined and compared and any other anomalous zones were not found in any of the sections. After this, it was concluded that there are no other mine-openings under the apartment building except for the earlier delineated three mining tunnels.

Another important purpose of the tomography was to establish the true underground model which was used for the rock engineering safety analysis. Based on the tomography images, the subsurface model for the numerical analyses of rock engineering was established, combining the results of geological survey and core logging. Figure 16 shows the procedure of the simplification of the derived subsurface model for the rock engineering analysis.

5.3 Survey results: ground safety analysis

Numerical analyses of rock engineering were performed to verify if the apartment building is affected by the presence of the old mine cavities or not. First, the behavior of the basement under the apartment was analyzed using FLAC code. In addition to this analysis assuming continuum model of FLAC code, another type of numerical modeling was performed using UDEC code which assumes the subsurface material as discontinuum since many faults and the accompanying joints and fractures have developed under the apartment in this site. In the model of FLAC code (Figure 17a), the detailed subsurface structure under the apartment was taken into account in the model geometry, while in UDEC model, it was impossible to do the same work due to the limitation in mesh generation because the established geological model was too complex.

An example of the analysis results obtained by FLAC is presented in Figure 17b which shows the distributions of yielding zones, where plastic deformation of the ground occurs. Although the mine cavities are located at deep depths, yielding zones occur in a much wider range than we expected before the analysis. This result is mainly due to the subsurface model for FLAC analysis which accounts for the weak geological formations beneath the apartment as well as the mining tunnels. As shown in Figure 17b, however, the yielding zones do not develop near the ground surface. Therefore, we can say that collapses might be expected around the cavity, while such an unstable state would not extend up to the ground surface. Similar results were obtained in UDEC analysis. The yielding zones from UDEC analysis, however, were much narrower than those from FLAC, and the displacement of the ground surface was much smaller. The main difference between two analyses by FLAC and UDEC codes was the subsurface model under consideration. In the FLAC analysis, subsurface model

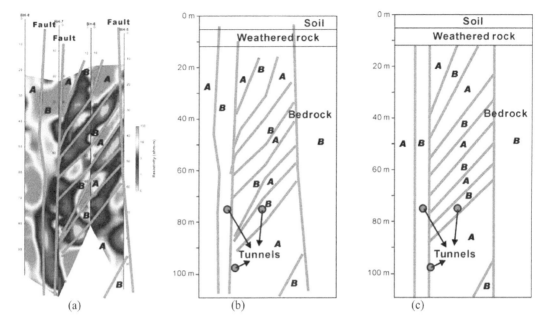

(a)   (b)   (c)

Figure 16. Interpreted geological structure presented on the merged tomogram (a) and its simplification for the numerical analysis of rock engineering (b and c). In these figures, A means the zone consisting of fairly competent rock, while B that of poorly competent rock.

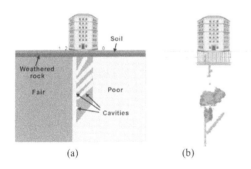

(a)   (b)

Figure 17. Result from the numerical analysis of rock engineering based on the continuum model of FLAC code. Using the subsurface model (a), yielding zones (b) were calculated.

reflects not only the cavities but also the subsurface structures such as faults and fractures, while that for UDEC considered the cavities only.

From these two analyses based on continuum as well as discontinuum models, it was concluded that there is no problem in the safety of the apartment, although the area around the mined cavities is unstable because of the weak geological formation. The second important conclusion is that for an accurate safety analysis, it is important to establish the underground model reflecting the subsurface structure precisely as in the case of FLAC analysis.

6 CONCLUSIONS

The essence of geophysical prospecting is to provide subsurface images in terms of material properties. Since the number of components defining the material property of anisotropic medium is multiple while that of isotropic medium single, we can get more information in principle when the subsurface material reveals anisotropy.

The geophysical imaging techniques of crosshole radar and electric resistivity have been discussed so far, addressing anisotropy: anisotropy effects, inversion algorithms and case histories. Among the four case histories of crosshole GPR and ERT methods discussed above, the first two refer to a situation where basement rock contains less fractures and/or altered minerals when the anisotropy ratio becomes high, while the third one of granite quarry is the opposite case. In the first and the last case histories, thanks to the anisotropy, it was possible to locate cavities more precisely. Understanding the origin of anisotropy in the host medium and the anisotropic characteristics of the prospecting target, we can interpret geophysical data more reliably and extract more information. A general conclusion of this paper is that anisotropic characteristic gives us additional and valuable information for understanding the internal status of basement rock, when the subsurface material is anisotropic. In particular, two anisotropy ratios of anisotropy radar and electric resistivity tomography defined in this paper are good measures to quantify the anisotropy characteristics of subsurface materials and useful to interpret the anisotropy crosshole tomograms.

ACKNOWLEDGEMENT

This research was supported by the Basic Research Project of the Korea Institute of Geoscience and Mineral Resources funded by the Ministry of Science and Technology of Korea.

REFERENCES

Daily, W. & Owen, E. 1991. Cross-borehole resistivity tomography. *Geophysics* 56: 1228–1235.

Daily, W. & Ramirez, A. 1995. Electrical resistance tomography during in-situ trichloroethylene remediation at the Savannah River Site. *Journal of Applied Geophysics* 33: 239–249.

Daily, W. & Ramirez, A.L. 2000. Electrical imaging of engineered hydraulic barriers. *Geophysics* 65: 83–94.

Herwanger, J.V., Pain, C.C., Binley, A., de Oliveira, C.R.E. & Worthington, M.H. 2004a. Anisotropy resistivity tomography. *Geophysical Journal International* 158: 409–425.

Herwanger, J.V., Worthington, M.H., Lubbe, R., Binley, A. & Khazanehdari, J. 2004b. A comparison of crosshole-electrical and seismic data in fractured rock. *Geophysical Prospecting* 52: 109–121.

Jung, Y. & Kim, J.-H. 1999. Application of anisotropic georadar tomography to monitor rock physical property changes. *Journal of Environmental and Engineering Geophysics* 4: 87–92.

Keller, G.V. & Frischknecht, C.F. 1966. *Electrical Methods in Geophysical Prospecting*. Oxford: Pergamon Press.

Kim, J.-H., Cho, S.-J. & Chung, S.-H. 1998. Three-dimensional imaging of fractures with direction finding antenna in borehole radar survey. *Proceedings of the 4'th SEGJ International Symposium – Fracture Imaging-*. 291–296. Tokyo, Japan,

Kim, J.-H., Cho, S.-J. & Yi, M.-J. 2004a. Borehole radar survey to explore limestone cavities for the construction of a highway bridge. *Exploration Geophysics* 35: 80–87.

Kim, J.-H., Cho, S.-J., Yi, M.-J. & Sato, M. 2006a. Application of anisotropy borehole radar tomography in Korea. *Near Surface Geophysics* 4: 13–18.

Kim, J.-H., Yi, M.-J. & Cho, S.-J. 2004b. Application of high-resolution geoelectric imaging techniques to geotechnical engineering in Korea. *Proceedings of the ISRM International Symposium 3rd ARMS*. Vol. 1: 191–196. Kyoto, Japan, Ohnishi, Y. and Aoki, K. (ed.).

Kim, J.-H., Yi, M.-J., Cho, S.-J., Son, J.-S. & Song, W.-K. 2006b. Anisotropic crosshole resistivity tomography for

ground safety analysis of a high-storied building over an abandoned mine. *Journal of Environmental and Engineering Geophysics* 11: 225–235.

LaBrecque, D.J. & Casale, D. 2002. Experience with anisotropic inversion for electrical resistivity tomography. *Symposium on the Application of Geophysics to Engineering and Environmental Problems (SAGEEP 2002)* 11ELE6. Las Vegas, U.S.A.

LaBrecque, D.J., Heath, G., Sharpe, R. & Versteeg, R. 2004. Autonomous monitoring of fluid movement using 3-D electrical resistivity tomography. *Journal of Environmental and Engineering Geophysics* 9: 167–176.

Lesmes, D.L., Decker, S.M. & R. D.C. 1997. A multiscale radar-stratigraphic analysis of fluvial acquifer heterogeneity. *Geophysics* 67: 1452–1464.

Maillol, J.M., Seguin, M.-K., Gupta, O.P., Akhauri, H.M. & Sen, N. 1999. Electrical resistivity tomography survey for delineating uncharted mine galleries in West Bengol, India. *Geophysical prospecting* 47: 103–116.

Parasnis, D.S. 1979. *Principles of Applied Geophysics*. 3rd ed. London, Chapman and Hall.

Peterson, J.E. Jr. 2001. Pre-inversion corrections and analysis of radar tomographic data. *Journal of Environmental and Engineering Geophysics* 6: 1–18.

Sasaki, Y. 1989. Two-dimensional joint inversion of magnetotelluric and dipole-dipole resistivity data. *Geophysics* 54: 254–262.

Sasaki, Y. 1992. Resolution of resistivity tomography inferred from numerical simulation. *Geophysical Prospecting* 40: 453–464.

Sénéchal, P., Hollender, F. & Bellefleur, G. 2000. GPR velocity and attenuation tomography corrected for artifacts due to media anisotropy, borehole trajectory error and instrumental drifts. *Proceedings of the 8th International Conference on Ground Penetrating Radar*, 402–407. Gold Coast, Australia.

Seol, S.J., Kim, J.-H., Cho, S.-J. & Chung, S.-H. 2004. A radar survey at a granite quarry to delineate fractures and estimate fracture density. *Journal of Environmental and Engineering Geophysics* 9: 53–62.

Shima, H. 1992. 2-D and 3-D resistivity image reconstruction using crosshole data. *Geophysics* 57: 1270–1281.

Slater, L., Binley, A.M., Daily, W. & Johnson, R. 2000. Crosshole electrical imaging of a controlled saline tracer injection. *Journal of Applied Geophysics* 44: 85–102.

Stratton, J.A. 1941. *Electromagnetic Theory*. New York and London: McGraw-Hill Book Company, Inc.

Tillard, S. 1994. Radar experiments in isotropic and anisotropic geological formations (granite and schists). *Geophysical Prospecting* 42: 615–636.

Vasco, D.W., Peterson, J.E., Jr. & Lee, K.H. 1997. Ground-penetrating radar velocity tomography in heterogeneous and anisotropic media. *Geophysics* 62: 758–1773.

Yi, M.-J., Kim, J.-H. & Chung, S.-H. 2003. Enhancing the resolving power of least-squares inversion with active constraint balancing. *Geophysics* 68: 931–941.

Geotechnical and Geophysical Site Characterization – Huang & Mayne (eds)
© 2008 Taylor & Francis Group, London, ISBN 978-0-415-46936-4

Characterization of deltaic deposits in the Nakdong River mouth, Busan

S.K. Kim
Dongguk University, Seoul, Korea

ABSTRACT: Sediments in the Nakdong River mouth were highly affected by the flow of the river stream, marine transgression and regression. The deposit is divided into four to five different layers depending on the depositional environment. It is featured that the upper silty clay layer is soft and thick, and the bottom sand layer lying directly on the bedrock is an excellent aquifer. Upward fresh water flow from the aquifer has leached salt and ions in the lower clay layer and a part of the upper clay layer. Effects of leaching on plasticity, sensitivity, shearing strength and OCR of the clay deposit are described with measurements in the field.

1 INTRODUCTION

A delta has been formed at the Nakdong River mouth as shown in Fig. 1. The longest river in Korea flowing from the north to the south gets divided into two

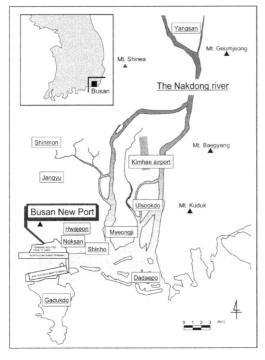

Figure 1. Delta formed at the Nakdong River mouth and locations of major residential and industrial complexes.

streams as it enters the delta. The delta area covers approximately 200 square km, and its south is open to sea while all other edges are surrounded by hills and mountains. Initially the land was a large rich farm producing plenty of agricultural products for feeding the citizens of neighboring Busan. Since late 1980s, the area was being developed to cope with industrialization of Korea. Gimhae International Airport is centered in the area and residential and industrial complexes were created in its north and the south. And now a new large port is under construction along the coastline (see Fig. 1). The area will be developed in the future as a key industrial belt in Korea.

At the initial stage of development, a heavy industrial complex was planned on the river mouth but the project was soon abandoned because of poor subsoil conditions. Local engineers are still facing several difficult problems because of an unusually thick soft clay deposit. Several failures such as collapse of a breakwater, tilting of bridge piers, unusual large settlements and so on were experienced. Those failures were mainly caused by insufficient knowledge of the deposit and characteristics of Busan clay.

2 DEPOSITIONAL CHARACTERISTICS OF THE DELTA AREA

2.1 Deposits in the deltaic valley

The old terrain of the delta area was a valley that was formed by right-lateral slip of Yangsan fault which runs parallel to the major stream of the Nakdong River (Cho, 1987). This is confirmed by a cross section of deltaic deposits of the Nakding River mouth, as presented in Fig. 2 (Dongil Consultants, 2003). The

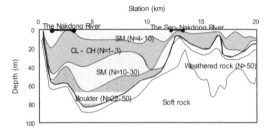

Figure 2. Soil profile of the delta area from Sasang to Gimhae (Dongil Consultants, 2003).

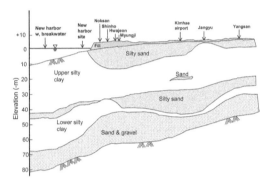

Figure 3. Longitudinal section of the deltaic deposits of the Nakdong River mouth.

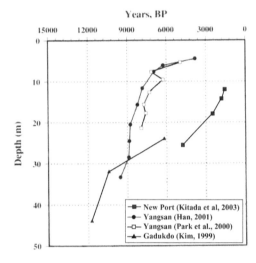

Figure 4. Radio-carbon dating for clays in the deltaic area (Baek, 1998; Kim, 2002).

right river bank of the river dropped up to 60 m by a vertical movement of Yangsan fault that occurred in the Cenozoic era, forming a deep basin in the east part. Judging from the boulder material that has been transported by the flow of the river stream, the river of the old terrain is believed to be very steep and wider than 10 km. However, the western basin is rather shallow and the deepest deposit is less coarse. Park et al. (1988) estimated that the Paleo-coastline of the Korean Peninsula extended to Kyushu, Japan, with erosion developing subsequently in the valley.

The sedimentation of sand and gravel in the valley was followed by deposits of different composition of materials, affected by marine transgression and regression through a long geological history. The deposits can be divided into five distinguished layers as can be seen in Fig. 3. The bottom gravel and sand layer is thick and an excellent aquifer holding fresh water flowing from the upper stream. The middle sand layer is sandwitched by the upper and lower clay layers but is missing or very thin in the coast and offshore.

The silty clay layer is divided into two, the upper and lower layers, based on the age of sedimentation and engineering properties. The upper clay layer became a top deposit on the sea side but on the land side the layer is covered by a deposit of silty sand, which has been transported by flooding after the sea level became

stable. The clay has been normally consolidated and is soft to medium in consistency in the upper layer. The lower clay layer is stiff to very stiff as the layer has been subjected to heavier overburden pressure. The dominant clay mineral is illite with some chlorite and kaolinite. Total thickness of the deposits ranges from 50 m to 80 m exhibiting thicker deposition on the sea side. The materials deposited in the two clay layers are, in general, referred to as Busan clay.

2.2 Ages of sediments and rate of sedimentation

Ages of clay sediments have been estimated by radio-carbon dating for clay samples taken at different sites in the area, as can be seen in Fig. 4. For example, a clay sediment at 44 m in depth in the breakwater site was probably deposited BP (before present) 12,000 years (Kim, 1999). This depth is within the lower clay layer. The upper clay layer was formed since BP 10,000 years, as shown in the upper two curves (Park et al., 2,000; Han, 2001). The lower two curves in the figure are the sedimentation ages for the clays on the sea side (Kidata et al., 2003), indicating that clay sediments on the seaside are younger than those on the inland.

In order to grasp the whole geological history of sedimentation of the estuarine deposits, the results of radio-carbon dating are combined with sea level curves studied by Kenny (1964); Fairbridge (1961) and Cho (1979), as presented in Fig. 5. As the sea level has become stable since approximately BP 6,000 years, according to geologists, the sedimentation of the upper sand layer on the land side seems to have occurred since then. As revealed already in Fig. 4, the upper clay layer was deposited during the rise of sea level from

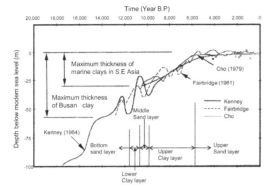

Figure 5. Changes of sea level and ages of the deltaic deposits.

BP 10,000 years to BP 6,000 years but sedimentation in offshore continued to take place even after BP 6,000 years.

The deposition of middle sand layer is estimated to occur during marine regression of BP 11,000 years to BP 10,000 years. It is featured that the depositional period is short (only 1,000 years), compared to the upper clay layer, resulting in shallow thickness on the offshore but much thicker at land side because the sedimentation was dominated by flooding. The deposition of lower clay layer took place during the rise of sea level between BP 12,000 years to BP 11,000 years.

It is noted that the upper silty clay layer is much thicker than the thickness of Southeast Asian marine clays or Osaka alluvium clay. The thickness of Busan clay is over 50m inclusive of the lower silty clay layer, even with the middle sand layer missing. This large thickness poses some difficulties for an improvement of the clay layer and design of foundations to satisfy settlement requirements among various foundation types.

The rate of sedimentation was rather fast, in particular, for the upper clay layer. The rate estimated during BP 10,000 years to BP 7,000 years was 1 cm/year. This rate may lead to an under-consolidation condition for the clay layer.

2.3 Depositional characteristics and effects on geotechnical properties

2.3.1 Depositional characteristics during sedimentation

The environmental characteristics of a deposit can be disclosed by an investigation of microorganisms that lived during sedimentation of soil particles. Some microorganisms can only live in salty water, while others cannot survive in salty marine water. For example, foraminifera, shell fragment, ostracods, pyrite, nanno fossil and marine diatom can survive only in salty water. The studies for Busan clay were conducted by

Jeong et al. (2001), Kitada et al. (2003), and Ryu (2003).

Table 1 presents a foraminiferal fauna investigation for the clay at Yangsan (Jeong et al., 2001). According to the investigation, there are no marine microfossils 16.42 m below ground level (GL). Instead, materials such as mica and floras that can survive in brackish water are abundant, suggesting that it was a brackish water deposit. An investigation for microfossils from ground level to the depth of 7.42 m also shows the same results. In the middle part between GL (−) 7.42 m and GL (−) 16.42 m, however, there exist abundant marine microfossils such as foraminifera, shell fragment, ostracods and sea urchin. This indicates that this clay has been deposited under marine environment.

Another investigation was conducted by Kitada et al. (2003) for the clay at New Port located on the sea side. Though the result was not shown herein, it was disclosed that marine microfossils appeared between GL (−) 8 m and GL (−) 30 m, which is much thicker than the thickness of the marine clay on the land side. Therefore, it can be concluded from the investigations that the clay was initially deposited in brackish water when the sea level started to rise, and then in a marine environment and once again in a brackish water environment. It is noted, however, that the change of the environment is a continuous process with salinity of water gradually increasing or decreasing. Because of such variable depositional environment, Busan clay can be termed as a typical of deltaic clay or estuarine clay.

2.4 Effect of depositional characteristics on geotechnical properties of clay

Depositional environment during sedimentation of soil particles affects geotechnical properties of clay. Fig. 6 illustrates the effects of depositional environment on physical behavior of the clay at Yangsan (Lim et al., 2003). It can be seen that the middle portion of the clay layer has higher water content, higher liquid limit, and higher void ratio and lower unit weight than those of the upper and lower portions. Suwa (2002) also carried out a similar investigation for the clay at New Port and obtained similar results including higher clay content in the middle portion. This is because flocculation causes sedimentation of clay particles under marine environment, as explained by Pusch (1973).

The depositional environment also affects the engineering properties of clay. Fig. 7 presents a variation of compression index, C_c, and compression ratio, $C_c/(1 + e_0)$, with depth for the clay at Shinho. The compression index and compression ratio, were higher in the middle portion of the upper clay layer with the maximum at a depth of approximately 30 m, which corresponds to the portion deposited in marine environment. It is noted that data is missing in the figure

Table 1. Investigation of microfossils for the Yangsan upper clay layer (Jeong et al., 2001).

Depth GL.-(m)	Foraminifera	Shell fragment	Ostracods	Sea urchin	Volcanic glass	Mica	Plant fragment	Pyrite	Nannofossils	Marine diatoms	Non-marine diatoms	Species of diatoms
5.20~5.21	–	–	–	–	+	+	–	–	–	+	–	Cyc.
5.41~5.42	–	++	–	–	+	+	++	–	–	–	–	Cos.
5.67~5.68	–	–	–	–	+	+	++	+	+	–	–	Cos.
6.21~6.22	–	–	–	–	+	++	++	+	+	–	+	
6.41~6.42	–	–	–	–	+	++	+++	–	+	–	–	
6.67~6.68	–	+	–	–	+	++	+++	++	+	–	–	Cos.
7.11~7.12	–	–	–	–	–	+++	+	+++	+	–	–	Cos.
7.42~7.43	++	++	+	–	–	+	+	+++	+	–	+	Cos., Ms., Nc., Epi.
7.68~7.69	+	+++	+	–	+	+	+	+++	++	–	+	Cos., Cs., Epi.
8.41~8.42	+++	+	+++	++	+	++	+	+++	+++	–	+	Cos., Epi., Cs.
8.67~8.68	+	+	++	–	++	+	+	++	+	–	–	Cos.
9.43~9.44	+++	+	–	–	++	+	+++	+	+	++	–	Cos., Sur., Nan., Nc.
9.69~9.70	+++	++	+++	–	+	+	+	++	++	–	–	Ds.
10.10~10.11	+++	++	+++	–	+	+	+	++	++	–	+	Nl.
10.41~10.42	+++	+++	++++	–	–	+	+	++++	++	–	+	Ds., Ms.
10.67~10.68	–	+++	+++	–	+	+	+	+++	+	–	–	Ms.
11.10~11.11	+++	++	–	+	+	+	+	+++	+++	–	+	
11.41~11.42	+++	++	+++	–	+	+	+	+++	+++	–	–	
11.67~11.68	+++	+++	+++	–	+	+	+	+++	+++	–	–	Dc.
12.10~12.11	+++	+++	+++	–	+	+	+	+++	+++	–	+	Nc.
12.41~12.42	+++	+++	+++	–	+	+	+	+++	+++	–	–	Cos.
12.67~12.68	+++	++	+++	–	+	+	–	+++	+++	–	–	
13.67~13.68	+++	+++	+++	–	+	+	+++	+++	+++	+	–	Cos.
14.10~14.11	+++	+++	++	–	+	+	+++	+	–	+	–	
14.41~14.42	+++	++++	++	–	+	+	–	+++	–	–	–	
14.67~14.68	++	++	++	–	+	+	++	+++	–	–	–	
15.10~15.11	+	++	–	–	+	+	++	+++	+++	+	–	Cos.
15.41~15.42	++	++++	+++	–	+	+	+	+++	+++	–	–	
15.67~15.68	+++	+++	+++	–	+	+	+	+++	+++	–	–	
16.10~16.11	+	++	++	–	–	+	+	+++	–	–	–	
16.41~16.42	+	++	++	–	+	+	+	++	–	–	–	
17.41~17.42	–	–	–	–	+	+	+	++	–	–	–	
17.67~17.68	–	–	–	–	+	+	+	+	–	+	–	
19.41~19.42	–	–	–	–	++	+	+	–	–	+	–	
19.57~19.58	–	–	–	–	+	+	–	+	–	–	–	
21.10~21.11	–	–	–	–	++	+	++	+	–	–	–	
21.41~21.42	–	–	–	–	+	+	+	+	–	–	–	
21.67~21.68	–	–	–	–	+	+	++	+	–	–	–	

– None, + very rare, ++ rare, +++ common, ++++ abundant, +++++ very abundant.

78

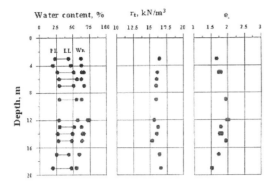

Figure 6. Variation of water content, unit weight, and void ratio with depth for the clay at Yangsan (Lim et al., 2003).

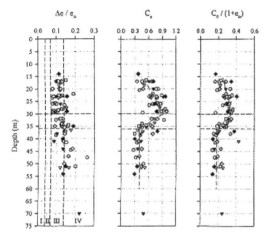

Figure 7. Variation of compression index and compression ratio with depth for the clay at Shinho (Industrial Research Institute, 2006).

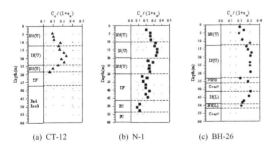

Figure 8. Variation of compression ratio with depth for the clay at New Port site (Huh, 2003).

above 15m depth because the deposit is the upper sand layer.

Fig. 8 presents compression ratio distribution with depth for the clay samples taken at different borings of the New Port (Huh, 2003). In this figure, a geological

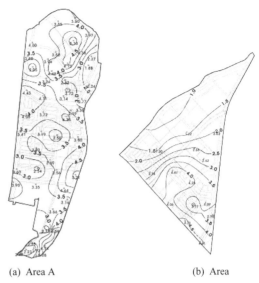

(a) Area A   (b) Area

Figure 9. Distribution of artesian pressure in the site of Yangsan (Han and Yoo, 1999).

term IS (inner shelf) indicates the marine environment, while NS (near shore) indicates the brackish environment. It is again confirmed that compression index of a clay layer deposited in the marine environment is higher. In this figure the value of compression ratio of the marine clay layer is 0.3 and the one for the brackish clay layer is 0.2 on average.

3 ARETESIAN PRESSURE

3.1 *Magnitude and distribution of artesian pressure*

Presence of artesian pressure was first known from a soil investigation for a residential complex at a site of Yangsan. Since then an extensive investigation has been carried out over the whole area by Han and Yoo (1999). Presented in Fig. 9 are contours of the pressure heads, which rose up to the height of EL 2.0 m to EL 4.5 m in Area A. Referring to ground level being at El. 0.0 to 0.8 m., the measurement reveals that most of the pressure heads rise above the ground level. The ground water table is less than 1m from the ground level. The pressure head at Area B is much lower compared to that of Area A. After the first finding of artesian pressure at Yangsan, investigations for the presence of artesian pressure have been extended for most parts of the delta area.

Presented in Fig. 10 is a measurement of artesian pressure with depth for a borehole in New Port site. It was revealed that an aquifer was located 57 m below ground level, at which the pressure measured was

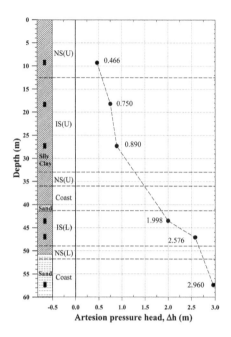

Figure 10. Artesian pressure measured at boring BH-26 of New Port site (Huh, 2003).

Table 2. Survey of artesian pressure head in the delta area.

Location	El. measured (−m)	Pressure head (m)	References
Yangsan	33	3.02	Han & Ryu (1998)
	30.5	3.52	
	37.8	4.04	
New Port	57	2.96	Huh (2003)
Kadukdo	40.3	1.57	Pusan Harbour (1998)
	26.5	1.35	
Boat race site	34.5	1.58	Daewoo Eng. (2001)
Shinho	34	1.11	Chung et al. (2003)

2.96 m/m^2 (Huh, 2003). The pressure decreases almost linearly with a decrease of depth.

Table 2 summarizes the results of artesian pressures measurements carried out at different locations. As can be seen in the table, higher pressures appear at locations in Yangsan in the north and in New Port in the south. Other locations show much lower values, however, it is obvious that the whole delta area is subjected to artesian pressure by fresh water flowing upward from an aquifer below. The aquifer that provides artesian pressure is the bottom sand and gravelly sand layer, which is also found in the middle sand layer in Jangyu (Daewoo engineering, 1998). The artesian pressure was caused by higher water tables of the

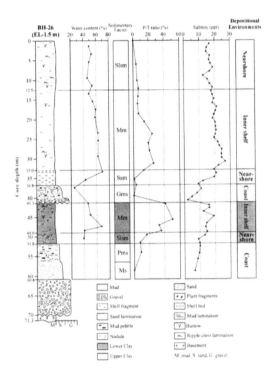

Figure 11. Variation of water content, P/T. and salinity with depth for borehole BH-26 of the site of New Port (Huh, 2003).

upper stream and surrounding mountains, which were associated with the aquifer in the delta area.

3.2 Leaching by artesian pressure

The upward flow of fresh water from an aquifer removes salt and cations from pore water of soils deposited in the marine and brackish water environments. If fresh water containing CO_2 permeates into salty clay, the flow lowers pH values and results in chemical weathering (Brenner et al., 1981).

Huh (2003) conducted a study on changes of depositional environment during and after sedimentation of soil particles as presented in Fig. 11. Column 3 in the figure represents a survey of the ratio (P/T) of marine plankton (P) to total foraminiferal fauna (T). The P/T ratio is an indicator of depositional environment, similar to method by Jeong et al. (2001) for representing depositional environment as mentioned earlier. However, the author uses geological terminology in this figure such as near-shore environment (NS) instead of brackish water and inner shelf environment (IS) instead of marine environment. It can be seen from the figure that values of P/T of the middle portions of the upper and lower clay layers are higher than the values for the other portions respectively, which indicate the marine deposit. The result agrees well with the

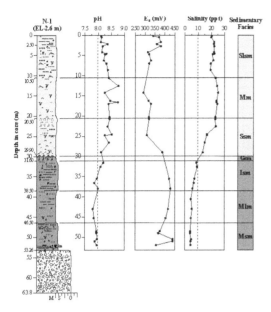

Figure 12. Geo-chemical analysis for a borehole (N-1) of Busan New port site (Huh, 2003).

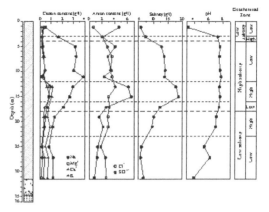

Figure 13. Geo-chemical analysis for clay at Jangju (Ryu, 2003).

previous description (see Fig. 11). It is noted that the survey was conducted at the same borehole where an artesian pressure measurement was made (see Fig. 10).

Soils deposited under marine and brackish waters have been subjected to changes of an environment through a long geological history. The last column indicates that the change of salinity occurred after sedimentation. Salinity holds maximum with a value of 27 g/l at depth of 30 m and decreases slightly with increase of depth. Low salinity is natural when soils have been deposited in fresh or brackish water as can be seen in the middle portion of the column. However, it is noted that the maximum salinity value (15 g/l) of the lower clay layer is much lower than that (27 g/l) of the upper clay layer. This clearly indicates leaching of salt has taken place.

A gradual decrease of salinity from the maximum value to the top of the upper clay layer can be noticed in the figure, which indicates a gradual change into the brackish water environment from the marine environment. Different from this gradual decrease of salinity, the sudden change of salinity in the lower portion is due to the upward flow of fresh water.

Fig. 12 details another survey for a different borehole (N-1) of the same area (Huh, 2003). The soil profile indicates that the upper and lower clay layers are almost connected with a thin sand layer in between. The lower portion of the clay profile exhibits lower pH and lower salinity. In particular, pH values were reduced below 8 and salt content get reduced gradually from a maximum of 24 g/l to 5 g/l. The value of the minimum may be the least for the deposit that has been

leached by fresh water for Busan clay. This is another typical example of leaching but a survey for artesian pressure was not conducted at this borehole.

Now consider exchange of ions that occurred after the sedimentation of clay. A geochemical investigation was carried out for a clay sample at Jangyu by Ryu (2003). The soil profile of Jangyu is composed of a silty clay layer and an underlying sand layer, as presented in Fig. 13. The crust has been desiccated because the ground surface is above the water table. The initial ionic composition of the pore water of a marine clay consists predominantly of Na^+, K^+, Mg^{2+} and Ca^+ ions, with Na^+ dominating (Brenner et al., 1981). The investigation reveals that Na^+ and SO_4^{2-} ions are abundant at a depth of 5 to 16 m, which is well consistent with high salinity. However, it can be seen that the ions were remarkably reduced below this depth together with a reduction of salinity, which may indicate that leaching removes Na^+ and SO_4^{2-} ions in pore water. Reductions of salinity, pH and Na^+ at a shallow depth of the crust were caused by percolation of rain water.

This experimental study, in which Na^+ ions have been removed from salty pore water by leaching, is different from leaching theory for Norwegian clays. Moun et al. (1971) described that leaching resulted in chemical weathering, by which K^+ ions disintegrated from feldspar and mica (including illite) were released into pore water. And then K^+ gets exchanged against Na^+ abundant in the pore water and bringing a decrease of the K^+/Na^+ ratio. However, a recent experimental study by Sridharan (2006) for a montmorillonite clay in Japan mentioned that leaching resulted in a decrease of Na^+, as can be seen in Fig. 16 (a), which agrees well with the experimental study for Busan clay. This problem needs further discussion but chemical weathering may be not active for Busan clay.

81

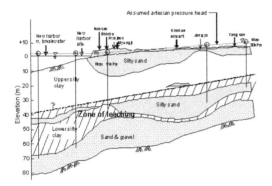

Figure 14.   Estimated zone of leaching.

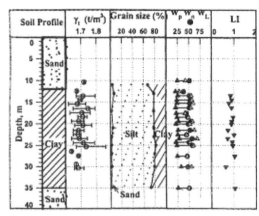

(a) Shinho

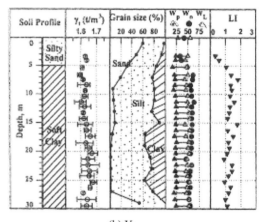

(b) Yangsan

Figure 15.   Variation of liquidity index with depth for clays (a) at Shinho and (b) Yangsan. (Chung et. al., 2001).

Judging from the previous discussions, it is obvious that sudden decreases of salinity, pH and cations were caused by leaching for Busan clay. The zone of leaching can be estimated as presented in Fig. 14.

4   EFFECTS OF LEACHING ON GEOTECHNICAL PROPERTIES OF BUSAN CLAY

It has been well known that leaching affects plasticity, strength and compressibility of soft clays. The following discussions will illustrate such effects in detail revealed through both laboratory and in-situ tests for Busan clay.

4.1   *Effect on plasticity*

Two typical examples for the measurements of liquid index are presented in Fig. 15. It can be seen from the figure that liquidity index is almost constant throughout the depth with a value of unity. Our main concern is that liquidity index in the lower portion of the clay layer is as high as unity even at a depth of 30 m. This is a significant behavior in engineering properties of Busan clay because the clay would be liquidized when it is completely disturbed. Of course, this is due to leaching, by which liquid limit decreases while the water content remains almost constant.

Bjerrum (1967) experimentally showed a reduction of liquid limit through leaching process, in which it reduced from 48% to 37% when salt concentration decreased to 1%. Another example is presented in Fig. 16, showing effects of cation concentration and salinity on liquid limit (Sridharan, 2006). This figure illustrates a linear reduction of liquid limit with a decrease of the logarithm of Na^+ ion or salinity respectively.

A measurement of liquid limit for the clay samples taken in the field was carried out together with salinity measurement by Ryu (2003). It is interesting to note that reduction of salinity corresponds well to those of liquid limit, as can be seen in Fig. 17. A decrease of salinity from 23 to 5 g/l resulted in a corresponding reduction of liquid limit from 70 to 50%. A linear relationship between liquid limit and the logarithm of salinity could not be established in this study.

4.2   *Effect of leaching on activity*

As it has been known that plastic limit of clay is almost unchanged during leaching, a reduction of liquid limit implies a decrease of plasticity index. Therefore, a reduction of liquid limit by leaching would result in a decrease of activity. Such examples are given in Fig. 18 (Chung et al., 2003), in which the variation of activities with depth appears clearly for two clays. Activities below 19 m in Jangyu are lower than those of the upper portion and those below 30 m in Shiho are also lower than those of the upper portion. The depths

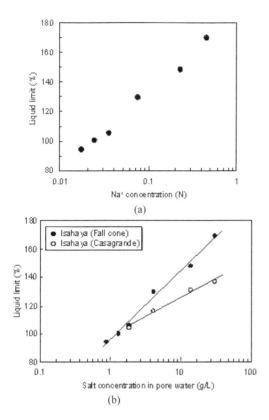

(a)

(b)

Figure 16. (a) Relationship between liquid limit and Na⁺ ion and (b) liquid limit and salt concentration for Isahaya clay (Sridharan, 2006).

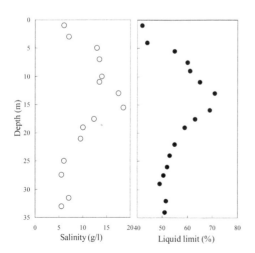

Figure 17. Relationship between salinity and liquid limit for a borehole located at Boembang-dong (Ryu, 2003).

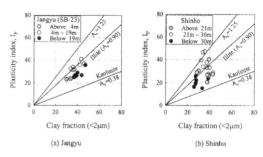

(a) Jangyu

(b) Shinho

Figure 18. Activities for clay in Jangyu and Shinho (Chung et al., 2003).

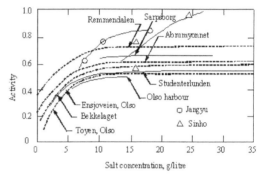

Figure 19. Effects of leaching on the activity of Norwegian clays and Busan clay (Bjerrum, 1954).

of low activities are within the zone of leaching. The measurements agree well with the theory.

Bjerrum (1954) conducted research on the effect of leaching on activity for Norwegian clays (see Fig. 19). From his leaching experiments, it is known that activities reduce when salt concentration drops below 10 g/l and a sharp drop occurs when it reduces below 5 to 6 g/l. The measurements of two clays shown in Fig. 18 are plotted in this figure for comparison. The activity of Busan clay is greater than that of Norwegian clays and more sensitive to reduction of salinity. The activity of Busan clay starts to reduce even at salinity decrease to 15 g/l.

4.3 Effect on sensitivity

Presented in Fig 20 are measurements of sensitivity for Busan clay conducted at different locations in the area. The measurements were carried out by means of both in-situ vane test (FV) and unconfined compression test (UC) (Baek, 2002). Values measured by the former are much lower than those of the latter. This may be due to incomplete disturbance of soil with the in-situ vane as the rotation of the vane would be confined within the soil mass. Neglecting the measurements by the vane, this figure clearly illustrates that sensitivities below 15 m to 20 m in depth are higher than the values at

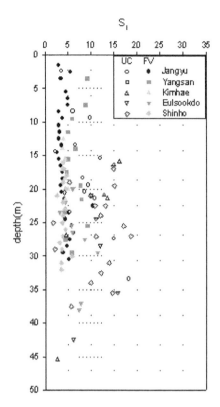

Figure 20. Measurements of sensitivity for different locations of the delta area by means of UC and FV (Baek, 2002).

Table 3. Classification of sensitivity.

Skempton & Northy (1952)	Rosenqvist (1963)
1.0: Insensitive clays	1.0: Insensitive clays
1–2: Clays of low sensitivity	1–2: Slightly sensitive clays
2–4: Clays of med. sensitivity	2–4: Medium sensitive clays
4–8: Sensitivity clays	4–8: Very sensitive clays
8–16: Extra-sensitivity clays	8–16: Slightly quick clays
>16: Quick clays	16–32: Medium quick clays
	32–64: Very quick clays
	>64: Extra quick clays

shallower depths, the depth agreeing with the upper boundary of the leached zone.

From those measurements, it is known that the sensitivity of the unleached zone for Busan clay is around 5 but the value increases to 10 to 15 in the leached zone. If Busan clay is classified according to Table 3, the sensitivity of the leached zone of Busan clay can be described as extra sensitive clay (Skempton and Northy, 1952) or slightly quick clay (Rosenqvist, 1963).

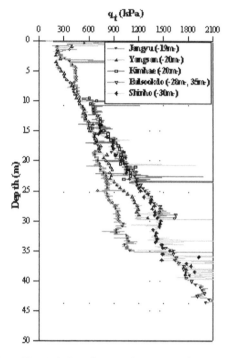

Figure 21. Variation of cone resistance, q_t, with depth for different locations in the delta area (Baek, 2002).

4.4 Effect on undrained shear strength

A large number of measurements of undrained shear strength have been carried out in the field as well as in the laboratory. Fig. 21 summarize measurements of cone resistance, q_t, for different locations of the delta area (Baek, 2002). The values of q_t, in general, increase linearly with depth as can be seen in the figure. Undrained strength measured by in-situ vane test also showed a linear increase with depth and shear strength ratio, s_u/σ'_{v0}, was 0.22 for clay deposits on the seaside (Chung et al., 2006). Considering plastic limit of around 25% for Busan clay, the result agrees well with behavior of young normally consolidated marine clay. As can be seen in Fig. 21, however, a deviation from the linear increase of shear strength with depth is found in all measurements, coinciding with the upper boundary of the leaching zone.

Fig. 22 illustrates clearly a decrease of shear strength due to desalination. For the clay at Jangyu, a linear increase of undrained shear strength was not obtained for salt concentrations below 10 g/l. In case of the clay at Shinho, an increase of shear strength was realized for salt concentration decreasing to 15 g/l. Salt concentration responsible to the reduction of shear strength is between 10 g/l and 15 g/l. After a decrease, the shear strength starts to increase with depth showing

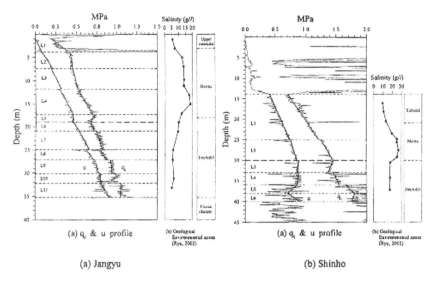

(a) Jangyu   (b) Shinho

Figure 22. Relationship between shear strength and desalination (Chung et al., 2003).

a slight decrease of the value of s_u/σ'_{v0}, as can be seen in the figures.

In his Rankine lecture, Bjerrum (1967) presented an experimental study illustrating well the effect of leaching on shear strength and compressibility for Drammen clay in Norway. Fig. 23 (a) illustrates that isostatic uplift resulted in artesian pore pressure and thereby vertical effective pressure was reduced. Leaching resulted in a decrease of void ratio and thus further loading did not follow the sedimentation curve. While leaching was taking place, undrained shear strength was reduced and the ratio, s_u/s'_{v0}, reduced from a value of 0.20 to a value of 0.12.

Comparing this experimental study to the in-situ shear strength behavior of Busan clay, a reduction of the shear strength ratio is negligible for the latter case. It is also noted that desalination has taken place until 5 g/l for Busan clay. It is not known whether desalination would proceed further through geological history. For Drammen clay, the dominant clay mineral was illite the same for Busan clay.

4.5 Effect on compressibility

Effect of leaching on compressibility has not been investigated either in the laboratory or in the field for Busan clay. Referring to the works of Bjerrum (1967) presented above and Kazi and Moun (1972), it is obvious that the settlement due to leaching has taken place even for Busan clay through a long geological history. As already presented in Fig. 24 (b) (Bjerum, 1967), the vertical strain increased from 3% to 4% during the leaching process, in which the salt concentration of the pore water had been reduced from 21 g/l to 1 g/l.

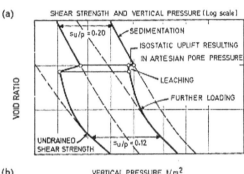

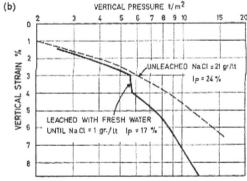

Figure 23. Change in compressibility and shear strength of marine clay caused by leaching: (a) general principle; (b) results of leaching tests in the laboratory with a lean clay from Drammen (Bjerrum, 1967).

Kazi and Moun (1972) conducted an interesting consolidation test with Drammen clay as presented in Fig. 24. According to their experiment, two identical undisturbed salty clay samples were gradually loaded

85

in oedometers until the vertical pressure reached to 5–6 t/m^2. One sample was leached until salt concentration dropped to 1 g/l but another sample was not leached. It is amazing to see the result that secondary settlement was 1.5% only but leaching settlement was 4.5%. The result demonstrated that the net 3% of settlement was contributed by leaching.

4.6 Effect on OCR

An artesian pressure uplifts a soil layer and results in a decrease of effective overburden pressure. Then the maximum past pressure estimated from an oedometer test or in the field becomes lower than the effective overburden pressure without artesian condition. This stress condition can be expressed by over-consolidation ratio (OCR).

Fig. 25 presents a variation of OCR with depth for the clay at Yangsan, which was determined by in-situ tests and the lab test (Lim et al., 2003). As direct measurements are not possible with in-situ equipments, interpretations were made using empirical formula such as Marchetti (1980), Lunne et al. (1997). The values of OCR vary depending on the equipment used but it is noted that all measurements indicate OCR gradually decreasing with depth from 7 m below the ground. .

Fig. 26 presents the distribution of OCR with depth for the clay at Hwajung. OCR was estimated from the results obtained from the oedometer test. As soil disturbance was greater when undisturbed samples were taken from deeper depths, Schmertmann's correction was applied for the determination of the maximum

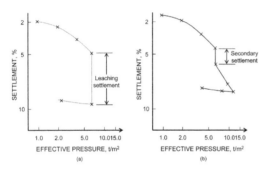

Figure 24. Comparison of settlements by (a) leaching and (b) by secondary settlement for Drammen clay in Norway (Kazi and Moum, 1972).

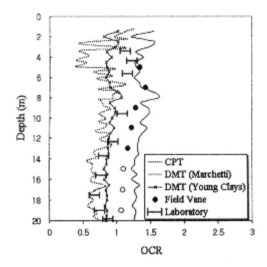

Figure 25. Measurement of OCR with in-situ tests for the clay at Yagsan (Lim et al., 2003).

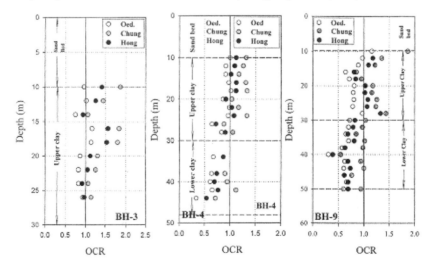

Figure 26. OCR measured in the lab for clay at Hwajung (Rao, 2004).

past pressure. The results show that the values of OCR are over unity above a depth of 30 m, while the values are less than unity for depths below. This is because the lower portion of the soil layer is subjected to artesian pressure.

5 CONCLUSIONS

The old terrain of the delta area was a deep valley formed by a vertical movement of Yangsan fault which runs parallel with the main stream of the river. Different layers have been deposited affecting by the river flow, the marine transgression and regression. The deposits are characterized by formation of distinguished different layers depending on the rise and fall of the sea level. The upper and lower silty clay layers are thick and normally consolidated. The clay, which is referred to Busan clay, has been accumulated under brackish water and marine environment during the rise of sea level. The bottom sand and gravel layer is an aquifer providing artesian pressure. The artesian pressure originates from high water level on the upstream and surrounding hills and mountains.

The lower portion of the upper clay layer, as well as the lower clay layer, has been leached by the upward flow of fresh water from the aquifer below. In the leached zone salt concentration dropped up to 5 g/l and Na^+ ions have been remarkably reduced. Leaching resulted in low plasticity, high sensitivity, low shear strength and low OCR. Sensitivity of Busan clay of the leached zone increases to a value of 10 to 15, which is classified as a very sensitive clay or slight quick clay. Leaching of Busan clay started much later geologically than quick clays in Norway and Canada which were produced by melting of glaciers in the Pleistocene glacial age. Leaching of Busan clay was gradual.

ACKNOWLEDGEMENTS

This paper is the summarized result of research works related to the deltaic deposit of the Nakdong River mouth and its geotechnical properties. A great number of researchers and engineers have contributed to this study. Their contribution is gratefully acknowledged. The final manuscript was reviewed by Prof. Madhav, M.R. and Prof. Kim, Y.T.

REFERENCES

Baek, K.J. 1998. Effects of depositional environment of the Nakdong delta on engineering properties of the deltaic deposit, Dissertation, Dr. of Engineering, Dongguk University, Seoul.

Baek, S.H. 2002. *Analysis of geotechnical characteristics with field measurements for Busan clay* (Korean).

Baek, S.H. 2005. *Investigation report on pore-water pressure measurements by piezometer for New Port Wharf site* (Korean), Samsung Mulsan, Seoul.

Bejerrum, L. 1954. Geotechnical properties of Norwegian marine clays, *Geotechnique*, 4: 49–69.

Bjerrum, L. 1967. Engineering geology of Norwegian normally-consolidated marine clays as related to settlements of buildings, *Geotechnique*, 17: 81–118.

Brenner, R.P., Chilingarian, G.V., and Roberson, Jr. J.O. 1981. Engineering geology on soft clay, *Soft Clay Engineering*, Brand E. W. and R. P. Brenner (eds): 159–238, Elsevier Scientific Publishing Co. Amsterdam.

Cho, H.R. 1987. *Alluvial plains in Korea* (Korean), Kyohak Yunggusa, Seoul.

Chung, S.G., Back, S.H., Ryu, C.K. and Kim, S.W. 2003. Geotechnical characterization of Pusan clays, *Prec., Korea-Japan Joint Workshop on Characterization of Thick Clay Deposits: Reclamation and Port Construction*: 3–44, ATC-7, Busan.

Chung, S.G., Kim, S.K., Kang, Y.J., Im, J.C. and Nagendra, P. 2006. Failure of a breakwater founded on a thick normally consolidated clay, *Geotechnique*, 56 (6): 393–409.

Daewoo Engineering. 2001. *An investigation of artesian pressure at the boat race site*. Seoul.

Dongil Consultants. 2003. *Advisory note on selection of pile foundation for Busan-Gimhae light rail electric car (Korean)*, Seoul.

Fairbridge, R.W. 1961. Eustatic changes in sea level, *Physics and Chemistry of the Earth*, 4: 99–186.

Industrial Research Institute. 2006. *Report on soil investigation for Myungji and Sinho residential complexes*, Dong-A University, Busan.

Jeong, K.H., Park, S.J., Park, J.H., Kim, C.H., Oh, K.T., Suwa, S. and Tanaka, H. 2001. Korea-Japan joint research on the soft clay deposit in Yangsan (Korean), *Proc. of ATC-7*: 165–181, Busan.

Han, Y.C. and Ryu, G.Y. 1999. A study on artesian pressure distribution in Yangsan-Mulgum area (Korean), *Proc. of Korean Geotechnical Society*, Seoul.

Huh, D.Y. 2003. *Research on geotechnical properties for Busan clay in the region of New Port* (Korean), Dissertation, Dr. of Engineering, Dong-A University, Busan.

Kazi, A. and Moum, J. 1972. Effect of leaching on the fabric of normally consolidated marine clays, *Proc. Int. Symp., Soil Structure*: 137–152, Gotothenburg.

Kenney, T.C. 1964. Sea level movements and the geologic histories of the post glacial marine soils at Boston, Nicolet, Ottawa, and Oslo, *Geotechnique*, 14: 203.

Kim, S.K. 1999. Relevance of foundation design to engineering characteristics of Kimhae clay, *Proc., Sang-Kyu Kim Symposium on Geotechnical Engineering* (Korean): 205–230, Seoul.

Kim, S.K. 2002. Engineering problems for thick clay deposit in the Nakdong estuarine delta, *Special lecture*, Kyushu Branch, The Japanese Geotechnical Society, Kyushu.

Kitada, N., Suwa, S., Saito, R., Iwaki, Shogaki, T., Nochigawa, Y., Jeoung, G.H. and Park, S.Z. 2003. Geological investigation of Pusan alluvial plain, *Proc. of Korea-Japan Joint Workshop, Characterization of Thick Clay Deposits: Reclamation and Port Construction*, ATC-7: 129–135.

Lim, H.D., Lee, C.H., and Lee, W.J. 2003. Geotechnical Characteristics of Yangsan Clay, *Proc. Korea-Japan*

Joint Workshop: Characterization of Thick Clay Deposit, Reclamation and Port Construction, ATC-7: 59–70.

Lunne, T., Robertson, P.K. and Powel, J.J. M. 1997. *Cone penetration testing in geotechnical practice*, Blacki Academic & Professional, an imprint of Chapman & Hall.

Marchetti, S. 1980. In-situ test by flat dilatometer, *Journal of Geotechnical Engineering*, ASCE, 106: 299–321.

Moun, J., Laken, T., and Torrence, J.K. 1971. A geotechnical investigation of the sensitivity of a normally consolidated clay from Drammen, Norway, *Geotechnique*, 21: 329–340.

Park, Y.A., Kim, B.K., and Zhao, S. 1998. Sea level fluctuation in the Yellow Sea basin, *Journal*, the Korean Society of Oceanography, 29(12): 42–49.

Pusch, R. 1973. Physico-chemical processes which affect soil structure and vice versa. *Proc. Int. Symp. Soil Structure*, Gothenberg, Appendix: 27–35.

Rosenqvist, I. Th. 1953. Considerations on the sensitivity of Norwegian quick-clays, *Geotechnique*, 3: 195–200.

Ryu, C.G. 2003. *Engineering geology for clayey deposits of Holocene epoch in the Nakdong river mouth* (Korean), Dissertation, Dr. of Science, Busan National University, Busan.

Skempton, A.W. and Northy, R.D. 1952. The sensitivity of clays, Geotechnique, 3: 30–53.

Sridharan, A. 2006. Engineering behavior of marine clays: some fundamental aspects, Special lecture, *Proc., 2006 ISLT*, Saga.

Suwa, S. 2002. Lecture notes, *Japan -Korea Seminar on Nakdong Delta in Busan and Osaka Bay*, Dong-A University, Busan.

88

Geotechnical and Geophysical Site Characterization – Huang & Mayne (eds)
© 2008 Taylor & Francis Group, London, ISBN 978-0-415-46936-4

Design parameters from in situ tests in soft ground – recent developments

M. Long
University College Dublin (UCD), Ireland

ABSTRACT: Geotechnical engineers have available a wide range of powerful tools for the purposes of soil characterisation. This work presented here confirms that design parameters in soft soils can be reliably obtained from in situ testing, used in conjunction with traditional sampling, provided engineers carefully assess the measured data and that appropriate correlations are used. CPTU pore pressure readings are particularly useful and such measurements should become standard. Pressuremeters, dilatometers and full flow probes all have a strong role to play, provided they are used correctly. In situ shear wave velocity can be measured easily and reliably by a variety of methods, independent of the technique used and of the operator.

1 INTRODUCTION

The purpose of this paper is to present recent developments in obtaining design parameters from in situ tests in soft ground. This is of course not the first time this task has been attempted and it is worth considering similar papers presented over the last 20 years. Some of these are listed on Table 1 below.

This paper will consider new developments in assessment of design parameters in soft soils over the past 4 years since the ISC'2 conference at Porto. The focus will be on practical routine engineering application for medium to important projects with emphasis on the piezocone (CPTU), full flow penetrometers, pressuremeters, dilatometer (DMT), field vane test (FVT), and using in situ geophysics techniques.

Many of the in situ techniques that will be discussed here have now been in use for some time and can be considered to have "come of age". The author has in mind the comments of Powell (2005) who said "many practitioners have felt that too often the capabilities of in situ tests have been over-sold or inappropriately applied" resulting in "dissatisfaction when the tests failed to deliver as promised". Powell (2005) rightly pointed out that consistency of both equipment and operation is essential. Therefore this paper will focus on the reliability, repeatability and accuracy of the various instruments, with respect to practical application.

Recently developed, research type, equipment will not generally be considered. Also emphasis will be on geotechnical parameters for building and infrastructure design and no consideration will be given to environmental parameters. Assessment of liquefaction potential (see Robertson, 2004) or issues relating to cyclic loading will in general not be covered.

Table 1.   Summary of previous keynote lectures on topic.

Date	Authors	Publication	Focus
1989	Lunne et al.	XII ICSMFE, Rio de Jan.	Design parameters from CPTU, DMT, FVT
1994	Jamiolkowski and Lo Presti	XIII ICSMFE, New Delhi	Young's modulus, K_0
1997	Tatsuoka et al.	XIV ICSMFE, Hamburg	Small strain stiffness
1998	Greenhouse et al. Fahey Houlsby	ISC'1 Atlanta	Geophysics Stiffness, K_0 Advanced interpretation
2001	Mayne Lunne	Both In Situ 2001, Bali	Empirical correlations Offshore application
2004	Stokoe et al. Yu Schnaid et al. Randolph Robertson	All ISC'2 Porto	Geophysical tests Theoretical Unusual materials Offshore application Liquefaction
2005	Schnaid	XVI ICSMGE, Osaka	Theoretical and empirical, sands and bonded materials
2006	Mayne	Natural Soils – 2 Singapore	Empirical, link to critical state

The paper will cover, in turn, issues associated with the equipment used, techniques for soil classification and finally recent developments in determination of soil parameters.

In the author's opinion, design parameters in soft ground should not be derived from in situ testing alone. In situ tests are ideally used in combination with physical logging and identification of the soil and laboratory tests on good quality samples.

2 EQUIPMENT

2.1 *Piezocone (CPTU)*

2.1.1 *General*
There seems little doubt but that the CPTU is now the most widely used in situ testing device for soft ground throughout the world. For a description of the equipment the reader is referred to the textbook by Lunne et al. (1997).

Up to the present the best official guideline for performing CPTU was the IRTP (International Reference Test Procedure) published by the International Society for Soil Mechanics and Geotechnical Engineering in 1999 (ISSMGE, 1999). A European Standard (ENISO 22476-1, 2007) has been completed and will be officially valid from late 2007 or early 2008. This document is an updated version of the IRTP, based on the same principles.

For the purposes of this paper it is assumed that, unless otherwise stated, the equipment used is the standard $10 \, cm^2$ cone, pushed at 2 cm/s with pore pressure measured in the u_2 position.

2.1.2 *Accuracy of transducers*
Some typical CPTU results for 2 well characterised research sites, namely the UK National soft clay research site at Bothkennar in Scotland and the Norwegian Geotechnical Institute (NGI) soft clay research site at Onsøy are shown on Figures 2a and 2b respectively. The location of these two sites together with others referred to in this paper is shown on Figure 1. The Bothkennar test was carried out by UCD and Lankelma Ltd. using a GeoPoint cone and that at Onsøy using an A.P van den Berg cone. Over the top 20 m at both sites cone resistance, corrected for pore pressure effects (q_t) is of the order of 0.4 MPa to 0.8 MPa, sleeve friction (f_s) is 5 to 15 kPa and pore pressure (u_2) is 0.2 MPa to 0.6 MPa.

Application Class 1 (for use in soft soils) of the European Standard requires that the "minimum allowable accuracy" should be the larger of:

q_t = 0.035 MPa or 5% of measured value
f_s = 5 kPa or 10% of measured value
u_2 = 0.01 MPa or 2% of measured value.

Although it is more instructive to compare these values to a range of measurements as will be done later, it can be seen that for both the Onsøy and Bothkennar tests, the measured q_t and u_2 values are significantly greater than the required accuracy. However the f_s

Figure 1. Location of test sites in Western Europe (not to scale).

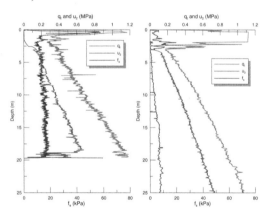

Figure 2. Typical CPTU results for soft clay sites (a) Bothkennar CPT002 (Boylan et al., 2007) and (b) Onsøy TestB5 (Lunne, 2005).

values are often less than the value required by the Standard. Measured sleeve friction will be influenced by (Lunne et al., 1986, 1997):

1. Pore pressure effects on the ends of the friction sleeve. This effect depends on the actual areas at each end of the sleeve and the difference in pore pressure at the top (u_3) and bottom (u_2). A correction should be made if possible.
2. Location of the sleeve relative to the cone tip and distribution of side friction and degree of remoulding behind the tip.

3. Interface friction between the soil and the sleeve (surface roughness).
4. From an instrumentation point of view f_s appears relatively less accurate compared to q_t and u_2 (e.g. measured value much less than full range of load cell).

A useful summary of these issues, with respect to soft soils, is given by Lunne and Andersen (2007). However further research work and equipment development, involving both geotechnical and instrumentation engineers, is needed in order to quantify these effects. Nonetheless care needs to be taken when using sleeve friction values for interpretation of soil strata or design parameters from the CPTU.

2.1.3 *Checks and corrections*

Perhaps the most important checks and corrections that need to be applied to CPTU results are:

i. Correction for pore pressure effects
ii. Correction for inclination
iii. Transducer drift (shift in zero values)
iv. Temperature.

It is vital that the user of the data checks that correction (i) is done correctly to the European Standard. In particular the user should check the correct area ratio (α) has been used. Ideally α should be determined experimentally in a simple calibration chamber (Lunne et al., 1997) as the actual α can vary considerably from the theoretical depending on how the seals are fabricated.

The influence of item (ii) is usually low, except in deep deposits of soft clay, but a simple correction can be made according to the European Standard.

Frequently issues (iii) and (iv) can have a very significant influence in the results. These two factors are interrelated as one main reason for the shift in reference values is temperature effects. However other factors such as load cell hysterisis can also alter the zero reference values. It is vital to check the load cell readings immediately on extraction, to assess whether there has been any drift. Peuchen et al. (2005) reviewed zero drift data for three CPT projects all with Fugro offshore systems and found that, for clays, the mean zero drift for q_c in 5 cm^2 cones (between start and completion of test) was about 0.03 MPa.

Work by Lunne et al. (1986) illustrated the importance of these effects. Boylan et al. (2008b) re-examined the problem for modern commercial cones. An example is shown for a typical Irish peat site at Loughrey on Figure 3a (Boylan and Long, 2006). Due to the soft highly compressible nature of the peat, q_t and f_s values are very low and occasionally become negative. Only the u_2 readings appear reliable. Similarly Boylan et al. (2008b) report CPTU data for the TU Delft A2 peat research site in the Netherlands, see

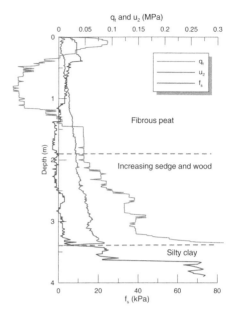

Figure 3a. CPTU results for a typical Irish peat site at Loughrey (Boylan and Long, 2006).

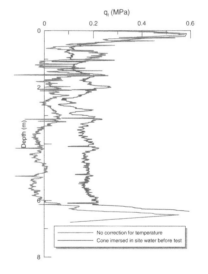

Figure 3b. Temperature effects on CPTU data from A2 peat research site in the Netherlands, Boylan et al. (2008b).

Figure 3b. Initially the cone was used directly as delivered to site, yielding the unreliable results. When the cone was immersed in a container of water from a dyke adjacent to the site for some time bore testing, a much more dependable profile resulted.

The main reason for the negative values can be seen in the results of a trial on the same GeoPoint cone used at Loughrey and another commercial cone produced by GeoMill given on Figure 4 (Boylan et al., 2008b). Output from the cones was initially equilibrated at room temperature of about 20°C and then the cones were immersed in water at a lower temperature so as to observe the effect of the differential temperature on the measured parameters. The temperature effect is very large on both q_c and f_s and confirms that (as required by the European Standard) prior to undertaking any test the cone should be immersed in a container of groundwater at the same temperature as that at the site until the readings become stable.

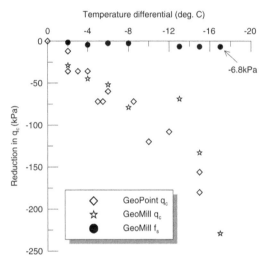

Figure 4. Influence of temperature on measured cone tip readings (Boylan et al., 2008b).

2.1.4 *Influence of different equipment*

Recent studies and experience show that even if the European Standard is followed the results can still vary significantly according to the equipment used. For example numerous CPT and CPTU tests have been carried out the Bothkennar site. The overall uniformity of the site has been confirmed by a large number of tests using the same cone by Hawkins et al. (1989). However tests in the same small area of the site with different cones can yield variable results. For example the recent UCD / Lankelma work (Figure 2a) using a GeoPoint cone is compared to that reported by Powell and Lunne (2005) for five different piezocones on Figure 5a to c. All of the Powell and Lunne (2005) tests were carried out by the same operator using the same data acquisition system. He had also carefully calibrated all of the cones so as to minimize any potential errors. The u_2 data shows least scatter. There is some scatter in the q_t data but it is significantly less than that of the f_s results. The GeoPoint q_t values are higher than those measured by the other cones. The f_s and u_2 values are very similar.

In contrast the UCD/Lankelma q_{net} ($=q_t - \sigma_{v0}$) data are more or less the same as those reported by Jacobs and Coutts (1992) for the Fugro – McClelland 10 cm^2 and 15 cm^2 cones, see Figure 5d.

The study by Tiggelman and Beukema (2006), at Almere in The Netherlands, included measurements of q_c and f_s by 11 different contractors in Dutch Holocene clay overlaying Pleistocene sand. They reported that the natural variability in the clay layer was relatively small, still q_c varies by a factor of 2 to 3 and the friction ratio deviates by a factor of 2 to 5, see Figure 6.

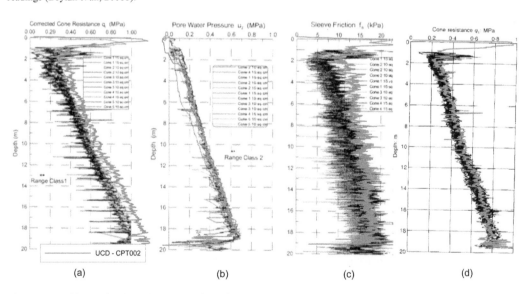

(a) (b) (c) (d)

Figure 5. Bothkennar CPTU results – comparison of UCD/Lankelma CPTU tests and those from (a to c) Powell and Lunne (2005) and (d) Jacobs and Coutts (1992).

Unfortunately pore pressure was not measured so the corrected cone resistance could not be computed. The scatter in q_t is likely to have been less than in q_c.

For both the Bothkennar and Almere sites the range in q_t (or q_c) and u_2 values was well in excess of the requirement of Application Class 1. However the range of f_s values were of the same order of magnitude.

A series of CPTU's were carried out at NGI's soft clay test site in Onsøy. Six organizations performed the tests together with 8 different CPTU devices (Lunne and Powell, 2008). The main objective of the work was to check the influence of equipment type and to evaluate if cone penetrometers used for commercial and research projects can meet the requirements of Application Class 1 of the European Standard. The results are shown on Figure 7 and it can be seen that:

- As for Almere, the range of q_t and u_2 values is well in excess of the requirement of Application Class 1. However the range of f_s values is again of the same order of magnitude.
- The scatter in u_2 values is relatively small (data for 3 devices showing clear lack of proper saturation omitted).
- Compared to u_2 the scatter in q_t values is relatively large.
- f_s values show relatively large variations.

Watson and Humpheson (2007) reported on the results of CPTU tests carried out using three different penetrometers at the site of the Yolla A platform offshore Australia. The site is underlain by normally consolidated silts and clays and sands. Significant differences, particularly in f_s, were noted for the three instruments used.

2.1.5 Conclusions for CPTU

The comments made in the previous sections are intended to make potential users aware of the pitfalls

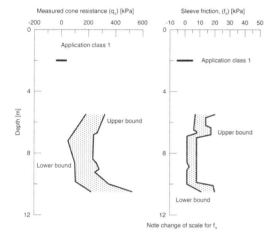

Figure 6. CPT data from Dutch Almere research site (Tiggelman and Beukema, 2006).

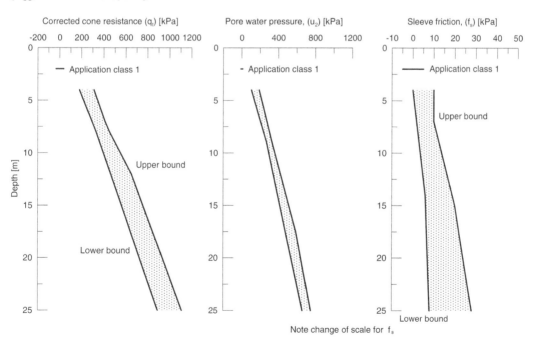

Figure 7. CPTU results from Onsøy (Lunne and Powell, 2008). Note Application Class 1 means that of ENISO 22476-1 (2007).

93

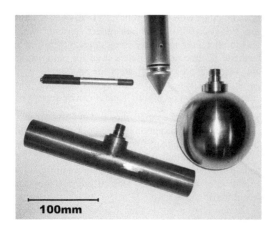

Figure 8. Full flow probes (Boylan and Long, 2006).

with the CPTU. They are not intended to be negative as in the author's opinion the tool is of great value if used properly. Good practice following the European Standard will reduce the variation from one test to another. If the pore pressure measurement system is sufficiently well saturated the measured pore pressure is the parameter that shows least variation from one type of CPTU equipment to another. The q_t values show somewhat more variation from one type of equipment to another as compared with the measured pore pressure.

Measured sleeve friction shows most variation from one type of equipment to another and these values should be treated with caution.

From the above it should be expected that derived soil parameters or soil classification based on pore pressure readings should be expected to be more reliable compared to classification based on sleeve friction measurements. This topic will be addressed later. Nonetheless it is recommended that pore pressure readings should be included in all routine investigations.

2.2 Full flow penetrometers

A feature of the 24 papers on in situ testing presented to the 2nd ISC Conference in Porto in 2004 was that four of them detail experience of full flow probes. These include the T-bar, ball (Figure 8) and plate penetrometers. Research in this area commenced in Australia in the late 1980's when researchers developed a miniature T-bar for use in the centrifuge so as to improve resistance measurements (Stewart and Randolph, 1991, 1994). These penetrometers were later scaled up for use in offshore site investigations (Randolph et al., 1998). The main motivation was to provide estimates of undrained and remoulded shear strength (s_u and s_{ur}) for very soft soils for both offshore and onshore use.

Only later, around 1997, did the Australian researchers become aware of work in the late 1930's at the Swedish Geotechnical Institute (SGI) by Kallstenius (1961) who describes the SGI Iskymeter, which is not unlike the T-bar penetrometer.

The ball penetrometer was later developed to reduce the chance of the load cell being subjected to bending moments induced from non-symmetric resistances along the T-bar (Watson et al., 1998). These penetrometers have several advantages over the standard cone (Randolph, 2004):

- The measured resistance requires minimal correction for pore pressure effects compared to possibly significant adjustment to the cone resistance.
- Improved resolution is obtained in soft soils due to the larger projected area – $100 \, cm^2$ compared to the $10 \, cm^2$ of the cone. This also results in reduced sensitivity to load cell drift.
- Measurements can be made during extraction as well as penetration thus giving additional useful data.
- Plasticity solutions based on simplified assumptions of soil behaviour exist which relate the net resistance to the shear strength of the soil (Randolph & Houlsby, 1984). More sophisticated solutions have also been developed (Randolph et al., 2000, Einav and Randolph, 2005, Martin & Randolph, 2006, Randolph and Anderson, 2006).

Recently more use has been made of the T-bar and ball probes rather than the plate as these are structurally more rigid. Furthermore the ball probe seems to be favoured over the T-bar due to bending effects on the T-bar load cell and some difficulties in the buckling of the driving rods when the T-bar is eccentrically loaded (see e.g. Long and Gudjonsson, 2004). The ball probe (possible of reduced diameter) is also more compatible with temporary steel casing used offshore (Peuchen et al., 2005).

A further recent development has been the inclusion of a pore pressure measuring element in the T-bar and ball (piezoball probe), see Kelleher and Randolph (2005), Peuchen et al. (2005) and Boylan and Long, (2006 and 2007). The balls used were developed by Benthic Geotech, Fugro and by Lankelma UK Ltd. respectively. Other piezoballs have been developed by Professor de Jong and his colleagues at University of Davis, California (see proceedings of this conference) and by Professor Randolph and his team at University of Western Australia.

A significant issue is that there is no standardisation in the design of these instruments, particularly with respect to the location of the pore water pressure transducer. For example the Benthic and Lankelma piezoballs are shown on Figure 9. The former has a diameter of 60 mm and comprises hardened smooth steel. The pore pressure transducer is at mid height.

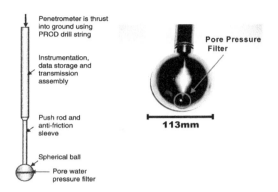

Figure 9. Piezoballs produced by (a) Benthic Geotech (Kelleher and Randolph, 2005) and (b) Lankelma, UK Ltd. (Boylan and Long, 2007).

Figure 10. Large arch Crozet Bridge (Monnet et al., 2006)

The Lankelma piezoball has a diameter of 113 mm with a lightly sandblasted surface and is capable of making pore water pressure readings through two 3.5 mm diameter porous elements at opposite sides of the ball, a third the way up from the tip. The Fugro ball has filters at its base and in the shaft just above the ball.

Kelleher and Randolph (2005) stated that the location of the probe on the Benthic ball reflects pore pressure change caused by shearing of the soil and is a very sensitive to changes in soil grading with depth and to the dilative response of the soil. Boylan and Long (2007) evaluated the performance of the Lankelma piezoball and found that the pore pressure readings tend to generally increase during a pause in penetration when a rod is being added and stay rather static during penetration. This may be due to the nature of soil flow around the ball but may also indicate a delayed pore pressure response caused by the relatively long channel which connects the filters on the bottom of the ball with the pore pressure element in the cone.

Peuchen et al. (2005) presented some data measured by pore pressure transducers on the axis of the T-bar one at its centre and one at its edge. T-bar pore pressure data is similar to that recorded by the piezoball. However Randolph et al. (1998) reported some pore pressure data from a T-bar with the pore pressure transducers in a similar location but the response is clearly not as good. Clearly there is a need for standardization of these useful tools.

2.3 *Pressuremeters*

After CPTU/full flow probes, perhaps the second most widely used of the modern (excluding SPT) in situ testing devices world-wide is the pressuremeter. The term pressuremeter is used in a broad sense and actually covers a wide variety of devices. A useful summary of recent developments in this topic can be found in the Proceedings of the International Symposium 50

years of pressuremeters (ISP5 – Pressio 2005) held in Marne-la-Valée, France in August, 2005. Papers submitted to the conference were from the following countries:

- France = 30 papers
- USA = 12 papers
- UK = 5 papers
- Algeria = 4 papers
- Tunisia = 3 papers
- Belgium, Germany, Brazil, Malaysia = 2 each
- Further 10 countries represented by 1 paper
- 29 National reports.

Although as expected most interest and use of the pressuremeter was from France and from French speaking countries there was also considerable interest in the use of the devices throughout the world. It is interesting to note the contribution of Lukas (2005) who summarised application of the pressuremeter in the US as follows:

- Devices being used in granular, residual and over-consolidated clayey soils and in improved soils.
- Settlement predictions are in reasonable agreement with measured values
- s_u from pressuremeter higher than for unconfined compression test but is more reliable.

The pressuremeter is widely used for foundation design in stiff clays. Baker (2005) reported that foundations for 9 of the sixteen highest buildings in the US were designed using the pressuremeter. He also detailed the case history of the Petronas Twin Towers in Kuala Lumpur, Malaysia. These were formerly the highest buildings in the world and the pressuremeter was used in the foundation design. The pressuremeter is also used in foundation design for soft and medium stiff clay, e.g. for the large arch Crozet Bridge near Grenoble in France, see Figure 10 (Monnet et al., 2006).

These techniques are also used in Taipei. For example Huang and Hsueh (2005) and Huang et al. (1999)

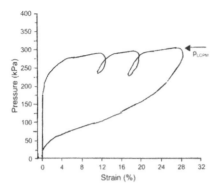

Figure 11. Typical CPM expansion curve (Powell, et al., 2005).

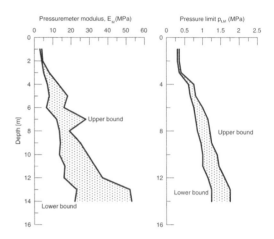

Figure 12. Variation in E_m and p_{LM} as derived from same data (Reiffsteck, 2005).

compare unload – reload shear modulus (G_{ur}) from a series of Cambridge high-pressure dilatometer (HPD) tests performed on soft sandstone for a super high-rise project in Taipei to G_{max} values from down-hole seismic tests.

The papers to ISP5 illustrate that perhaps the greatest additional benefit of pressuremeters over other in situ devices is that:

1. The full stiffness/strain plot (stiffness decay curve) can be obtained from unload/load cycles. An example is shown on Figure 11. Indeed Powell and Shields (1995) pointed out that similar stiffness decay curves will be obtained from the cone pressuremeter (CPM) and the self boring pressuremeter in clays where no significant structural breakdown occurs.
2. Semi empirical design rules can be applied with confidence, though several of the papers to the conference suggested that current practice using the original Ménard rules may be somewhat conservative.

A prediction competition was also held at the ISP5 conference. Participants were given data recorded during Ménard pressuremeter testing and asked to make pile bearing capacity and settlement predictions.

One worrying aspect of the results was that there were significant variations in the basic pressuremeter modulus (E_m) and the pressure limit (p_{LM}) derived by the 9 participants from the same set of input data, see Figure 12 (Reiffsteck, 2005). The variation in the derived parameters increased significantly with depth.

It would seem that similar conclusions for the pressuremeter could be made as for the CPT above. Clearly these are excellent devices and can give valuable and useful information but the tests and the subsequent data analysis need to be undertaken with care to a standard set of procedures so as to ensure correlations and empirical designs are based on sound datasets. Powell (2005) highlighted that in some cases pressuremeter

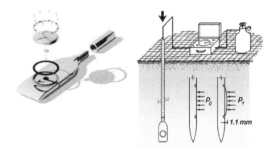

Figure 13. Flat dilatometer.

correlations have been applied in an inappropriate way different from the original derivative.

It also seems clear that there is a future role for cone pressuremeters given the potential usefulness of both devices, though some research work is needed into the derivation of a better relationship of horizontal effective stress (σ'_h) for clays from the CPM.

2.4 Flat dilatometer/seismic dilatometer (DMT/SDMT)

2.4.1 General

Perhaps the third most widely used modern test worldwide for the purposes of soil characterisation and determination of soil parameters is the flat dilatometer (DMT) (Marchetti, 1980). A good summary of recent developments in this topic is given in the 2nd International Flat Dilatometer Conference held in Washington DC in April 2006. The conference proceedings contain 51 papers from seven different countries.

A dilatometer test consists of pushing a flat blade located at the end of a series of rods into the ground, Figure 13. Once at the testing depth, a circular steel membrane located on one side of the blade is expanded horizontally into the soil. The pressure is recorded at

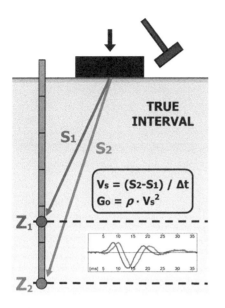

Figure 14. Seismic dilatometer (Monaco and Marchetti, 2007).

specific moments during the test (p_0 and p_1 on Figure 13). The blade is then advanced to the next test depth. Various soil parameters can then be derived from these measurements together with knowledge of the in situ effective stress and pore water pressure.

Recently (Marchetti et al., 2008) have introduced seismic piezocone technology into the DMT, by the inclusion of geophones, to form the seismic dilatometer (SDMT), Figure 14. According to Monaco and Marchetti (2007) the "true-interval" two-receiver test configuration results in a high degree of repeatability of the shear wave velocity (V_s) measurements. This system comprises a cylindrical element placed above the DMT blade, equipped with two receivers located at 0.5 m distance. The signal is amplified and digitized at depth. The shear wave source at the surface is a pendulum hammer which hits horizontally a rectangular steel base pressed vertically against the soil and oriented with its long axis parallel to the axis of the receivers, so that they can offer the highest sensitivity to the generated shear wave.

2.4.2 Repeatability, reliability and accuracy

Readings from the DMT/SDMT have a good degree of repeatability. Although the tests at Onsøy (Figure 15a) were taken some time ago in 1981 they are still useful as they illustrate the near perfect repeatability, except in the top 1 m of desiccated crust. According to Lacasse et al. (1982) and Marchetti (1981), the reduced scatter is particularly encouraging as the tests were carried out by 4 different operators (S. Marchetti, F. Cestari, S. Lacasse and T. Lunne) on different days under different weather conditions.

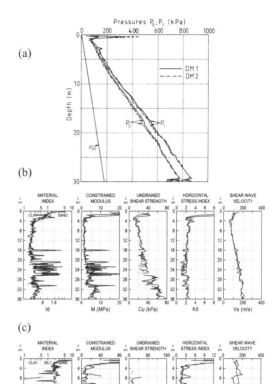

Figure 15. Some examples of repeatability of DMT tests for (a) Onsøy (Lacasse and Lunne, 1982), (b) Fiumicino, Italy and (c) Puerto de Barcelona (Marchetti et al., 2008)

More recent testing in 2005 and 2007 at Fiumicino, Italy (Figure 15b) and at Puerto de Barcelona (Figure 15c) also show excellent repeatabilty. The sites are underlain by different materials (clay and silt) and there were different operators (Diego Marchetti and Norma Perez of IGEOTEST Girona Spain).

Mylanarek et al. (2006) described some work carried out at the Zelazny Most tailings pond in Poland and showed that the SDMT gives highly repeatable V_s values with a difference of at most 1 m/s to 2 m/s between readings. Similarly Marchetti et al. (2008) reported difference of 1 m/s or less for V_s measured by several SDMT tests at the Italian National Research site at Fucino.

Some possible contributions to the reliability/repeatability of the DMT are:

• Once it is set and calibrated, the 1.1 mm centre displacement of the dilatometer blade cannot be

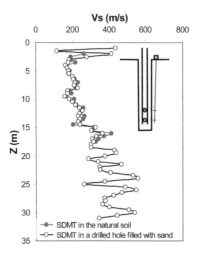

Figure 16. SDMT tests in non-penetrable ground (Marchetti et al., 2008).

Figure 17. ENVI MemoVane II (www.envi.se).

adjusted. It is mechanically fixed in the instrument and is not operator dependant.
• Measurements are by two pressure gauges at ground level. These can be easily calibrated and can have a variety of pressure ranges.
• Pressure transducer precision is 1 about kPa.

2.4.3 Tests in non penetrable soils

Marchetti et al. (2008) reported some work carried out in soils which are not penetrable by the DMT in its normal mode of operation. Here a borehole is drilled and filled with loose sand and the seismic portion of the test is carried out in the normal manner. As can be seen from Figure 16, the results of the tests (V_s only) in virgin soil and from the sand filled borehole are more or less identical. This application could have advantages in soil with a high portion of gravel or at site which is overlain by a considerable thickness of fill or stiff or dense soils.

2.4.4 Summary for DMT/SDMT

Clearly the DMT/SDMT is a reliable and adaptable tool and should be considered for use in any project involving soft ground. Like for the pressuremeter it is important to apply appropriate correlations only. A possible weakness of the DMT is that derivation of geotechnical parameters involves the use of empirical correlations, which were developed some time ago, mainly for Italian soils. This is an area which well warrants research.

2.5 In situ vane testing

Recent developments of in situ vane testing techniques have mostly made use of improved data gathering and recording systems. Advanced vane testing systems record torque down-hole rather than at the surface and torque and time are recorded continuously, typically at 1 Hz (Peuchen and Mayne, 2007). These systems are most commonly used offshore (see for example www.fugro.com) but are also used for advanced onshore application (www.envi.se or www.geotech.se).

For example in the "MemoVane II" produced by the Swedish company ENVI (www.envi.se) measurement takes place immediately above the vane, thus avoiding the influence of friction on the rods. Data is sent acoustically to a laptop at the surface. This makes it possible to use the vane at great depths. The vane is shown in Figure 17 and can be either 55 mm × 110 mm or 60 mm × 120 mm. An aluminum shoe protects the vane during penetration and cleans the vane after each test level. ENVI are currently developing a MemoVane III, which will have the gearbox in the vane unit so the rotation takes place down the hole.

Data is recorded as stress versus time rather than a simple peak and residual value, thus giving additional valuable data on the material. An example is shown for the Onsøy site on Figure 18b.

On Figure 18a, intact undrained shear strength (s_u) values for Onsøy using both a standard Geonor vane (Lunne et al., 2006) and the MemoVane II are compared. The agreement is very good.

A disadvantage of the vane test is if the approach required by regular standards, such as ISO (2003), is followed the test is relatively slow. Typically measurement of intact and remoulded shear strength at one depth takes about 20 minutes. Peuchen and Mayne (2007) presented an approach for rapid in situ vane testing where the measured data needs to be corrected for rate effects. Testing is carried out at limited locations at variable rates, so as rate correction can be applied to rapid tests.

2.6 Standard penetration testing

Despite its well known limitations the standard penetration test (SPT) is often used in the investigation of soft soils. An example of some SPT data in Taipei soft clay is shown on Figure 19.

Undoubtedly the test has some application in assessment of liquefaction potential, in the appraisal of ground improvement or in a regional study where the test is locally well calibrated. However due to

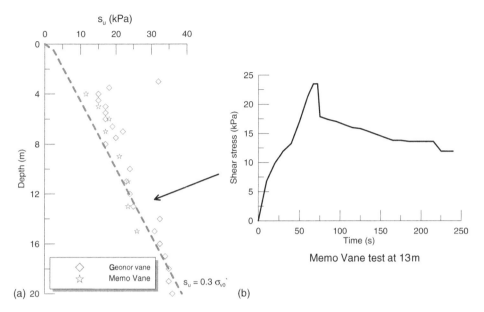

Figure 18. In situ vane data for Onsøy: (a) comparison of Geonor and MemoVane II data and (b) example of stress v time data for MemoVane II.

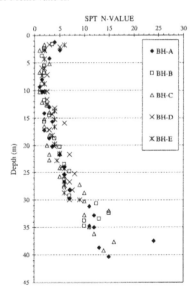

Figure 19. Typical SPT-N profile Taipei soft clay (Chin et al., 2007)

uncertainties over the imparted energy, the influence of the equipment used and the operators the results should be treated with great caution. In the authors opinion the test has little use in the general characterisation of the engineering properties of soft soils and the investment would be better made on a more advanced tests such as the CPTU or DMT.

2.7 Piezoprobe

Onshore in situ pore pressure can normally be measured relatively easily using piezometers. Offshore it can be important to determine whether the pore pressure deviates from the hydrostatic. An alternative is to carry out dissipation tests during a CPTU test. Due to time and cost limitations, this is not always possible and this has led to the development of small diameter piezoprobes, e.g. work by Whittle et al. (2001a and b) at MIT. A good review of this topic is given by Lunne (2001). An important finding of the MIT work was that the pore pressure at the piezo-probe element was influenced by the larger diameter cone or push rods located at the end of the tapered section.

Piezoprobes can also be use for high resolution stratigraphic profiling. This could be useful in thinly layered or laminated soils. An example of such a piezoprobe is shown on Figure 20 (De Jong et al., 2007).

2.8 Geophysical techniques – surface waves

2.8.1 Surface wave techniques

The use of geophysical surface wave techniques has found an increasing application in geotechnical engineering practice. According to Stokoe et al. (2004), who gave a good review of the topic, the recent great interest in this method arises from the non-intrusive nature of the technique combined with the capability of imaging softer layers beneath stiffer material and the

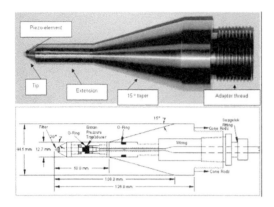

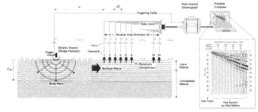

Figure 21. MASW surface wave technique (www. kgs.ku.edu)

Figure 20. Piezoprobe (De Jong et al., 2007).

ability to test large areas rapidly and cost effectively. The four most common techniques are:

- Continuous surface waves (CSW)
- Spectral analysis of surface waves (SASW)
- Multi channel analysis of surface waves (MASW)
- Frequency wave number (f-k) spectrum method.

The steady state Rayleigh wave/Continuous Surface Wave (CSW) technique was introduced by Jones (1958) into the field of geotechnical engineering. It was subsequently developed by others, such as Tokimatsu et al. (1991) and Mathews et al. (1996). The CSW method uses an energy source such as vibrator to produce surface waves.

In the early 1980's the widely used Spectral Analysis of Surface Waves (SASW) method was developed by Heisey et al. (1982) and by Nazarian and Stokoe (1984). The SASW method uses a single pair of receivers that are placed collinear with an impulsive source (e.g. a sledgehammer). The test is repeated a number of times for different geometrical configurations. Crice (2005) acknowledged the usefulness of SASW but suggests that solutions are neither unique nor trivial and that an expert user is required for interpretation.

The MASW technique was introduced in the late 1990's by the Kansas Geological Survey (Park et al., 1999) in order to address the problems associated with SASW. A schematic of this technique is shown on Figure 21. The MASW method exploits multichannel recording and processing techniques that are similar to those used in conventional seismic reflection surveys. Donohue et al. (2004) demonstrated the reliable use of MASW at several soft soil sites in Ireland.

The MASW method has improved production in the field due to multiple transducers, and improved characterisation of dispersion relationship by sampling spatial wave-field with multiple receivers. Advantages of this method include the need for only one-shot gather and its ability to identify and separate fundamental and higher mode surface waves. Crice (2005)

suggests that "MASW is the wave of the future" because of the usefulness and interpretability of the data and the potential for dramatically higher productivity. He illustrates how MASW survey data can be reliably interpreted by computer software without human intervention. Long and Donohue (2007) reported that this is only accurate for simple soil profiles. Significant user experience and intervention is required for more complex profiles as the inversion formulation in MASW can suffer the same uniqueness problems as in SASW. In the view of the author an informed user is certainly important for MASW data analysis.

2.8.2 *Reliability of surface wave techniques*

At least 5 investigations have been carried out at the Bothkennar research site for the purposes of determining shear wave velocity (V_s) and this comprehensive database allows an assessment of the reliability of the various techniques used. These include two surface wave techniques and the investigations were carried out by:

1. University of North Wales (Hepton, 1988): SCPT and SDMT
2. UK Building Research Establishment (BRE) (Powell and Butcher, 1991, Powell, 2001, Hight et al., 2003): cross hole and SCPT
3. Surrey University (SU) (Hope et al., 1999, Sutton, 1999): cross hole
4. GDS Instruments Ltd. (Sutton, 1999): CSW
5. UCD: MASW.

All of the available data are shown on Figure 22. In Figure 22a a comparison is made between the two sets of SCPT data and the UNW SDMT results. The agreement is very good. Figure 22b shows the crosshole data from BRE and SU. The subscripts refer to the directions of propagation and wave polarisation respectively.

The BRE work was carried out using conventional down-hole equipment, whereas the SU investigation included a novel technique for the determination of V_{hh} where the source was at the surface. A clear implication of the data on Figure 22b is that the natural anisotropy of stiffness of Bothkennar clay is very low.

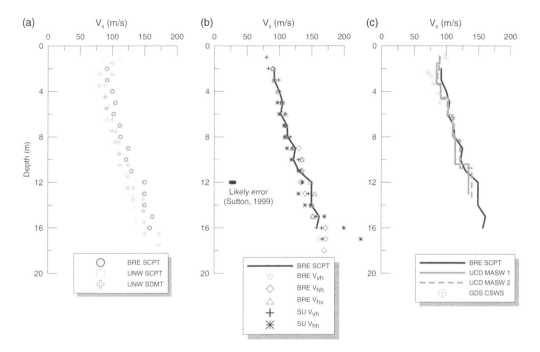

Figure 22. V_s data from Bothkennar: (a) BRE and UNW SCPT and SDMT, (b) BRE and SU cross-hole and (c) surface wave techniques.

This has recently been confirmed by multi directional bender element tests by Bristol University on high quality block samples of the clay (Nash et al., 2006 and Sukolrat, 2007).

Also shown in Figure 22b is the error of ±8% associated with the cross-hole work suggested by Sutton (1999). It can be seen that the agreement between the various sets of data is good and the scatter is generally of the same order of magnitude as the expected error.

Finally on Figure 22c the UCD MASW data and the GDS Instruments CSW data are compared with the BRE SCPT results. Again the agreement is excellent. A limitation of both surface wave techniques, especially the CSW, is that the range of penetration is limited. Note that for the CSW work, GDS used a lightweight vibrator ($\approx$14 kg) and not the 60 kg unit used by them today. The lighter unit has a lower bound frequency range of 12 Hz to 14 Hz compared to 6 Hz to 7 Hz for the heavier one, meaning its penetration range is more limited.

MASW profiles taken at the NGI Onsøy site (Long and Donohue, 2007) are compared to three SCPT tests on Figure 23. Again the agreement is very good. These data also reflect the uniformity of this site.

MASW (Long and Donohue, 2007) and SASW data (taken by GDS for this study) for the medium stiff clay site at Glava near Trondheim, Norway is shown on Figure 24.

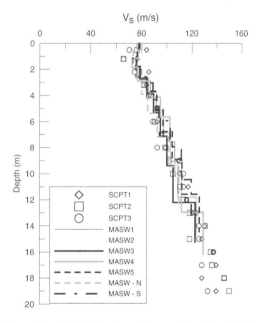

Figure 23. V_s data from Onsøy (Long and Donohue, 2007).

Glava clay has been investigated by researchers at the Geotechnics Division of the Norwegian University of Science and Technology (NTNU formerly NTH) since the mid 1980's (e.g. Sandven, 1990,

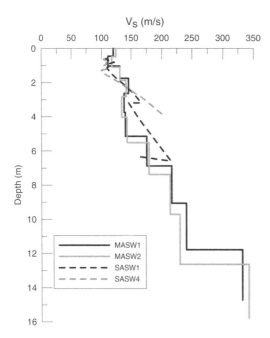

Figure 24. MASW (Long and Donohue, 2007) and SASW data from NTNU Glava research site.

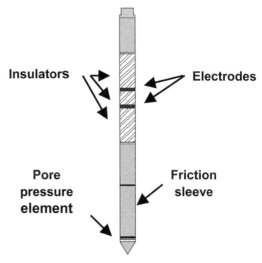

Figure 25. Soil moisture probe (Boylan et al., 2008a).

Sandven and Sjursen, 1998). The agreement between the measured values is relatively good. The penetration depth for the MASW technique is about twice that of the SASW. Perhaps this is not surprisingly given the number of transducers used and the difference sin the data interpretation routine.

Lo Presti et al. (2003) and Soccodato (2003) also compared V_s derived from SASW with that obtained from other techniques for Pisa clay and Fucino clayey soil respectively. Reasonable agreement was found in both cases.

2.8.3 Conclusions for surface wave techniques

The important implication of the results presented above for practicing engineers is that, for relatively homogenous soft clays, in situ shear wave velocity can be measured easily and reliably by a variety of methods. The results, for the examples shown, seem to be relatively independent of the technique used and of the operator. In turn the in situ shear stiffness at small strain (G_{max}) can be determined from the measured V_s data and material density (ρ) and the formula:

$$G_{max} = \rho V_s^2 \qquad (1)$$

Care needs to be taken in materials exhibiting significant natural material anisotropy (e.g. see Powell and Butcher, 1991).

2.9 Other new developments

The literature contains reference to a good number of other in situ tools, particularly those which can be used in conjunction with standard CPT testing. Examples are resistivity probes, heat flow measurements, radioisotope measurements, acoustic measurements (Lunne et al., 2007). Cones can also be fitted with special cameras so a visual image of the strata can be obtained (several papers in ISC'1 and ISC'2 conferences).

Despite all these developments the use of these instruments in routine practice seems limited. A particular issue seems to be the calibration of the equipment to produce useful engineering parameters. For example the paper by Boylan et al. (2008a) to this conference, on the soil moisture probe (Figure 25) pointed out that significant errors can be obtained if calibration relationships suggested in the literature are used. However in the authors opinion such probes can very useful (see also Gardiner, 2005) and more effort is required to develop calibrations for local soil conditions.

2.10 Developments in data acquisition systems

Cone penetrometers or other similar devices are only part of a larger entity dependant on the data acquisition system. Developments in data acquisition systems have improved the accuracy and reliability of geotechnical testing equipment. For example (www.apvdberg.com, www.envi.se or www.geotech.se):

- With the introduction of digital CPT cones and digital data transfer, there is no influence of cables or connectors on the measured data.

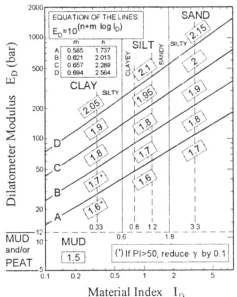

SOIL DESCRIPTION and ESTIMATED γ/γ_w

EQUATION OF THE LINES:
$E_D = 10^{(n+m \log I_D)}$

	m	n
A	0.585	1.737
B	0.621	2.013
C	0.657	2.289
D	0.694	2.564

(*) If PI>50, reduce γ by 0.1

Material Index I_D

Figure 26. Chart for estimating soil type with DMT (Marchetti and Crapps, 1981).

- Cables and connectors of smaller diameter can be used. Calibration data can be stored in the cone, thus minimising operator error.
- Measured data can also be stored in the cone as back-up.
- Cordless CPT units use radio waves to transfer the data to the surface, thus increasing sampling rate and eliminating the need for a cable.
- Optical transmission of data has also been considered (www.apvdberg.com).

3 SOIL CLASSIFICATION BY IN SITU TESTS

3.1 Soil classification by DMT

Though most use is made of the CPT for this purpose it is possible to use DMT data to classify soils as shown on Figure 26 (Marchetti and Crapps). ISSMGE (2001) suggested that this chart would give a good average for "normal" soils. The author is not aware of any recent detailed assessment of the reliability of the chart but such an exercise would be very useful.

3.2 Soil classification by CPT/CPTU

3.2.1 Introduction

Use of the CPTU for classifying soil has now gained world-wide acceptance. A number of well established soil classification or soil behaviour charts exist. Generally these charts use a combination of corrected cone resistance (q_t), sleeve friction (f_s) and pore water pressure (u_2) data or derived parameters such as pore pressure parameter (B_q) and friction ratio (R_f). Some examples are:

- Schmertmann (1978): for CPT based on q_t and R_f (Figure 27a)
- Robertson et al. (1986): for CPTU based on q_t, B_q and R_f (Figure 27b)
- Senneset et al. (1989): based on q_t and B_q (Figure 27c)
- Robertson (1990): similar to above but parameters normalized (Figure 27d)
- Robertson et al. (1995): based on normalized q_t and V_s
- Olsen and Mitchell (1995): based on normalised q_t and R_f (Figure 27e)
- Eslami and Fellenius (1997): based on effective cone resistance (q_e) and f_s (Figure 27f), where:

$$R_f = \left(\frac{f_s}{q_t}\right) x100\% \qquad (2)$$

$$B_q = \text{pore pressure parameter} = \frac{u_2 - u_0}{q_t - \sigma_{v0}} \qquad (3)$$

u_0 = in situ pore pressure
σ_{v0} = total overburden stress
q_e = effective cone resistance = $q_t - u_2$ (4)

3.2.2 Use of the charts

It is interesting to review the proceedings of the various conferences to see which of the CPTU classification charts have been used in practice. A summary for some selected major conferences on soil characterisation and in situ testing is given on Table 2.

Undoubtedly over this period (1998 – 2006), the Robertson et al. (1986) and Robertson (1990) charts remained the most popular and were widely used.

3.2.3 Reliability of CPT/CPTU based charts

Mollé (2005) carried out a very useful review of the reliability of the CPT/CPTU based charts. He used data from the literature to compare the chart predictions to the actual soil type as determined from laboratory classification tests. A summary of his findings is given on Table 3. It should be pointed out that the data used was from a wide range of geological settings. Nonetheless some useful findings can be made from this summary:

- The Robertson et al. (1986) and Robertson (1990) charts yield reasonable to very good results.
- There appears to be no advantage to using the normalized version of these charts (1990) over the non normalized ones (1986).

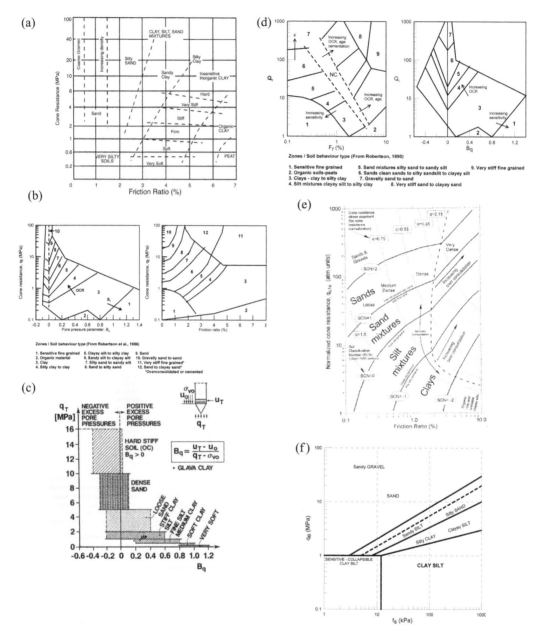

Figure 27. Some commonly used CPTU/CPT classification charts: (a) Schmertmann (1978), (b) Robertson et al. (1986), (c) Senneset et al. (1989), (d) Robertson et al. (1990), (e) Olsen and Mitchell (1995) and (f) Eslami and Fellenius (1997).

- The Jeffries and Davies (1991) and Ramsey (2002) charts also yield good results but both of these charts are based closely on that of Robertson (1990).
- The Eslami and Fellenius (1997) chart was also used successfully but only by the authors themselves.

- The Senneset et al. (1989) and Jones and Rust (1983) charts performed poorly.

The work of Mollé (2005) contributed significantly to the confidence that exists in the geotechnical

Table 2. Summary of classification charts used by authors various conferences (Table gives number of papers in which chart used).

Chart	ISC'1			ISC'2	
	Atlanta (1998)	Bali (2001)	S'pore (2002)	Porto (2004)	S'pore (2006)
Schmertmann (1978)		1			
Robertson et al. (1986)	1	1		4	
Robertson (1990)	7	1	1	4	3
Robertson et al. (1995)			1	1	
Olsen (1984) or Olsen and Mitchell (1995)	2				
Eslami and Fellenius (1997)				1	
Senneset et al. (1989)				1	
Others		1			

Table 3a. Success rate for CPT charts (Mollé, 2005).

Classification chart	Pijpers (2002)	Berry et al. (1998)	Bennert et al. (2002)
Schmertmann (1978)		63%	
Robertson et al. (1986) $q_t - R_f$	55%	80%	83%
Robertson et al. (1986) $q_t - B_q$			78
Robertson (1990) $Q_t - F_r$	72%	77%	61%
Robertson (1990) $Q_t - B_q$			50%
Eslami and Fellenius (1997)	58%		
Olsen and Mitchell (1995)		68%	
Senneset et al. (1989)			
Jones and Rust (1983)			
Ramsey (2002)			
Jefferies and Davies (1991)			78%
No. of data points	107	51 (min.)	200

Table 3b. Success rate for CPT charts (Mollé, 2005).

Classification chart	Chenghou et al. (1990)	Powell et al. (1988)	Fellenius & Eslami (2000)	Ramsey (2002)
Schmertmann (1978)				
Robertson et al. (1986) $q_t - R_f$	41%			
Robertson et al. (1986) $q_t - B_q$	80%			
Robertson (1990) $Q_t - F_r$			100%	
Robertson (1990) $Q_t - B_q$				
Eslami and Fellenius (1997)			100%	
Olsen and Mitchell (1995)				
Senneset et al. (1989)	2%	30%		
Jones and Rust (1983)	30%	58%		
Ramsey (2002)				75–80%
Jefferies and Davies (1991)				
No. of data points	40 (min.)	12 (min.)	18	

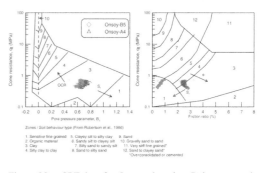

Figure 28. CPT data for Onsøy plotted on Robertson et al. (1986) charts.

3.2.4 Application of Robertson et al. (1986) chars to various soil types

In this section the application of the Robertson et al. (1986) charts to various soil types is explored. For onshore and relative shallow offshore conditions there appears to be no need to use the normalised 1990 version of the charts. However for deep strata both onshore and offshore the normalised charts may be more appropriate.

3.2.4.1 Clay

Data for the Onsøy site, between 7 m and 15 m (so as to avoid the upper dry crust), are shown on the Robertson et al. (1986) charts on Figure 28. As can be seen from Figures 2, 6, 18 and 23, this site is highly uniform. The

community in the use of the Robertson charts. It's not clear however as to which soil types the charts are most applicable and this will be explored in the next section.

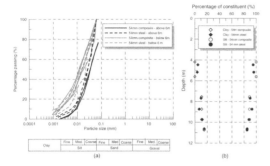

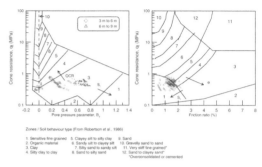

Figure 29. Particle size distribution tests for Os, Bergen (Long and Gudjonsson, 2005).

Figure 30. CPT data for Os plotted on Robertson et al. (1986) charts.

clay deposit has the following properties (Lunne et al., 2003a):

- Water content (w): 61%–65%,
- Clay content: 40% to 60%,
- Plasticity index (I_P): 30% to 50%,
- OCR: 1.4 to 1.9,
- Sensitivity (S_t) (fall cone): 3–7.

The tests chosen were carried out by two different organizations both using a $10\,cm^2$ cone. These were at NGI grid locations B5 and A4, which are 2.1 m apart (Lunne and Powell, 2008).

Test A4 suggests the material is in Behaviour Zone 3 (clay) which is correct. According to the q_t/B_q chart Test B5 also places the material in Zone 3. However the q_t/R_f chart for this test places the material in Zone 1 (sensitive fine grained). Although the Onsøy site is highly uniform this clearly highlights the difficulties in measuring f_s in soft clays and in classifying soil based on R_f values only.

3.2.4.2 Clay/silt

The Norwegian Public Roads Administration (Statens vegvesen) Os research site, which is just south of Bergen, has been developed for the purposes of studying the behaviour of silty soils.

As can be seen from Figures 29 the soils can be sub-divided into two layers:

- A more silty soil between 3 m and 6 m (also has relatively high OCR and low sensitivity),
- A more clayey soil below 6 m (with relatively low OCR and high sensitivity).

The data were sub-divided into these two zones and plotted on the Robertson et al. (1986) charts on Figure 30. The B_q chart in particular appears to work well for these materials. According to the chart the soil becomes more clayey with depth and has decreasing OCR and increasing sensitivity. This is consistent with the index properties. The q_t v R_f chart does not give any additional useful results and in fact plots the two layers closer together.

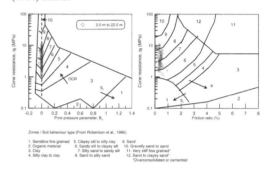

Figure 31. CPT data for Holmen, Drammen plotted on Robertson et al. (1986) charts.

3.2.4.3 Sand

Lunne et al. (2003b) sub-divided the different strata at the Holmen, Drammen research site into different layers depending on their geological history and material properties. The layer between 2 m and 22 m is a fluvial medium to coarse grained sand and CPTU data for this layer are shown on the Robertson et al. (1986) charts on Figure 31.

Both charts broadly characterise the material correctly, though as the B_q values are so small the q_t v R_f plot provides the most detailed information.

3.2.4.4 Sandy silt

The UCD Sligo research site in Western Ireland (Long, 2007) is underlain by about 2 m of peat over sandy silt. Between these 2 layers, there is an approximately 2 m intermediate/transition zone of peaty silt. The sandy silt layer comprises about 35% silt and 55% sand.

The Sligo CPTU data plotted on the charts of Robertson et al. (1986) on Figures 32 and it can be seen that the charts seem to give reasonable classification for the sandy silt but do not work for the peat.

3.2.4.5 Organic clays and peat

Long et al. (2007) explored the use of CPTU tests in classifying organic clays and peats at three test sites in the UK and Ireland. Index test data for the Crayford marsh site is shown on Figure 33. This site is located on

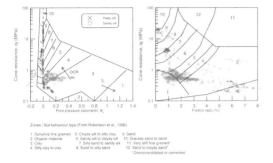

Figure 32. CPT data for Sligo plotted on Robertson et al. (1986) charts.

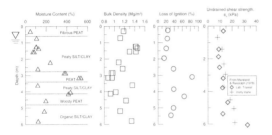

Figure 33. Crayford marsh site UK (Long et al., 2007).

the southern banks of the River Thames, 23 km east of central London, and consists of recent marsh deposits overlying alluvial silts and clays.

These comprise 6 m of varying layers of peaty and organic silt/clay overlying the Thames gravels. Loss of ignition at 440°C range from as low as 14% in the organic silt/clay to 84% in the peat. Moisture contents range from 100% in the organic silt/clay to 500% in the peat. Bulk densities vary between 1 Mg/m³ to 1.4 Mg/m³ with lowest values found in the peat. I_p values range from 35% to 260% with higher values found with increasing levels of organics.

Strength data for the site published by Marsland and Randolph (1978) show s_u to vary between 8 to 32 kPa from laboratory undrained triaxial tests and between 8 and 22 kPa from in situ vane tests.

CPTU data for the site is plotted on the Robertson (1990) charts on Figure 34. The strata were not well defined by the Robertson et al. (1986) charts and it was hoped that normalisation by in situ effective stress would be beneficial for these relatively shallow deposits with low unit weight. The normalized parameters are defined as:

$$Q_t = \frac{q_t - \sigma_{v0}}{\sigma_{v0}{}'} \tag{5}$$

$$F_r = \frac{f_s}{q_t - \sigma_{vo}} \tag{6}$$

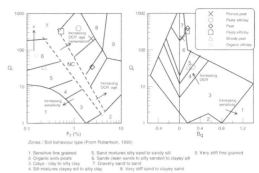

Figure 34. CPT data for Crayford plotted on Robertson et al. (1986) charts.

As can be seen the resulting high Q_t value places all of the data within the sand and sand and gravel behaviour zones of the charts. It seems these charts (or the original non-normalised versions) are not suitable for reliably classifying organic material.

Hamza et al. (2005) presented a case history of a study of the Nile delta deposits which comprise a complicated stratigraphy of silty sand, layered clay, organic silt and sand. They showed how well the Robertson (1990) charts identify the changing stratigraphy except for the organic silts which are missed on the chart.

3.2.4.6 Summary

The Robertson et al. (1986) q_t v B_q and q_t v R_f or Robertson (1990) Q_t v B_q and Q_t v F_r charts are probably adequate to reasonably accurately characterise uniform soft to medium stiff clay and uniform sand sites. Output from these charts seems consistent with the material index properties.

However, more importantly, it seems that for intermediate soils such as silty clay or clayey silt and sandy silt these charts also work well.

There seems to be difficulties with the use of the characterising peat and organic clay soils. In the author's experience the charts also do not readily identify laminated soils.

Work presented here again identifies the importance of reliable f_s measurements. Inaccuracies in f_s measurements can decrease the reliability of the Robertson et al. (1986) and Robertson (1990) q_t v R_f and Q_t v F_r chart.

3.3 Development of new charts

3.3.1 Charts based on V_s and u_2

As has been detailed above it is clear that CPTU u_2 and shear wave velocity (V_s) are two of the more reliable and accurate parameters that can be obtained from in situ testing. It seems logical then to attempt to use them in combination for the purposes of classifying soils. Robertson et al. (1995) have previously

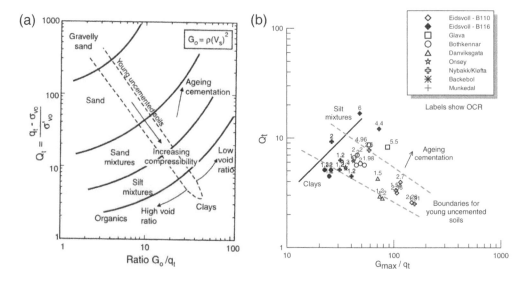

Figure 35. (a) Soil classification chart based on G_{max} and q_t (Robertson et al (1995) and (b) data from sites in Norway and Sweden plotted in Robertson et al. format.

suggested a chart based on normalized cone resistance (Q_t) and the ratio of small strain shear modulus (G_0) to cone resistance (G_0/q_t). G_0 is calculated directly from V_s using Formula 1. This chart is shown on Figure 35a.

It was intended for identifying "unusual" soils such as highly compressible sands, cemented and aged soils and clays with either high or low void ratio (see Lunne et al., 1997).

Data for nine soft clay sites in Norway and Sweden (Lunne et al., 2006, Larsson and Mulabdić, 1991a and 1991b, Long and Donohue, 2007) are plotted in the Robertson et al. (1995) format on Figure 35b. Note that a portion of the chart only is shown and the boundaries for young uncemented soils have been changed. This method of analysing the data does clearly separate the young uncemented soils, with OCR less than 2, from the moderately overconsolidated soils with OCR of 4 to 6.

The data are re-plotted in the form of normalized shear wave velocity V_{s1} and normalized pore pressure B_E on Figure 36, where:

$$V_{s1} = \frac{V_s}{\left(\dfrac{\sigma_{v0}'}{p_{atm}}\right)} \quad \text{(Mayne et al., 1998)} \qquad (7)$$

$$B_E = \frac{u_2 - u_0}{u_0} \quad \text{(Eslami and Fellenius, 1996)} \qquad (8)$$

Again it appears there is some promise in this method of plotting data as it clearly separates the two groups.

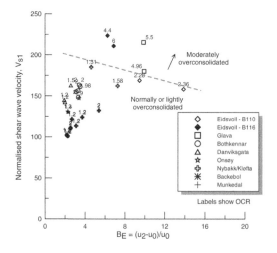

Figure 36. Soil classification chart based V_{s1} and B_E.

3.3.2 Charts using new normalised parameters

In a paper to this conference (Schneider et al., 2008) propose classification charts based on Q and $\Delta u_2/\sigma_{v0}'$. Q is normalized cone resistance, calculated in the same way as Q_t (Formula 5) but the authors feel that the subscript on Q_t is inappropriate. These parameters were chosen so as to separate the influences of yield stress ratio and partial consolidation on normalized CPTU parameters. The charts were developed from a combination of large strain finite element analyses of undrained cone penetration, cavity expansion modeling of pore pressure generation, data from variable rate CPTU tests together with a relatively

108

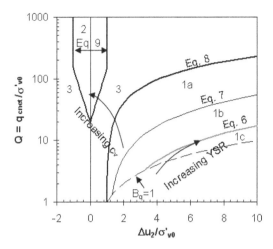

Figure 37. CPTU classification chart (Schneider et al., 2008). 1a: silts and low I_r clays, 1b: undrained clays, 1c: sensitive clays, 2 drained sands and 3 transitional soils.

large database of field testing. The resulting proposed chart in semi log Q-$\Delta u_2/\sigma'_{v0}$ space is shown on Figure 37.

4 RECENT DEVELOPMENTS IN SOIL PARAMETER INTERPRETATION FROM IN SITU TESTS

4.1 Introduction

In this section it is only intended to report on significant developments in this area since the previous conferences and reports listed in Table 1. Three significant developments, at least in the author's experience, have been in:

1. Interpretation of s_u from CPTU
2. Interpretation of s_u from full flow probe tests
3. Interpretation of remoulded shear strength from full flow probe tests.

4.2 Interpretation of s_u from CPTU

The undrained shear strength (s_u) of soil is normally determined by dividing the net cone resistance (q_{net}) by a cone factor (N_{kt}) using Equation 9.

$$s_u = \frac{q_{net}}{N_{kt}} = \frac{(q_t - \sigma_{vo})}{N_{kt}} \qquad (9)$$

An alternative approach is to use the excess pore water pressure and another empirical factor $N_{\Delta u}$:

$$s_u = \frac{\Delta u}{N_{\Delta u}} = \frac{u_2 - u_0}{N_{\Delta u}} \qquad (10)$$

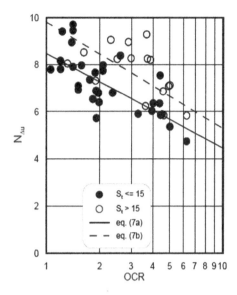

Figure 38. Relationship between $N_{\Delta u}$ and OCR, S_t and I_p (Karlsrud et al., 2005). *Note*:
Eq7(a): If $S_t < 15$ then $N_{\Delta u} = 6.9 - 4.0 \log OCR + 0.07 I_p$
Eq7(b): If $S_t > 15$ then $N_{\Delta u} = 9.8 - 4.5 \log OCR$

A third approach is to use the effective cone resistance and the empirical factor N_{ke}:

$$s_u = \frac{q_e}{N_{ke}} = \frac{(q_t - u_2)}{N_{ke}} \qquad (11)$$

Although theoretical solutions for N_{kt}, $N_{\Delta u}$ and N_{ke} exist conventional practice is to use empirically derived factors. Karlsrud et al. (2005) have developed CPTU correlations for clays based on carefully executed CPTU tests and laboratory triaxial tests on high quality Sherbrooke block samples. With respect to the determination of s_u they concluded:

- $N_{\Delta u}$ gives the best and most consistent results and reliability in $N_{\Delta u}$ determined strengths ($\pm 10\%$) might be higher than others.
- N_{kt} data is somewhat more scattered.
- N_{ke} lies somewhere between N_{kt} and $N_{\Delta u}$ when it comes to variability.
- They recommend the use of $N_{\Delta u}$ for the purposes of determining s_u (corresponding to peak triaxial compressive strength).

Karlsrud et al. (2005) presented various charts for the determination of the empirical parameters. It is important to note that Karlsrud et al.'s correlations are related to triaxial anisotropically consolidated undrained shear strength (CAUC). An example, which relates $N_{\Delta u}$ to OCR, S_t and I_p is shown on Figure 38.

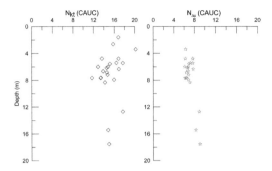

Figure 39. N_{kt} and $N_{\Delta u}$ factors for Bothkennar – related to triaxial compression CAUC tests (Boylan et al., 2007).

Note that this work is strictly speaking valid only for relatively shallow lightly overconsolidated Norwegian marine clays.

Boylan et al. (2007) examined the range of N_{kt} and $N_{\Delta u}$ factors determined for the Bothkennar research site based on triaxial test data from high quality Sherbrooke block and Laval samples. N_{kt} values lie largely within the range suggested by Karlsrud et al. (2005), see Figure 39. $N_{\Delta u}$ factors (related to anisotropically consolidated triaxial compression tests – CAUC) occupy a significantly narrower deviation band than N_{kt}, reinforcing the findings of Karlsrud et al. (2005).

4.3 Interpretation of s_u from full flow probe tests

In terms of research effort on interpretation of in situ tests since 2004, arguably most work has been invested in full flow penetrometers. Developments in equipment and procedures has been summarised in Section 2.2. Advances on theoretical developments have been reported by Randolph and Andersen (2006), Einav and Randolph (2005, 2006) and Chung et al. (2006).

Practical application of full flow probes have been reported by Boylan and Long (2006) for peat, by Boylan et al. (2007) for Bothkennar clay, by Yafrate and DeJong (2006) for sites in Norway, Australia, Canada and the US, by Weemees et al. (2006) for soft clays and silts in British Columbia and by Lunne et al. (2005) for two onshore and one offshore clay in Norway and Australia. There seems to be particular interest in the use of full flow penetrometers in the offshore environment (Randolph et al., 2005, Peuchen et al., 2005, Kelleher and Randolph, 2005, Teh et al., 2007, Gaudin et al., 2007, Randolph et al., 2007 and Lunne and Andersen, 2007).

Special applications of full flow probes for determining remoulded shear strength have been reported by Yafrate and DeJong (2005, 2006) and Yafrate et al. (2007) and for the determination of coefficient of consolidation by Low et al. (2007).

Net resistances for the T-bar and Ball is normally calculated using (Randolph, 2004):

$$q_{\text{T-bar}} \text{ or } q_{\text{Ball}} = q_m - [\sigma_{vo} \text{-} (1-a)u_0]\frac{A_s}{A_p} \quad (12)$$

where A_s/A_p is the ratio of shaft area to the bearing area of the probe, which in this case is 0.1 for both probes and q_m is the measured resistance. Randolph et al. (2007) suggested that this equation is not quiet correct and a more precise solution would involve u_2 and σ'_v (parameters which are not always available)

The undrained shear strength is then determined for the T-bar and for the Ball by:

$$s_u = \frac{q_{\text{Tbar}}}{N_{\text{Tbar}}} \quad (13)$$

$$s_u = \frac{q_{\text{Ball}}}{N_{\text{Ball}}} \quad (14)$$

where N_{Tbar} and N_{Ball} are bearing capacity factors for the T-bar and Ball.

Plasticity solutions based on simplified assumptions of soil behaviour were initially used for determination of N_{Tbar} and N_{Ball}. Typically values of 10.5 and 12 were used in design and these seemed to fit reasonably well with field observations.

Recent research has drawn attention to the fact that all in situ tests (CPTU, vane, T-bar, Ball etc.) impose strain rates that are 3 to 5 orders of magnitude higher than in a careful laboratory test. This needs to be considered in data interpretation and may be easier for the T-bar and Ball because of the simpler geometry. In addition one needs to consider the effects of anisotropy and gradual strain softening of the soil as the T-bar passes a given horizon (Randolph & Anderson, 2006).

Randolph and Andersen (2006) showed the effect of anisotropy was relatively small (less than 5%) but the effects of rate and strain softening could be very significant. These two latter effects may in fact be compensating, thus leading to field values of N_{Tbar} and N_{Ball} being similar to those determined from simple plasticity solutions.

Clearly further research effort is required to satisfactorily derive bearing capacity factors for full flow probes, particularly as current practice of empirical test interpretation and theoretical design is inconsistent (Randolph et al., 2007)

Nonetheless in current practice, it is essential to make empirical correlations for the determination of undrained shear strength, similar to practice for the CPTU. In the absence of local data Lunne et al. (2005) recommend the following T-bar N factors (related to CAUC triaxial strength and average CAUC, CAUE and DSS laboratory strength), see Table 4.

However Yafrate and DeJong (2006) reported an N_{Tbar} range of 5.9 to 11.7 for sites in Norway, Canada, Australia and the US, the lower values being

Table 4. T-bar factors recommended by Lunne et al. (2005).

In situ test	Empirical factor	Undrained shear strength	Recommended range
T-bar	N_{Tbar}	s_u^{CAUC}	8–11
		s_u^{av}	10–13

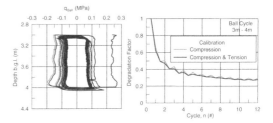

Figure 40. Cyclic ball testing in Bothkennar clay (Boylan et al., 2007).

associated with high sensitivity clays. Boylan et al. (2007) reported values in the range 10.3 to 14.7 for Bothkennar clay. There are fewer reported N_{Ball} values. Yafrate and DeJong (2006) and Boylan et al. (2007) determined values in the range 6.7 to 12.2 and 9.9 to 14.7 respectively.

All of these values are based on CAUC tests on high quality samples. Nonetheless some further research effort is required to provide practicing engineers with reliable design guidelines.

4.4 Interpretation of remoulded shear strength from full flow probe tests.

Remoulded shear strength is assessed by cycling the penetrometer continually up and down a number of times (typically 10) within a defined layer (Yafrate and DeJong, 2005 and Yafrate et al., 2007). While a cycle region of 0.2 m would be suitable (Yafrate and DeJong, 2005) a larger test region is usually chosen.

Some typical results for Bothkennar clay (Boylan et al., 2007) are presented in Figure 40. The instruments used in this study were calibrated both in compression and tension to evaluate if the 'saw-tooth' degradation profile regularly occurring in cyclic penetrometer tests (e.g. Boylan & Long, 2006) is due to the differences in the compression and tension calibrations of the cones. The use of the appropriate calibrations caused a marked change to the degradation curve. However this was not the case for all the tests reported and further work is required to investigate these effects.

The remoulded shear strength (s_{ur}) is then determined in a similar way to s_u by dividing the final steady state penetration resistance by a suitable N factor.

Theoretical considerations suggest that N factors related to s_{ur} should be greater than for s_u. This is because for intact conditions there are compensating rate effects and softening effects, whereas for remoulded conditions there are only rate effects, since all softening has occurred. Field studies have also shown this to be the case (Yafrate and DeJong, 2006). Lunne and Andersen (2007) suggested a N_{Tbar} of about 13.5 (related to field vane and fall cone tests) may be appropriate for determining s_{ur} in typical soft clays. Yafrate et al. (2007) reported higher values (>20) for Connecticut Valley varved clay (again related to field vane) and values in the range 7.2 to 39.2 for clays in

Norway, Canada, Australia and the US. Clearly further work is required in this area.

5 CONCLUSIONS

As has been detailed in this paper and as pointed out by Powell (2005) geotechnical engineers have available a wide range of powerful tools for the purposes of soil characterisation. These tools need to be used consistently and the available standards need to be carefully followed. It is apparent that a combination of in situ tests can be the most effective means of characterisation of the soils. It is also clear that the in situ test should be used in conjunction with traditional physical logging and sampling, using the best available sampling system.

Some specific findings of this work are as follows:

1. CPTU sleeve friction (f_s) measurements are less reliable than cone resistance (q_t), which in turn are less reliable than pore pressure (u_2).
2. It follows that undrained shear strength (s_u) determined from CPTU pore pressure data ($N_{\Delta u}$) and soil classification based on pore pressure (B_q) are more dependable than other techniques, provided the system is adequately saturated.
3. Pore pressure measurements (i.e. CPTU) should become standard with all CPT testing.
4. Geotechnical engineers need to put more effort into routine checks such as filter saturation, temperature effects and transducer drift.
5. Full flow probes such as T-bar and ball tests are undoubtedly useful in the determination of intact and remoulded shear strength, particularly in very soft soils and in deepwater situations. Further theoretical work is required to provide engineers with reliable bearing capacity factors for intact and remolded shear strength.
6. In the meantime full flow probe data needs to be empirically correlated to strength values as for the CPTU.
7. Pressuremeter and dilatometer tests will continue to provide useful data. Further effort is required to provide consistent correlations for design parameters for a variety of soft soils, particularly for the

111

dilatometer. Similar to the CPTU engineers need to take care in the determination of the fundamental instrument parameters which are used in the correlations. This is particularly the case for the pressuremeter modulus and limit pressure.

8. It also seems clear that there is a future role for cone pressuremeters, though some research work is needed into the derivation of a better relationship of horizontal effective stress (σ'_h) for clays from the CPM.

9. It seems that in situ shear wave velocity can be measured easily and reliably by a variety of methods. The results seem to be relatively independent of the technique used (having accounted for natural material anisotropy) and of the operator.

10. The well known Robertson et al. (1986) and Robertson (1990) charts for soil classification using CPTU data seem to work well in clays, clayey silts, silty sands and sands. There are difficulties with using these charts in organic soils and peat and in complicated stratigraphies.

ACKNOWLEDGEMENTS

Much of the findings presented and opinions expressed in this paper come from the work and from discussions with friends and colleagues of the author. He is especially grateful to Noel Boylan and Shane Donohue, UCD, Tom Lunne of NGI, John Powell of BRE, Mark Randolph of COFS and to Prof. Silvano Marchetti, University of L'Aquila who provided ideas and input and reviewed the paper. Several organizations contributed with in situ testing data which has been included in this paper. In particular assistance has been provided by Darren Ward, Andy Barwise and Brian Georgious of Gardline Lankelma, UK, Peter O'Connor of APEX Geoservices, Gorey, Ireland and Chris Jones and Jerry Sutton of GDS, UK.

REFERENCES

Baker, C.N. 2005. The use of the Ménard pressuremeter in innovative foundation design from Chicago to Kuala Lumpur. *Proc. ISP – Pressio 2005. 50 years of pressuremeters, Gambin, Magnan and Mested (eds.), l'ENPC/LCPC, Paris,* 2: 63–102.

Bennert, T., Maher, A. and Gucunski, N. 2002. Evaluation of geotechnical design parameters using the seismic piezocone, *New Jersey Department of Transportation,* FHWA NJ 2001-032, 2002.

Berry, K.M., Olson, S.M. and Lamie, M. 1998. Cone penetration testing in the Mid-Mississippi river valley. *Proceedings of the 1st. International Conference on Geotechnical Site Characterization,* ISC'1, Atlanta, 2: 983–987.

Boylan, N. and Long, M. 2006. Characterisation of peat using full flow penetrometers. In *Proceedings 4th Int. Conf.*

on Soft Soil Engineering (ICSSE), Vancouver, October. Chan & Law (Eds.): 403–414. London: Taylor & Francis Group.

Boylan, N., Long, M., Ward, D., Barwise, A. and Georgious, B. 2007. Full flow penetrometer testing in Bothkennar clay". In *Proceedings 6th International Conference, Society for Underwater Technology, Offshore Site Investigation and Geotechnics (SUT-OSIG),* London, September: 177–186.

Boylan, N., Ward, D. and Long, M. 2008a. Characterisation of soft clay with a soil moisture probe (SMP). *Proceedings 3rd International Conference on Site Characterisation (ISC'3),* Taipei, April.

Boylan, N., Mathihssen, F., Long, M. and Molenkamp, F. 2008b. Cone penetration testing of organic soils. *Proc. Baltic Geotechnical Conference,* Gdansk, Poland, September.

Cheng-hou, Z., Greeuw, G., Jekel, J. and Rosenbrand, W. 1990. A new classification chart for soft soils using the piezocone. *Engineering Geology,* 29: 31–47.

Chin, C.-T., Chen J.-R., Hu, I.-C., Yao, D.T.C. and Chao, H-C. 2007. Engineering characteristics of Taipei clay. In *Proceedings International Workshop on Characterisation and Engineering Properties of Natural Soils ("Natural Soils 2006").* NUS Singapore, December. Tan, T.S. et al. eds. 3: 1755–1804. London, Taylor and Francis.

Chung, S.F., Randolph, M.F. and Schneider, J.A. 2006. Effect of penetration rate on penetrometer resistance in clay. *Jnl of Geotech and Geoenvironmental Eng., ASCE,* 132(9): 1188–1196.

Crice, D. 2005. MASW, The wave of the future. *Journal of Environmental and Engineering Geophysics,* 10 (2): 77–79.

De Jong, J.T., Yafrate, N.J.. and De Groot, D.J. 2007. Design of a miniature piezoprobe for high resolution stratigraphic profiling. *ASTM Geotechnical Testing Journal,* 30 (4).

Donohue, S., Long, M, O'Connor, P. and Gavin, K. 2004. "Use of multichannel analysis of surface waves in determining Gmax for soft clay". In *Proc 2nd. Int. Conf on Geotechnical Site Characterisation, ISC'2,* Porto, September: 1: 459–466. Rotterdam, MillPress.

Einav, I. & Randolph, M.F. 2005. Combining upper bound and strain path methods for evaluating penetration resistance. *Int. J. Numer. Meth. Engng,* 63(14): 1991–2016.

Einav, I. & Randolph, M.F. 2006. Effect of strain rate on mobilized strength and thickness of curved shear bands. *Géotechnique,* 56(7): 504–510.

ENISO 22476-1. 2005. *Geotechnical investigation and testing – Field testing – Part 1: Electrical cone and piezocone penetration tests.* Geneva: ISO / CEN.

Eslami, A. and Fellenius, B.H. 1997. Pile capacity by direct CPT and CPTU methods. *Can Geo. Jnl.* 24 (6): 880–898.

Fahey, M. 1998. Deformation and in situ stress measurement. In *Proceedings First International Conference on Geotechnical Site Characterisation, Atlanta (eds. Robertson and Mayne).* 1: 49–68. Rotterdam, Balkema.

Fellenius, B.H. and Eslami, A. 2000. Soil profile interpreted from CPTU data. *Year 2000 Geotechnics,* Geotechnical Engineering Conference, AIT, Bangkok, Nov. 18p.

Gardiner, R. 2005. Moisture monitored. *Ground Engineering,* February: 22–23.

Gaudin, C., Cassidy, M.J. and Donovan, T. 2007. Spudcan reinstallation near existing footprints. In *Proceedings*

6th International Conference, Society for Underwater Technology, Offshore Site Investigation and Geotechnics (SUT-OSIG), London, September: 285–292.

Greenhouse, J., Pehme, P., Coulter, D. and Yarie, Q. 1998. Trends in geophysical site characterization. In *Proceedings First International Conference on Geotechnical Site Characterisation, Atlanta (eds. Robertson and Mayne).* 1: 23–34. Rotterdam, Balkema.

Hamza, M.M., Shmahien, M.M. and Ibrahim, M.H. 2005. Characterisation and undrained shear strength of Nile delta soft deposits using piezocone. In *Proc. 16th International Conference on Soil Mechanics and Geotechnical engineering, Osaka,* September, Vol. 2.

Hawkins, A.B., Larnach, W.J., Lloyd, I.M. and Nash, D.F.T. 1989. Selecting the location and the initial investigation of the SERC soft clay test bed site. *Quarterly Journal of Engineering Geology,* 22(4): 281–316.

Heisey, J.S., Stokoe, K.H and Meyer, A.H. 1982. Moduli of pavement systems from Spectral Analysis of Surface Waves. *Transportation Research Record,* 852: 22–31. Washington D.C..

Hepton, P. 1988. Shear wave velocity measurements during penetration testing. *In Proceedings Penetration testing in the UK.* Geotechnology Conference, Birmingham, July: 275–278. London, Thomas Telford.

Hight, D.W, Paul, M.A, Barras, B.F, Powell, J.J.M, Nash, D.F.T, Smith, P.R, Jardine, R.J and Edwards, D.H. 2003. Characterisation of the Bothkennar clay. In *Proceedings International Workshop on Characterisation and Engineering Properties of Natural Soils ("Natural Soils 2002").* NUS Singapore, December. Tan, T.S. et al. eds. 2: 543–597.

Hope, V.S., Clayton, C.R.I. and Butcher, A.P. 1999. In situ determination of G_{hh} at Bothkennar using a novel seismic method Quarterly Journal of Engineering Geology and Hydrogeology. 32(2): 97–105.

Houlsby, G.T. 1998. Advanced interpretation of field tests. In *Proceedings First International Conference on Geotechnical Site Characterisation, Atlanta (eds. Robertson and Mayne).* 1: 99–114. Rotterdam, Balkema.

Huang, A.B., Pan, Y.W., Wang, C.H., Liao, J.J and Hsieh, S.Y. 1999. Pressuremeter tests in poorly cemented weak rocks. 37th US Rock Mechanics Symposium, Balkema, 1: 247–252.

Huang, A.B. and Hsueh, C.K. 2005. Pressuremeter tests in Taiwan. *Proc. ISP – Pressio 2005. 50 years of pressuremeters, Gambin, Magnan and Mested (eds.), l'ENPC/LCPC, Paris,* 2: 369–374.

International Society for Soil Mechanics and Geotechnical Engineering, ISSMGE. 1999. *International reference test procedure for the CPTU.* In Proceedings XIIth ECSMFE, Amsterdam. Rotterdam, Balkema.

International Society for Soil Mechanics and Geotechnical Engineering, ISSMGE. 2001. The flat dilatometer test for soil investigations. Report of the ISSMGE TC16, In *Proceedings International Conference on In Situ Measurement of Soil Properties and Case Histories, In Situ 2001.* (Rahardjo and Lunne eds.). Bali, Indonesia.

International Organisation for Standardisation (ISO) (2003) ISO/WD 22476-9, 2003: *Geotechnical investigation and testing – Field testing – Part 9: Field vane test.* Geneva ISO.

Jacobs, P.A. and Coutts, J.S. 1992. Comparison of electric piezocone tips at the Bothkennar test site. *Géotechnique,* 42(2): 369–375.

Jamiolkowski, M and Lo Presti, D.C.F. 1994. Validity of in situ tests related to real behaviour. In *Proceedings XIII ICSMFE, New Delhi,* 5: 51–55. New Delhi: Oxford Publishing.

Jeffries, M.G. and Davies, M.P. 1991. Discussion on Robertson (1990) – Soil classification by the cone penetration test. *Can. Geo. Jnl.,* 28: 173–176.

Jones, R. 1958. In-situ measurements of the dynamic properties of soil by vibration methods. Géotechnique, 8(1): 1–21.

Jones, G.A. and Rust, E. 1982. Piezometer penetration testing CUPT. *Proc. 2nd European Symposium on Penetration Testing, ESOPT-2,* Amsterdam, 2: 607–614.

Kallstenius, T. 1961. Development of two modern continuous sounding methods. *Proc.5th ICSMFE,* Paris, 1: 475–480

Karlsrud, K., Lunne, T., Kort, D.A. and Strandvik, S. 2005. CPTU correlations for clays. In *Proc. 16th International Conference on Soil Mechanics and Geotechnical engineering, Osaka,* September, 2: 693–702.

Kelleher, P.J. and Randolph, M.F. 2005. Seabed geotechnical characterisation with a ball penetrometer deployed from the Portable Remotely Operated Drill. In *Proc., Int. Symp. on Frontiers in Offshore Geotechnics (ISFOG),* Perth, Australia: 365–371.

Lacasse, S. and Lunne, T. 1982. Dilatometer tests in two soft marine clays. In *Proceedings on Updating Subsurface Sampling of Soils and Rocks and their In Situ Testing,* Santa Barbara, California, January.

Larsson, R. and Mulabdić, M. 1991a. Shear moduli in Scandinavian clays. *Swedish Geotechnical Institute Report* No. 40, Linköping.

Larsson, R. and Mulabdić, M. 1991b. Piezocone tests in clay. *Swedish Geotechnical Institute Report* No. 42, Linköping.

Long, M. and Gudjonsson, G. 2004. T-bar testing in Irish soft soils. In *Proc 2nd. Int. Conf on Geotechnical Site Characterisation, ISC'2,* Porto, September: 1: 719–726. Rotterdam, MillPress.

Long, M. and Gudjonsson, G. 2005. FoU – prosjekt. Parameterbestemmelser for siltige materialer. Delrapport C. Interpretative Report. *Statens Vegvesen Intern Rapport Nr. 2383,* October 2005.

Long, M and Donohue, S. 2007. In situ shear wave velocity from multichannel analysis of surface waves (MASW) tests at eight Norwegian research sites. *Canadian Geotechnical Journal,* 44 (5): 533–544.

Long, M. 2007. Engineering characterisation of estuarine silts. *Quarterly Journal of Engineering Geology and Hydrogeology (QJEGH),* 40: 147–161.

Long, M., Boylan, N. and O'Connor, S. 2007. The application of CPTU to soft soils and alternative techniques. In *Proceedings SGE – 2007, Soft Ground Engineering Conference,* Athlone, Ireland, February 2007, ISBN 1 898 012 83 0, paper 1.1.

Lo Presti, D.C.F., Jamiolkowski, M. and Pepe, M. 2003. Geotechnical characterization of the subsoil of Pisa Tower. In *Proceedings International Workshop on Characterisation and Engineering Properties of Natural Soils ("Natural Soils 2002").* NUS Singapore, December. Tan, T.S. et al. eds. 2: 909–946.

Low, H.E., Randolph, M.F. and Kelleher, P. Comparison of pore pressure generation and dissipation rates from cone and ball penetrometers. In *Proceedings 6th International Conference, Society for Underwater Technology, Offshore Site Investigation and Geotechnics (SUT-OSIG)*, London, September: 177–186.

Lukas, R.G. 2005. The pressuremeter in consulting engineering practice in the USA. *Proc. ISP – Pressio 2005. 50 years of pressuremeters, Gambin, Magnan and Mested (eds.), l'ENPC/LCPC, Paris*, 2: 329–336.

Lunne, T., Eidsmoen, T., Gillespie, D. and Howland, J.D. 1986. Laboratory and field evaluation of cone penetrometers. *Proceedings of the ASCE Specialty Conference In Situ '86: Use of In Situ Tests in Geotechnical Engineering*, Blacksburg, American Society of Civil Engineers: 714–729.

Lunne, T., Lacasse, S. and Rad, N.S. 1989. General report/Discussion session 2: SPT, CPT, pressuremeter testing and recent developments in situ testing – Part 1: All tests except SPT, In *Proceedings XII ICSMFE, Rio de Janeiro*, 4: 2339–2403. Rotterdam, Balkema.

Lunne, T., Robertson, P.K and Powell, J.J.M. 1997. *Cone penetration testing in geotechnical practice*. London: Spon Press.

Lunne, T. 2001. In situ testing in offshore geotechnical investigations. In *Proceedings International Conference on In Situ Measurement of Soil Properties and Case Histories, In Situ 2001*. (Rahardjo and Lunne eds.). Bali, Indonesia: 61–78.

Lunne, T., Long, M. and Forsberg, C.F. 2003a. Characterisation and engineering properties of Onsøy clay. In *Proceedings International Workshop on Characterisation and Engineering Properties of Natural Soils* ("Natural Soils 2002"). NUS Singapore, December. Eds. Tan, T.S. et al. Published by Balkema, 1: 395–428.

Lunne, T., Long, M. and Forsberg, C.F. 2003b. Characterisation and engineering properties of Holmen, Drammen sand. In *Proceedings International Workshop on Characterisation and Engineering Properties of Natural Soils ("Natural Soils 2002")*. NUS Singapore, December. Eds. Tan, T.S. et al. Published by Balkema, 2: 1121–1148.

Lunne, T. 2005. *Personal communication to author*.

Lunne, T., Randolph, M.F., Chung, S.F., Andersen, K.H., and Sjursen, M. 2005. Comparison of cone and T-bar factors in two onshore and one offshore clay sediments. In *Proc., Int. Symp. on Frontiers in Offshore Geotechnics (ISFOG)*, Perth, Australia: 981–989.

Lunne, T., Berre, T., Andersen, K.H., Strandvik, S. and Sjursen, M. 2006. Effects of sample disturbance on measured shear strength of soft marine Norwegian clays. *Canadian Geotechnical Journal*. 43: 726–750.

Lunne, T. and Andersen, K.H. 2007. Soft clay shear strength parameters for deepwater geotechnical design. In *Proceedings 6th International Conference, Society for Underwater Technology, Offshore Site Investigation and Geotechnics (SUT-OSIG)*, London, September: 151–176.

Lunne, T. and Powell, J.J.M. 2008. Comparative testing of piezocones at the Onsøy test site in Norway. *Submitted for publication*.

Marsland, A. and Randolph, M.F. 1978. A study of the variation and effects of water pressures in the pervious strata underlying Crayford Marshes. *Géotechnique*, 28(4): 435–464.

Mathews, M.C., Hope, V.S. and Clayton, C.R.I. 1996. The use of surface waves in the determination of ground stiffness profiles. *Proc. Institution of Civil Engineers Geotechnical. Engineering*, 119 (Apr.): 84–95.

Marchetti, S. 1980. In Situ Tests by Flat Dilatometer. *Journal of the Geotechnical Engineering Division, ASCE*, 106 (GT3): 299–321.

Marchetti, S. 1980. In Situ Tests by Flat Dilatometer; closure to discussion. *Journal of the Geotechnical Engineering Division, ASCE*, 107 (GT6): 832–837.

Marchetti, S. and Crapps, D.K. 1981. Flat dilatometer manual. *Internal Report of G.P.E. Inc.*

Marchetti, S., Monaco, P., Totani, G. and Marchetti, D. 2008. In situ tests by seismic dilatometer. *From Research to Practice in Geotechnical Engineering, A Geotechnical Special Publication Honoring John H. Schmertmann*, ASCE GeoInstitute.

Martin, C.M. and Randolph, M.F. 2006. Upper-bound analysis of lateral pile capacity in cohesive soil. *Géotechnique*, 56(2): 141–145.

Mayne, P.W., Robertson, P.K. and Lunne, T. 1998. Clay stress history evaluated from seismic piezocone tests. In *Proceedings of the 1st. International Conference on Geotechnical Site Characterization*, ISC'1, Atlanta, 2: 1113–1118.

Mayne, P.W. 2001. Stress-strain-strength-flow parameters from enhanced in situ tests. In *Proceedings International Conference on In Situ Measurement of Soil Properties and Case Histories, In Situ 2001*. (Rahardjo and Lunne eds.). Bali, Indonesia: 27–48.

Mayne, P.W. 2007. In situ test calibrations for evaluating soil parameters. In *Proceedings International Workshop on Characterisation and Engineering Properties of Natural Soils ("Natural Soils 2006")*. NUS Singapore, December. Tan, T.S. et al. eds. 3: 1601–1652. London, Taylor and Francis.

Mlynarek, Z., Gogolik, S., Marchetti, S. and Marchetti, D. 2006. Suitability of SDMT test to assess geotechnical parameters of post – flotation sediment. In *Proceedings of 2nd International Flat Dilatometer Conference:* 148–153. Washington DC, June.

Mollé, J. 2005. The accuracy of the interpretation of CPT-based soil classification methods for soft soils. *MSc Thesis* Section for Engineering Geology, Department of Applied Earth Sciences, *Delft University of Technology*. Memoirs of the Centre of Engineering Geology in the Netherlands, No. 242. Report AES/IG/05-25, December.

Monaco, P. and Marchetti, S. 2007. Evaluating liquefaction potential by seismic dilatometer (SDMT) accounting for aging and stress history. *In Proceedings 4th International Conference on Earthquake Geotechnical Engineering, ICEGE Thessaloniki, Greece, June, Paper 1626.*

Monnet, J., Allagnat, D., Teston, J., Billet, P. and Baguelin, F. 2006. Foundation design for a large arch bridge on alluvial soils. *Proc. Inst. Civ. Eng., Geotech Eng.*, 159, 1: 19–28, Thomas Telford, London.

Nash, D.F.T., Lings, M.L., Benahmed, N., Sukolrat, J. and Muir Wood, D. 2006. The effects of controlled destructuring on the small strain shear stiffness G_0 of Bothkennar clay. In *The Tatsuoka Geotechnical Symposium*. Rome.

Nazarian, S. and Stokoe, K.H. 1984. In situ shear wave velocities from spectral analysis of surface waves. In *Proceedings 8th world conf. on earthquake engineering.* 3: 31–38.

Olsen, R.S. and Mitchell, J.K. 1995. CPT stress normalisation and prediction of soil classification. In *Proceedings of the International Symposium on Cone Penetration Testing, CPT'95*, Lingköping, Sweden, Swedish Geotechnical Society, 2: 257–262.

Park, C.B., Miller, D.M., and Xia, J. 1999. Multichannel Analysis of surface waves. *Geophysics*, 64 (3): 800–808.

Peuchen, J., Adrichem, J. and Hefer, P.A. 2005. Practice notes on push-in penetrometers for offshore geotechnical investigation. In *Proc., Int. Symp. on Frontiers in Offshore Geotechnics (ISFOG)*, Perth, Australia: 973–979.

Peuchen, J. and Mayne, P. 2007. Rate effects in vane shear testing. In *Proceedings 6th International Conference, Society for Underwater Technology, Offshore Site Investigation and Geotechnics (SUT-OSIG):* 259–266. London, September: 187–194.

Pijpers, B. 2002. Sonderen, van Begemann tot Hryciw, een drieluik over de interpretatie van sonderingen, twee 'speciale' milieusondes en de camerasonde. *GeoDelft Internal Report.*

Powell, J.J.M., Quarterman, R.S.T. and Lunne, T. 1988. Interpretation and use of the piezocone test in UK clays, *Proc. Penetration testing in the UK*, Thomas Telford, London: 51–156.

Powell, J.J.M. and Butcher, A.P. 1991. Assessment of ground stiffness from field and laboratory tests. In Proceedings Xth ECSMFE, Fierenze. 1: 153–156. Rotterdam, Balkema.

Powell, J.J.M. and Shields, C.H. 1995. Field studies of the full displacement pressuremeter in clays. In Proc. 4th Int. Symposium on Pressuremeters (ISP4), Sherbrooke, Canada, 239–248.

Powell, 2001. In situ testing and its value in characterising the UK National soft clay testbed site, Bothkennar. In *Proceedings International Conference on In Situ Measurement of Soil Properties and Case Histories, In Situ 2001.* (Rahardjo and Lunne eds.) Bali, Indonesia.

Powell, J.J.M. 2005. Technical session 1c: In situ testing. In *Proc. 16th International Conference on Soil Mechanics and Geotechnical engineering, Osaka*, September, Gen Reps. Vol.: 51–61.

Powell, J.J.M., Shields, C.H., Frank, R., Dupla, J.-C. and Mokkelbost, K.H. 2005. A cone pressuremeter method for design of axially loaded piles in clay soils. *Proc. ISP – Pressio 2005. 50 years of pressuremeters, Gambin, Magnan and Mested (eds.), l'ENPC/LCPC, Paris*, 1: 547–561.

Powell, J.J.M. and Lunne, T. 2005. A comparison of different sized piezocones in UK clays. In *Proc. 16th International Conference on Soil Mechanics and Geotechnical engineering, Osaka*, September, 1: 729–734.

Ramsey, N. 2002. A calibrated model for the interpretation of cone penetration tests (CPT's) in North Sea Quaternary soils. *Proc. Int. Conf. Offshore SI and Geotechnics, SUT* 2002, London, Nov.: 341–356.

Randolph, M.F. and Houlsby, G. 1984. The limiting pressure on a circular pile loaded laterally in cohesive soil. *Géotechnique*, 34 (4): 613–623.

Randolph, M.F., Hefer, P.A., Geise, J.M. and Watson, P.G. 1998. Improved seabed strength profiling using T-bar penetrometer. In *Proc. Int. Conf. Offshore Site Investigation and Foundation Behaviour, SUT, London:* 221–236.

Randolph, M.F., Martin, C.M. Hu, Y. 2000. Limiting resistance of a spherical penetrometer in cohesive material. *Géotechnique*, 50(5), pp. 573–582.

Randolph, M.F. 2004. Characterisation of soft sediments for offshore applications. In *Proc. of 2nd Int. Conf. on Geotechnical and Geophysical Site Characterization – ISC'2*, Porto, 1: 209–232. Rotterdam, Millpress.

Randolph, M.F., Cassidy, M., Gourvenec, S. and Ebrich, C. 2005. Challenges of offshore geotechnical engineering. In *Proc. 16th International Conference on Soil Mechanics and Geotechnical engineering, Osaka*, September, 2: 123–133.

Randolph, M.F. Anderson, K.H. 2006. Numerical Analysis of T-bar Penetration in soft clay. *Int. J. Geomechanics, ASCE*, 6(6): 411–420.

Randolph, M.F., Low, H.E. and Zhou, H. 2007. In situ testing for the design of pipelines and anchoring systems. In *Proceedings 6th International Conference, Society for Underwater Technology, Offshore Site Investigation and Geotechnics (SUT-OSIG)*, London, September: 251–262.

Reiffsteck, P. 2005. Bearing capacity and settlement of a deep foundation – presentation of the results of the prediction exercises. *Proc. ISP – Pressio 2005. 50 years of pressuremeters, Gambin, Magnan and Mested (eds.), l'ENPC/LCPC, Paris* 2: 521–536.

Robertson, P.K., Campanella, R.G., Gillespie, D. and Greig, J. 1986. Use of piezometer cone data. In *Proc. ASCE Speciality Conf. In Situ '86: Use of In Situ Tests in Geotechnical Engineering*, Blacksburg: 1263–1280.

Robertson, P.K. 1990. Soil classification using the cone penetration test. *Can Geo. Jnl.*, 27(1): 151–158.

Robertson, P.K., Sasitharan, S., Cunning, J.C. and Segs, D.C. 1995. Shear wave velocity to evaluate flow liquefaction. *Journal of Geo. Eng., ASCE*, 121 (3): 262–273.

Robertson, P.K. 2004. Evaluating soil liquefaction and post-earthquake deformations using the CPT. In *Proc. of 2nd Int. Conf. on Geotechnical and Geophysical Site Characterization – ISC'2*, Porto, 1: 233-252. Rotterdam, Millpress.

Sandven, R. 1990. Strength and deformation properties of fine grained soils obtained from piezocone tests. *PhD thesis* NTH, Trondheim.

Sandven, R.B. and Sjursen, M. 1998. Sample disturbance in soils – results from an overconsolidated marine clay. In *Proc. 1st Int. Conf. On Site Characterisation – ISC '98*, Atlanta, 1: 409–417.

Schmertmann, J.H. 1978. Guidelines for cone tests, performance and design. *Federal Highway Administration, Report* FHWA-TS-78209, Washington, 145p.

Schnaid, F., Lehane, B.M. and Fahey, M. 2004. In situ test characterization of unusual geomaterials. In *Proc. of 2nd Int. Conf. on Geotechnical and Geophysical Site Characterization – ISC'2*, Porto, 1: 49–74. Rotterdam, Millpress.

Schnaid, F. 2005. Geo-characterisation and properties of natural soils by in situ tests. In *Proceedings XVI ICSMGE, Osaka, 1: 3–45.* Rotterdam: Millpress.

Schneider, J.A., Randolph, M.F., Mayne, P.W. and Ramsey, N. 2008. Influence of partial consolidation during penetration on normalised soil classification by piezo-cone. *Proceedings 3rd International Conference on Site Characterisation (ISC'3), Taipei, April.*

Senneset, K., Sandven, R. and Janbu, N. 1989. The evaluation of soil parameters from piezocone tests. In *Proc. Research Council, Trans. Research Board, In Situ Testing of Soil Properties for Transportation Facilities,* Washington D.C..

Soccodato, F.M. 2003. Geotechnical properties of Fucino clayey soil. In *Proceedings International Workshop on Characterisation and Engineering Properties of Natural Soils ("Natural Soils 2002").* NUS Singapore, December. Tan, T.S. et al. eds. 1: 791–807, Rotterdam, Balkema.

Sukolrat, J. 2007. Structure and destructuration of Bothkennar clay. *PhD Thesis,* University of Bristol.

Stewart, D.P. and Randolph, M.F. 1991. A new site investigation tool for the centrifuge. *Proc. Int. Conf. on Centrifuge Modelling Centrifuge 91,* Boulder Colorado: 531–538.

Stewart, D.P. and Randolph, M.F. 1994. T-bar penetration testing in soft clay. *ASCE Journal of Geotechnical Engineering, 120(12):* 2230–2235.

Stokoe, K.H. II, Joh, S-H and Woods, R.D. 2004. The contributions of in situ geophysical measurement to solving geotechnical engineering problems. In *Proc. of 2nd Int. Conf. on Geotechnical and Geophysical Site Characterization – ISC'2,* Porto, 1: 97–132. Rotterdam, Millpress.

Sutton, J.A. 1999. Engineering seismic geophysics at Bothkennar. *M.Phil Thesis.* University of Surrey.

Tatsuoka, F., Jardine, R.J., Lo Presti, D., DiBenedetto, H. and Kodaka, T. 1997. Theme lecture: Characterising the prefailure deformation properties of geomaterials. Proc. 14th ICSMFE Hamburg, 4: 2129–2175.

Teh, K.L., Leung, C.F. and Chow, Y.K. 2007. Some considerations for predicting spudcan penetration resistance in two layered soil using miniature penetrometer. In *Proceedings 6th International Conference, Society for Underwater Technology, Offshore Site Investigation and Geotechnics (SUT-OSIG),* London, September: 279–284.

Tokimatsu, K., Kuwayama, S., Tamura, S. and Miyadera, Y. 1991. V_s determination from steady state Rayleigh wave method. *Soils and Foundations,* 31(2): 153–163.

Watson, P.G., Newson, T.A. and Randolph, M.F. 1998. Strength profiling in soft offshore soils. In *Proc. 1st Int. Conf. On Site Characterisation – ISC '98,* Atlanta, 2: 1389–1394.

Watson, P.G. and Humpheson, C. 2007. Foundation design and installation of the Yolla A platform. In *Proceedings 6th International Conference, Society for Underwater Technology, Offshore Site Investigation and Geotechnics (SUT-OSIG),* London, September: 399–412.

Weemees, I, Howie, J., Woeller D., Sharp, J., Cargill, E. and Greig, J. 2006. Improved techniques for in situ determination of undrained shear strength in soft clays. In *Proc. Sea to Sky Geotechnics:* 89–95.

Yafrate, N.J. and DeJong, J.T. 2005. Considerations in evaluating the remoulded undrained shear strength from full flow penetrometer cycling. In *Proc., Int. Symp. on Frontiers in Offshore Geotechnics (ISFOG),* Perth, Australia: 991–997.

Yafrate, N.J. and DeJong, J.T. 2006. Interpretation of sensitivity and remoulded undrained shear strength with full flow penetrometers. In *Proc. Conference of Int. Soc. For Offshore and Polar Engineering (ISOPE-06),* San Francisco: 572–577.

Yafrate, N.J., DeJong, J.T. and DeGroot, D.J. 2007. The influence of full flow penetrometer area ratio on penetration resistance and undrained and remoulded shear strength. In *Proceedings 6th International Conference, Society for Underwater Technology, Offshore Site Investigation and Geotechnics (SUT-OSIG),* London, September: 461–468.

Whittle, A.J., Sutabutr, T., Germaine, J.T. and Varney, A. 2001a. Prediction and interpretation of pore pressure dissipation of a tapered piezoprobe. *Géotechnique,* 51, 7: 601–617.

Whittle, A.J., Sutabutr, T., Germaine, J.T. and Varney, A. 2001a. Prediction and measurement of pore pressure dissipation for a tapered piezoprobe. *Proc. Offshore Technology Conf., Houston,* Paper 13155.

Yu, H.-S. 2004. In situ soil testing: from mechanics to interpretation. In *Proc. of 2nd Int. Conf. on Geotechnical and Geophysical Site Characterization – ISC'2,* Porto, 1: 3–38. Rotterdam, Millpress.

Recompacted, natural and in-situ properties of unsaturated decomposed geomaterials

C.W.W. Ng & R. Chen

Department of Civil Engineering, The Hong Kong University of Science and Technology, Hong Kong

ABSTRACT: The behaviour and properties of unsaturated soils have attracted increasing attention world-wide. To improve our understanding of unsaturated soil behaviour and properties, this paper investigates and summarises some recent findings on the soil-water characteristics, small strain stiffness and shear strength of unsaturated soils. Measured results obtained from laboratory tests on recompacted and natural specimens and in-situ tests are compared and discussed. This paper focuses on decomposed geomaterials, which are relatively not well studied but relevant to many parts of the world.

1 INTRODUCTION

Civil engineers build on or in the earth's surface (Ng & Menzies 2007). A large portion of the earth surface is found in semi-arid and arid regions where the groundwater table is deep because the annual evaporation from the ground surface in these regions exceeds the annual amount of precipitation (Fredlund & Rahardjo 1993). The land surface in semi-arid and arid regions may be comprised of notoriously hazardous geomaterials called "unsaturated soils". These soils are hazard to slopes, earth structures and earth-supported structures because on wetting by rain or other means, they can expand or collapse; on drying by evaporation or other means, they can desiccate and crack with serious consequences for safety and high costs. Among unsaturated soils, some are particularly problematic for engineers, for example, expansive plastic clays commonly found in Colorado and Texas, USA, in Hubei and Shandong, China and in Madrid, Spain (Ng & Menzies 2007); loess soils commonly found in Missouri and Wisconsin, USA, Ningxia and Shanxi, China (Liu 1988) and in Kent, Sussex and Hampshire, United Kingdom (Jefferson et al. 2001); and residual and saprolitic soils located above the ground water table, particularly at many hillsides in Brazil, Portugal and in the Far East such as Hong Kong and Malaysia.

The location of the groundwater table and hence the state of the soil (saturated or unsaturated) are strongly influenced by climatic conditions in a region. If the region is arid or semi-arid, the groundwater table is slowly lowered with time. If the climate is temperate or humid, the ground water table may remain quite close to the ground surface. As shown in Figure 1, it is the difference between the downward flux (i.e.

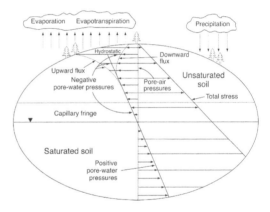

Figure 1. A visualization of soil mechanics showing the role of the surface flux boundary condition (Fredlund 1996).

precipitation) and the upward flux (i.e. evaporation and evapotranspiration) that determines the location of the groundwater table and the state of the soils. Accordingly, the geotechnical world is divided by a horizontal line representing the groundwater table (Fredlund 1996). Below the water table, the pore-water pressures will be positive and the soil will, in general, be saturated. Above the water table, the pore-water pressure will, in general, be negative with respect to the atmospheric pressure (i.e. the gauge pressure). The entire soil zone above the water table is called the vadose zone (see Fig. 2). Immediately above the water table there is a zone called the capillary fringe where the degree of saturation approaches 100%. This zone may range from less than one metre to approximately 10 metres in thickness, depending upon the soil type

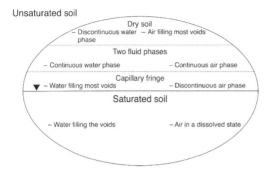

Figure 2. A visualization of saturated/unsaturated soil mechanics based on the nature of the fluid phases (Fredlund 1996).

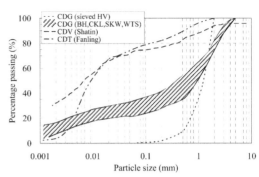

Figure 3. Particle size distributions of CDG, CDV and CDT.

(Fredlund 1996). Inside this capillary zone, the water phase may be assumed to be continuous whereas the air phase is generally discontinuous. Above this capillary zone, a two-phase zone may be identified in which both the water and air phases may be idealised as continuous. Inside this zone, the degree of saturation may vary from about 20% to 90%, depending on soil type and soil state. Above this two-phase zone, the soil becomes dryer and the water phase will be discontinuous whereas the air phase will remain continuous (Fredlund & Rahardjo 1993, Fredlund 1996). It is generally recognised that the soil located in the dry and in the capillary fringe zones may be described and modelled adequately by using Terzaghi's effective stress principle. However, the effective stress principle runs into some difficulties and complications when the soil is located in the two-phase zone. To avoid unnecessary complication in this paper, it may be reasonable to assume that unsaturated soil behaviour and properties are essentially governed by two independent stress state variables, net normal stress $(\sigma - u_a)$ and matric suction $(u_a - u_w)$, where σ, u_a and u_w are the total normal stress, pore air and pore water pressures, respectively (Fredlund & Rahardjo 1993).

To improve our understanding of unsaturated soil behaviour, this paper investigates three fundamental and important aspects of unsaturated soil properties: soil-water characteristics, small strain stiffness and shear strength obtained from laboratory tests on recompacted and natural specimens and in-situ tests. Natural soils differ from the corresponding recompacted soils both with respect to fabric and bonding (soil structure) (Burland 1990). Measured results obtained from laboratory and in-situ tests are compared and discussed. The influences of the drying and wetting cycle and the stress state on soil-water characteristics are investigated in this paper. In addition, suction effects on small strain stiffness and shear strength are studied. This paper mainly focuses on tests on unsaturated decomposed geomaterials carried out by the unsaturated soil research group at the

Hong Kong University of Science and Technology (HKUST), which was established in 1996 and since then, a PG course called "Unsaturated Soil Mechanics", has been developed and taught.

2 TEST MATERIALS

Three typical types of unsaturated decomposed material from Hong Kong are tested here. They are completely decomposed volcanic soil (CDV), completely decomposed granite (CDG) and completely decomposed tuff (CDT). The particle size distributions (PSD) of CDV, CDG and CDT are shown in Figure 3. According to the GCO (1988), CDV can be classified as slightly sandy silt–clay with high plasticity. CDG is described as silty sand whereas CDT is described as slightly sandy silt with intermediate plasticity. In the figure, HV, BH, CKL, SKW and WTS denote the sampling locations at Happy Valley, Beacon Hill, Cha Kwo Ling, Shau Kei Wan and Wong Tai Sin in Hong Kong, respectively.

To prepare recompacted specimens, the soil was first oven-dried for 48 hours. After removal from the oven, those particles with one dimension larger than 2 mm were discarded by dry sieving. A sieved PSD of HV is shown in Figure 3 as an example. Then water was added to achieve the desired water content. After mixing, the soil was kept in a plastic bag for moisture equalization in a temperature- and moisture-controlled room. The soil was then compacted in layers to a desired dry density in a mould with desired dimensions by the static or dynamic compaction method (Ng et al. 1998, Ng & Chiu 2001).

Natural specimens were obtained from Mazier (GCO 1987) and block samples taken from the field. Each natural sample was carefully hand trimmed to create a natural specimen with suitable dimensions for laboratory testing. The trimming was conducted in a temperature- and humidity-controlled room to decrease water loss during specimen preparation (Ng et al. 2004).

3 SOIL-WATER CHARACTERISTICS

To study transient seepage in unsaturated soils, it is essential to know the relationship between the amount of water in a soil and the soil suction (i.e. soil-water characteristics). There have been different terms used to describe the relationship in different disciplines. Fredlund et al. (2001) recommended that the term "soil-water characteristic curve (SWCC)" should be used in civil engineering-related disciplines. SWCC is a key hydraulic property of an unsaturated soil. It represents the water retention capability of the unsaturated soil. SWCC can be determined from both laboratory and in-situ tests by controlling or measuring both the water content and soil suction.

3.1 Soil-water characteristics of recompacted CDV and CDG specimens

SWCC of a given soil can be influenced by many factors, e.g. initial specimen properties (including initial water content, initial dry density and compaction method), soil structure (recompacted or natural), drying and wetting cycle, stress history and current stress state (Vanapalli et al. 1999, Ng & Pang 2000a, b).

This section discusses the influence of drying and wetting cycles and the stress level on SWCCs of recompacted specimens of CDV (Ng & Pang 2000a, b) and CDG (Ho et al. 2006, 2007, Tse 2007).

3.1.1 Apparatuses

Although it is generally recognised that unsaturated soil behaviour and properties are governed by the two independent stress state variables (i.e. matric suction and net stress), effects of net stress on SWCCs have been ignored for many years. At HKUST, two pressure plate systems have been developed for measuring stress-dependent soil-water characteristic curves (SDSWCCs) under one-dimensional (1D) (Ng & Pang 2000a) and triaxial stress conditions (Ng et al. 2001). In both systems, the axis-translation technique (Hilf 1956) is adopted to control matric suction in an unsaturated specimen.

Figure 4 shows the one-dimensional stress-controllable pressure-plate extractor for measuring SDSWCC under 1D stress conditions (Ng & Pang 2000a). This apparatus was designed to overcome some limitations of conventional volumetric pressure plate extractors. An oedometer ring equipped with a high air-entry value (AEV) ceramic plate at its base is located inside an airtight chamber. Vertical stress is applied through a loading frame to a soil specimen inside the oedometer ring. Because radial deformation is zero for the K_0 stress condition, the change in total volume of the specimen is measured from the vertical displacement of the soil specimen using a dial gauge. The axis-translation technique is adopted to impose a matric suction on the soil specimen. The

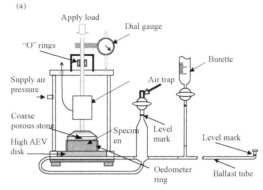

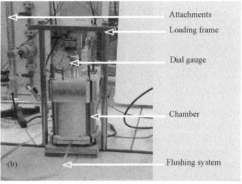

Figure 4. The one-dimensional stress-controllable pressure plate extractor developed at HKUST (Ng & Pang 2000a): (a) Schematic diagram; (b) Photograph.

u_a is controlled through a coarse porous stone together with a coarse geotextile located at the top of the specimen. The u_w is controlled at the atmospheric pressure through the high AEV ceramic plate mounted at the base of the specimen. A set of hysteresis attachments, including an air trap, a ballast tube and a burette, are used to measure the water volume change during the drying and wetting processes. The air trap is attached to collect air that may diffuse through the high AEV ceramic disc. The ballast tube serves as horizontal storage for water flowing in or out of the soil specimen. The burette is used to store or supply water and to measure the water volume change in the soil specimen. This type of 1D apparatus has been commercialised as Fredlund cell very recently.

Figure 5 shows the modified triaxial pressure plate system at HKUST for measuring SDSWCC under triaxial stress conditions. This system was modified from the original design by Ng et al. (2001). It consists of five main components, namely a net normal stress control, a suction control, a flushing system, a measuring system for water volume change and a new measuring system for total volume change. The cell pressure is applied from the top of the outer cell. Because of the

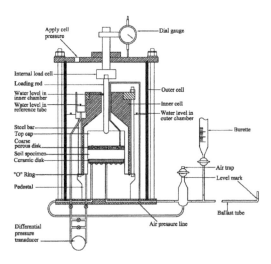

Figure 5. A schematic diagram of the modified triaxial pressure plate system at HKUST (Lai 2004, Ng & Menzies 2007).

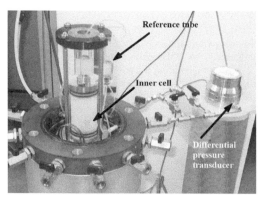

Figure 6. A new double-cell total volume change measuring system for unsaturated soils (Ng et al. 2002).

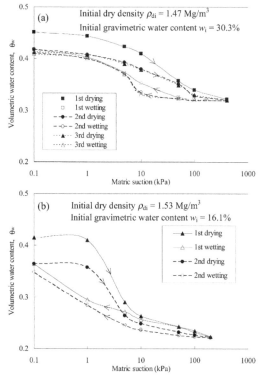

Figure 7. The influence of drying and wetting cycles on SWCCs of (a) CDV from Shatin (Ng & Pang 2000b) and (b) CDG from HV specimens (Ho et al. 2007).

open-ended design of the inner cell, the same pressure can be applied to both the inner and the outer cells (Ng et al. 2002). Axial force, if required, is exerted on the specimen through the loading ram. An internal load cell and a linear variable differential transformer (LVDT) are attached to the loading ram to measure the axial force exerted on and the axial displacement of the test specimen, respectively. The axis-translation technique is adopted. The u_a is controlled through a coarse porous disc at the top of the specimen and the u_w is maintained at the atmospheric pressure through a saturated high AEV ceramic disc at the base of the specimen. As in the 1D stress-controllable pressure plate extractor, a set of hysteresis attachments are used to measure the water volume change during the drying and wetting processes. The total volume change of the test specimen is measured by adopting a new simple device as shown in Figure 6.

Figure 6 shows the setup of the new total volume change measuring system. The basic principle is that the total volume change of a specimen is measured by recording differential pressures between the water inside an inner cell and the water inside a reference tube. The differential pressures are measured by a highly accurate differential pressure transducer (DPT). Several important steps were taken to improve the accuracy and the sensitivity of the open-ended, bottle-shaped inner cell measuring device over some existing double-cell systems. Detailed calibrations were carried out to account for apparent volume changes due to changes in the cell pressure, fluctuation in the ambient temperatures, creep in the inner cell wall and relative movement between the loading ram and the inner cell. The estimated accuracy for total volume changes measured by this device is about 0.03% of the

volumetric strain of a specimen with 70 mm diameter and 19 mm height (Ng et al. 2002).

3.1.2 Influence of drying and wetting cycles

Figure 7a shows measured SWCCs of a recompacted CDV specimen subjected to three repeated drying and wetting cycles. The specimen was compacted to an initial dry density (ρ_{di}) of 1.47 Mg/m3 at the initial gravimetric water content (w_i) of 30.3%. In each cycle, a marked hysteresis loop between the drying and wetting paths can be observed. This is mainly attributed to the geometric non-uniformity of individual pores, resulting in the so-called "ink-bottle" effect (Hillel 1998). The size of the hysteresis loop is the largest in the first cycle but seems to become independent of the subsequent cycles. The desorption characteristics are dependent on the drying and wetting history. The rate of desorption is relatively high during the first cycle compared with that during the second and third cycles. This may be due to the presence of relatively large voids initially. During the first wetting process, a significant volume change is likely to take place and this results in a smaller void due to collapse of soil structures of the virgin soil (Ng et al. 1998). Thus, a smaller rate of desorption for the second and third drying and wetting cycles is observed. The adsorption characteristics of the first wetting process also seem to be different from that of the subsequent wetting processes. The value of the matric suction at which the soil starts to absorb water significantly is higher during the first wetting process (i.e. about 50 kPa) than during the subsequent cycles (i.e. about 10 kPa). The rates of adsorption are substantially different for the first and subsequent drying and wetting cycles at suctions ranging from 10 to 50 kPa. This might be caused by some soil structure changes after the first drying and wetting cycle. According to Bell (1992), drying initiates cementation by aggregation formation, leading to some relatively large inter-pores formed between the aggregated soil lumps. These large inter-pores reduce the specimen's rate of absorption along a certain range of the wetting path.

Figure 7b shows the influence of drying and wetting cycles on the SWCCs of a recompacted CDG specimen. Two cycles of drying and wetting were applied to this specimen. The major observed difference between the two cycles is a reduction in the size of hysteresis loop in the second drying and wetting cycle. Although the shape of the two loops is surprisingly similar, the difference between the initial and final volumetric water contents of each cycle is also greatly reduced in the second drying and wetting cycles, similar to that observed in CDV (see Fig. 7a).

3.1.3 Influence of stress state

Figure 8a shows the influence of 1D stress on the soil-water characteristics of CDV specimens recompacted to an almost identical initial ρ_{di} at the same w_i. A drying and wetting cycle was applied on three specimens at different net vertical stresses. In a specimen subjected to a higher vertical stress, the AEV of an SDSWCC appears to be larger and the rate of desorption is lower, whereas the size of the hysteresis loop is smaller. These observations are likely to be caused by

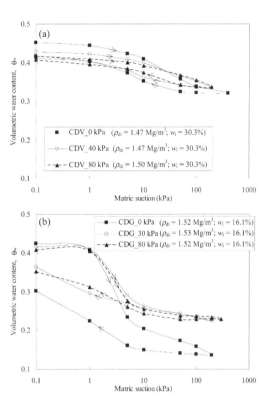

Figure 8. The influence of 1D stress on the soil-water characteristics of (a) CDV from Shatin (Ng & Pang 2000b) and (b) CDG from HV (Ho et al. 2006).

the presence of a smaller average pore-size distribution in the soil specimen under higher applied vertical stress.

Measured SDSWCCs of recompacted CDG specimens are shown in Figure 8b. The three specimens were recompacted to an almost identical ρ_{di} at the same w_i. A drying and wetting cycle was applied on three CDG specimens at different net vertical stresses. There is a distinct difference between the SWCC (at zero stress) and the two SDSWCCs loaded to 30 kPa and 80 kPa. When the applied net vertical stress is increased from 0 to 30 kPa, the desorption rate and the size of hysteresis loop significantly decrease but the AEV and the absorption rate remain almost unchanged. For the specimens under 30 and 80 kPa, the difference in SDSWCCs is not very significant.

Figure 9 shows SDSWCCs of CDG specimens compacted to the same initial ρ_{di} at the same w_i but tested under different stress conditions (1D, isotropic and deviatoric (DEV) stress states) but with the same net mean stress of 20 kPa. The stress ratios applied by the isotropic stress state, 1D stress state and deviatoric stress state are 0, 0.75 and 1.2, respectively (Tse 2007). The stress ratio is defined as $q/(p - u_a)$, where q and $(p - u_a)$ are the deviator stress and net mean stress,

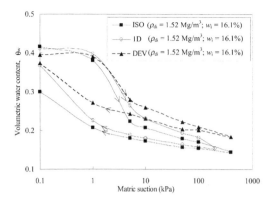

Figure 9. The influence of the stress ratio on the soil-water characteristics of CDG from HV (Tse 2007).

respectively. It can be seen that the SDSWCCs shift upward from isotropic to 1D, and from 1D to deviatoric, illustrating that the specimen under a higher stress ratio has a higher water retention ability at a given suction. At a given higher stress ratio, the SDSWCC shows a lower desorption rate and higher residual water content. A smaller hysteresis loop and a smaller amount of air entrapment after a drying and wetting cycle are also observed under a higher stress ratio condition. The AEV of each specimen does not seem to be affected. When a soil specimen is subject to a higher stress ratio, the deviatoric stress reduces the void ratio (i.e. shear-induced volume change) inside the specimen more effectively (Tse 2007). The water retention ability of the soil is thus increased, resulting in an upward shift in the SDSWCC, a decrease in the desorption rate and an increase in the residual water content. Consequently, the ink-bottle effect is less pronounced as the void ratio is reduced, leading to the shrinkage of the size of the hysteresis loop and reduction in the amount of air entrapment after a drying and wetting cycle.

3.2 Comparison between recompacted and natural CDV and CDG specimens

Figure 10a compares the SDSWCCs of a recompacted and a natural CDV specimen. The recompacted specimen was compacted to the same initial ρ_{di} at the same w_i as that of the natural specimen. Both specimens were loaded to the same net vertical stress of 40 kPa. The recompacted specimen seems to have a higher AEV than that of the natural one and the size of the hysteresis loop in the former is considerably larger than that in the latter. According to Ng & Pang (2000b), the recompacted specimen, which is recompacted on the wet side of optimum, is generally believed to be more homogenous, whereas the natural specimen has relatively non-uniform pore size distributions due to various geological processes, such

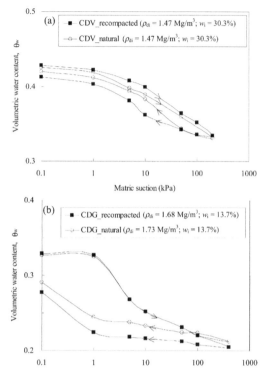

Figure 10. Comparison of the SDSWCCs between recompacted and natural specimens of (a) CDV from Shatin (Ng & Pang 2000b) and (b) CDG from HV (Tse 2007).

as leaching, in the field. As the two specimens have the same initial density, it is reasonable to postulate that the natural specimen would have some larger pores than the recompacted specimen would have, at least statistically. Thus, the natural specimen has a slightly lower AEV and a higher rate of desorption than the recompacted specimen for suctions up to 50 kPa, as shown in Figure 10a. The rates of desorption of the two soil specimens appear to be the same for high suctions. On the other hand, the rates of adsorption for the two specimens are considerably different. The rate of the wetting curve obtained from the natural specimen is substantially higher than that of the recompacted specimen. This observed behaviour may be explained by the difference in pore size distributions between the natural and recompacted specimens. As the natural soil specimen has a non-uniform pore-size distribution in which some relatively large pores exist with some relatively small pores, the presence of these small pores would facilitate the ingress of water to the specimen as the soil suction reduces. On the contrary, the lack of small pores and the presence of the relatively uniform, medium-sized pores in the recompacted specimen would slow the rate of water

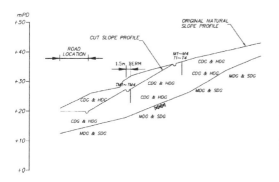

Figure 11. Cross-section and instrumentation locations in a CDG/HDG slope at Ma On Shan (Li et al. 2005).

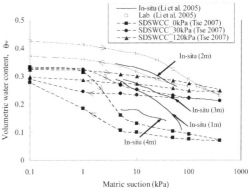

Figure 12. Comparison of SDSWCCs from in-situ and laboratory tests from Ma On Shan (Li et al. 2005) and laboratory tests on CDG from HV in Hong Kong (Tse 2007).

entering the soil specimen. The rates of adsorption of the two soil specimens appear to be the same at small suctions (less than 5 kPa).

Figure 10b shows the SDSWCCs of a natural and a recompacted CDG specimen. Both specimens have similar initial dry densities and the same initial water contents subjected to the same stress level of 30 kPa. The major difference between the two curves is the smaller size of the hysteresis loop for the natural specimen than for the recompacted specimen, consistent with results from CDV. This may be due to the non-uniform pore size distribution in the natural specimen as discussed earlier. The AEV and the desorption and adsorption rates are similar for both specimens.

3.3 Comparison between laboratory and in-situ SDSWCCs of CDG and CDT

In-situ SWCCs of a natural soil can be obtained by field measurements of both water content and soil suction based on the so-called instantaneous profile method (Watson 1966). As pointed out by Basile et al. (2003), various surveys and comparisons of in-situ and laboratory measurements of SWCC have been suggested. Differences between laboratory and field measurements have long been noted. The differences may be related to sample size, sample disturbance, lack of consideration of stress state, measurement error and different hysteretic paths. Up to now, there are still great difficulties in providing quantitative explanations for and formalization of differences between laboratory and in-situ hydraulic properties, including SWCCs.

Li et al. (2005) conducted a full-scale field experiment in an instrumented cut slope which consists of CDG and highly decomposed granite (HDG) at Ma On Shan. Figure 11 shows the cross-section and the locations of instrumentation in the slope. At the crest of the slope, four time-domain reflectometer (TDR) moisture probes were installed to measure the volumetric water contents in boreholes M1-M4 at depths

of 1, 2, 3 and 4 m, respectively. Moreover, four vibrating wire tensiometers were installed to measure matric suction in boreholes T1-T4 at depths of 1, 2, 3 and 4 m, respectively. Based on the measurements of water content and matric suction, in-situ SDSWCCs at different locations were determined. The measured SDSWCCs include several wetting-drying cycles due to rainfall infiltration and evaporation for about one year.

Figure 12 shows the measured in-situ SDSWCCs at 1 m, 2 m, 3 m and 4 m depth at the crest of the cut slope. Each in-situ SDSWCC shows negligible hysteresis during repeated wetting-drying cycles in the field. The measured SDSWCCs are different at different depths. No clear trend can be identified.

Li et al. (2005) compared the in-situ SDSWCC at 2 m depth with a laboratory SWCC on a natural Mazier specimen taken from the same location. The laboratory SWCC for both drying and wetting paths in the natural specimen was determined by HKUST using a conventional pressure plate extractor at zero net normal stress. A clear hysteresis loop can be seen between the drying and wetting paths. As compared in the figure, the in-situ SDSWCC at 2 m depth appears to be consistent with the wetting path of the SWCC determined in the laboratory. This seems to suggest that the natural specimen may have similar density (i.e. void ratio) as the in-situ one.

To further study the in-situ measurements of SDSWCCs, laboratory measurements of natural CDG samples taken from another site at Happy Valley and loaded at different stresses are included in the figure. It is clear that the measured soil-water characteristics are stress-dependent and hysteretic. The higher the stress applied on the specimen (i.e. the smaller the void ratio), the flatter and higher location of the SDSWCC, suggesting the greater ability of the specimen to retain water at a given suction. Similar laboratory results from other soils are also reported (Ng & Pang 2000b). The observed discrepancies between the in-situ and

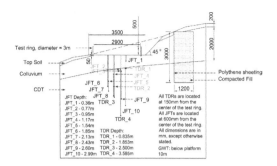

Figure 13. Cross-section and instrumentation locations at Tung Chung (Tse 2008).

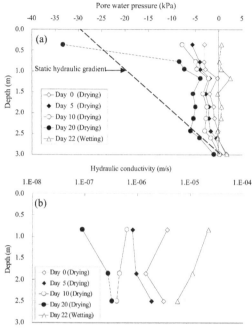

Figure 14. Profiles of (a) pore water pressure and (b) hydraulic conductivity during the first drying and wetting cycle in CDT at Tung Chung (Tse 2008).

laboratory SDSWCCs may be worth to be further investigated.

Supported by the Geotechnical Engineering Office of the Civil Engineering and Development Department of the HKSAR, a joint government-industry-university full-scale field monitoring program has been conducted on CDT at Tung Chung recently by HKUST and Ove Arup and Partners (HK) Ltd in Hong Kong (Tse 2008). Various instruments including tensiometers, TDRs, earth pressure cells, inclinometers and a rain gauge were installed. Figure 13 shows a cross-section and the instrumentation locations for in-situ one-dimensional water permeability tests using the instantaneous profile method (Watson 1966). The field permeability tests were conducted within a test ring of 3 m in diameter. Ten jet-filled tensiomters (JFTs) were installed at depths from 0.36 to 2.99 m to monitor changes in matric suction. All these tensiometers were located at 0.6 m from the centre of the test ring. Four TDR probes were installed at depths from 0.835 to 3.585 m to measure volumetric water content. All these TDR probes were located at 0.15 m from the centre of the test ring. Data were recorded at 5 minute intervals. In this paper, only some monitoring data associated with the first drying and wetting cycle for the permeability tests are interpreted preliminarily and presented here. The results of the second drying and wetting cycle and other field data such as earth pressures and ground deformations will be interpreted and reported in detail by Tse (2008).

During the installation of field instruments in September, eight out of the 30 days were raining over the site. These eight raining days were spreading fairly uniformly in September and hence the ground was quite wet before the 1D permeability test. At the beginning of the test in early October, a 0.1 m deep water pond had been maintained for about five days to soak the ground before the drying process was permitted. The initial rainfall and ponding resulted in the measured low suction profile (small negative pore water pressures) at the first day (Day 0) of the first drying and wetting cycle as shown in Figure 14a. It can be seen

from the figure that the measured negative pore water pressures were generally small along the depth (i.e. less than 10 kPa) and the measured positive pressure gradients gradually becomes less positive from Day 0 to Day 10. This indicates that there was a downward flux before the first drying and wetting cycle due to the initial rainfall and ponding. The trend of the downward flux gradually reduced and eventually turned into an upward flux near the ground surface (within the top metre). During the first wetting between Day 20 and Day 22, negative pore water pressure (suction) decreased significantly and rapidly at different depths within 2 days. At Day 22, zero and small positive pore water pressures were measured at the top 3 metres. The hydraulic gradient was approximately equal to 1.0 suggesting that the downward flux was equal to the water permeability of the soil. (see Fig. 14b).

Figure 14b shows the measured hydraulic conductivity profiles. The changes of hydraulic conductivity are consistent with the trend of the measured pore water pressures during the first drying and wetting cycle. The hydraulic conductivity decreased with an increase in soil suction during the drying process but it increased with an increase in pore water pressure during wetting. The variations of the measured hydraulic conductivity ranging from about 1×10^{-5} to 1×10^{-7} m/s appear to be very significant (an order of 2) during the first drying and wetting cycle. The rapid responses

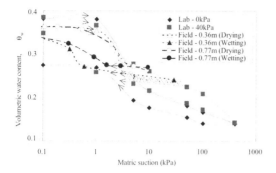

Figure 15. Comparisons of SDSWCCs from in-situ and laboratory measurements in CDT at Tung Chung (Tse 2008).

in the pore water pressure and hydraulic conductivity may be attributed to the presence of fissures and cracks in the ground.

Figure 15 shows measured in-situ SDSWCCs at depths of 0.36 m and 0.77 m. Hydraulic hysteresis can be clearly seen at the two depths. This is opposite to those observed at Ma On Shan (Li et al. 2005). For comparison purposes, two laboratory measured SDSWCCs of two natural specimens taken at the same site (i.e. Tung Chung) are also included in the figure. The laboratory tests were carried out at net vertical stresses of 0 kPa and 40 kPa. Consistent agreements can be seen between the in-situ and laboratory measured SDSWCCs.

4 SMALL-STRAIN STIFFNESS

The shear modulus at very small strains (0.001% or less), G_0, is important for predicting ground deformations and dynamic responses associated with many engineering structures. G_0 can be determined from in-situ seismic tests and laboratory measurements using resonant column and bender elements.

4.1 Laboratory testing apparatus

Figure 16 shows a computer-controlled triaxial system has been developed for determining anisotropic G_0 in both saturated and unsaturated soils at HKUST. The axis translation technique is adopted to control matric suction in an unsaturated soil specimen. The u_a is controlled through a coarse low AEV corundum disc placed on top of the specimen, while the u_w is controlled through a saturated high AEV ceramic disc sealed to the pedestal of the triaxial apparatus.

Figure 17 shows a modified base pedestal for embedding a high AEV ceramic disc and a bender element. A spiral-shaped drainage groove of 3 mm wide and 3 mm deep connected to the water drainage system is carved on the surface of the modified pedestal. It

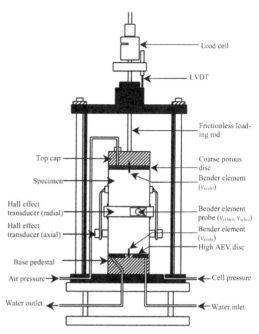

Figure 16. Schematic diagram of a triaxial apparatus with multidirectional shear wave velocity and local strain measurements for testing unsaturated soils (Ng & Yung 2007).

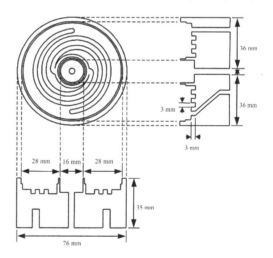

Figure 17. A modified base pedestal (Ng & Yung 2007).

serves as a water channel for flushing air bubbles that may be trapped or accumulated beneath the high AEV ceramic disc as a result of air diffusion during long periods of unsaturated soil testing. More details of the modified triaxial testing system are given by Ng & Yung (2007) and Ng & Menzies (2007).

In order to determine the shear wave velocities, $v_{s(ij)}$, and hence the shear moduli, $G_{o(ij)}$, in different planes

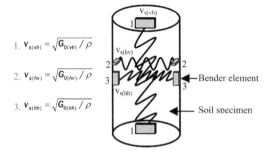

1. $v_{s(vh)} = \sqrt{G_{0(vh)} / \rho}$

2. $v_{s(hv)} = \sqrt{G_{0(hv)} / \rho}$

3. $v_{s(hh)} = \sqrt{G_{0(hh)} / \rho}$

← Bender element

— Soil specimen

Figure 18. Schematic diagram showing the arrangement of the bender elements (Ng & Yung 2007).

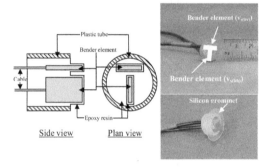

Figure 19. Details of the bender element probe (Ng et al. 2004).

of a soil specimen, three pairs of bender elements are used as illustrated in Figure 18. The shear modulus in the vertical plane of each soil specimen, $G_{0(vh)}$, is determined by using a pair of bender elements incorporated in the top cap and base pedestal. The shear moduli in the vertical and horizontal planes of each soil specimen, $G_{0(hv)}$ and $G_{0(hh)}$, are measured by a pair of bender element probes as shown in Figure 19. The bender element probes are inserted into the mid-height of each soil specimen with the aid of water-proof silicon grommets.

Two axial and one radial Hall-effect transducers (Clayton & Khatrush 1986) are used to measure local axial and radial displacements of each soil specimen. Based on the local strain measurements, the current tip-to-tip travelling distance of the shear waves and the volume change of each soil specimen can be determined throughout a test (Ng et al. 2004).

4.2 Stiffness characteristics of saturated CDT and CDG

4.2.1 Inherent stiffness anisotropy of saturated CDT

Figure 20 shows the measured shear moduli ($G_{0(hh)}$, $G_{0(hv)}$ and $G_{0(vh)}$) of the saturated block and Mazier specimens of CDT under isotropic stress conditions. The CDT shows stiffness anisotropy illustrating that

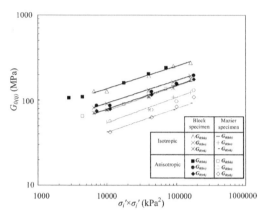

Figure 20. Variations of $G_{0(ij)}$ with $\sigma'_i \times \sigma'_j$ in saturated CDT from Fanling (Ng & Leung 2006).

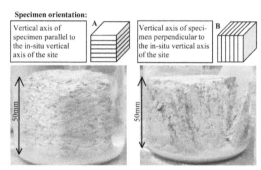

Figure 21. Soaking tests on block specimens of CDT from Fanling (diameter of beaker: 100 mm) (Ng & Leung 2007).

$G_{0(hh)}$ is higher than both $G_{0(hv)}$ and $G_{0(vh)}$. This is due to the horizontal layering structure of CDT, which can be demonstrated by two soaking tests as shown in Figure 21. Two 50 mm³ block specimens of CDT were soaked in distilled water for three months (Ng & Leung 2007). Specimen A was extracted and soaked with its vertical axis parallel to the in situ vertical axis of the site. Specimen B was withdrawn with its vertical axis perpendicular to the in situ vertical axis of the site. The two specimens show different degrees of disintegration during soaking. The foliation pattern is visible in the specimens. Specimen A did not disintegrate whereas Specimen B disintegrated at some places along the joints between the foliation planes. The soaking test results show that CDT has a horizontal layering structure that enhances wave propagation and results in higher $G_{0(hh)}$. The horizontal layering structure possibly resulted from various geologic processes.

As shown in Figure 20, $G_{0(hh)}$, $G_{0(hv)}$ and $G_{0(vh)}$, of the block specimen are consistently higher than those of the Mazier specimen. It may be attributed to different degrees of sample disturbance. Sample disturbance

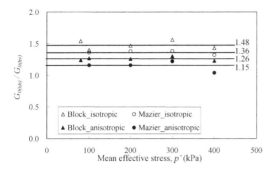

Figure 22. Variations in the shear modulus ratio $G_{0(hh)}/G_{0(hv)}$ with mean effective stress in saturated natural CDT specimens from Fanling (data from Ng & Leung 2006).

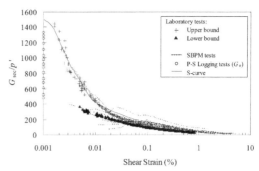

Figure 23. Comparisons between in-situ and laboratory measurements on CDG at Kowloon Bay (Ng et al. 2000).

in the Mazier specimen is considered larger than that in the block specimen (Ng & Leung 2007).

The degree of stiffness anisotropy may be expressed in terms of the ratio of the shear modulus in the horizontal plane ($G_{0(hh)}$) to that in the vertical plane ($G_{0(hv)}$) in this study. To eliminate the differences in frequency and boundary effects, $G_{0(hh)}$ and $G_{0(hv)}$ were chosen for comparisons since they were generated by the same bender element probe and they had the same travelling distance and boundary conditions (Ng et al. 2004). Figure 22 shows the degree of stiffness anisotropy $G_{0(hh)}/G_{0(hv)}$ of the block and Mazier specimens. The average values of $G_{0(hh)}/G_{0(hv)}$ of the block and Mazier specimens were 1.48 and 1.36, respectively. Based on the theoretical derivations by Ng & Leung (2007), any measured ratio of $G_{0(hh)}/G_{0(hv)}$ under isotropic stress states is mainly attributed to the soil fabric. At any given p', the Mazier specimen shows a lower degree of anisotropy than those of the block specimen. This may be also due to sample disturbance (Ng & Leung 2007). The natural anisotropic structure of CDT may be damaged by the Mazier sampling process, resulting in the lower degree of anisotropy and smaller shear moduli in the Mazier specimen.

4.2.2 Stress-induced stiffness anisotropy of saturated CDT

Figure 20 also shows the measured $G_{0(hh)}$, $G_{0(hv)}$ and $G_{0(vh)}$ of the block and Mazier specimens of CDT under anisotropic stress conditions (effective stress ratio $q/p' = 1.0$). The shear moduli determined at the anisotropic stress state are generally consistent with the data obtained at the isotropic stress state. This implies that the shear wave velocities depend on the two effective stresses in the shear planes as expected.

As shown in Figure 20, similar to isotropic compression tests, measured $G_{0(hh)}$ is higher than $G_{0(hv)}$ and $G_{0(vh)}$ in both block and Mazier specimens at the anisotropic stress state. This means that the effect of a stronger layering structure in the horizontal plane

on the shear modulus (i.e. $G_{0(hh)} > G_{0(hv)}$) prevails even under a higher vertical effective stress condition (i.e. $q/p' = 1.0$). The shear moduli of the block specimens are again consistently higher than those of the Mazier specimen due to the lower degree of sample disturbance of the block specimens (Ng & Leung 2007).

As shown in Figure 22, the average values of $G_{0(hh)}/G_{0(hv)}$ at the anisotropic states are 1.26 and 1.15 for block and Mazier specimens, respectively. For a given type of soil specimen (i.e. block or Mazier), the measured values of the degree of stiffness anisotropy at the anisotropic stress state are consistently lower than those measured at the isotropic states. This can be explained theoretically to be caused by the stress-induced anisotropy (in the opposite sense) along the anisotropic stress path applied to the specimen (Ng & Leung 2007).

4.2.3 Comparison of shear modulus-shear strain relationship of CDG determined in triaxial and in-situ tests

In-situ G_0 can be determined by using geophysical methods. The methods used in Hong Kong include suspension P-S (compression and shear wave) velocity logging (Ng et al. 2000), crosshole (Ng & Wang 2001) and downhole (Wong et al. 1998) seismic measurements. In-situ measurements of the shear modulus and the stress-strain relationships at small strains can be obtained from self-boring pressuremeter (SBPM) tests using, for example, the Cambridge-type SBPM (Clarke 1995). The self-boring mechanism of the pressuremeter minimizes disturbance to the surrounding soil during installation.

Figure 23 shows a comparison between the relationships of normalized secant shear stiffness and shear strain of CDG obtained in the field using the Cambridge SBPM and the suspension method, and in the laboratory using a triaxial apparatus equipped with internal strain measuring devices (Ng et al. 2000). For plotting the data from the P-S logging method, it is assumed that the shear strain mobilized in the soil is on the order of 0.001% and that the CDG has an initial K_o of 0.4. Despite some scatter in the measured values, reasonable consistency can be seen between the measurements obtained using the SBPM and the triaxial apparatus for shear strains greater than 0.01%. Because the recent stress history (Wang & Ng 2005) and stress path for the soil elements around the SBPM and the soil specimens tested in the triaxial apparatus are very different, the consistency between the two sets of test results may be fortuitous. Effects of the recent stress history on the small-strain stiffness of CDG are reported by Wang & Ng (2005). Assuming that the shear moduli obtained from the suspension method represent the state of the soil at strains on the order of 0.001% or smaller in the figure, the determined shear moduli are in reasonable agreement with the laboratory measurements. However, significant differences in laboratory and in-situ measurements of shear stiffness were reported by Viana Da Fonseca et al. (1997), who compared the shear stiffness values of granitic saprolites in Portugal obtained using the cross-hole technique and those obtained in a triaxial apparatus instrumented with local strain measuring devices. They suggested that sample disturbance was the cause of a threefold higher measured stiffness using the cross-hole technique than in the tests in the laboratory. This does not seem to be the case in Hong Kong.

4.3 Stiffness characteristics of unsaturated CDT

The influence of suction (or water content) on G_0 of unsaturated soils has been studied using different equipment and approaches (Wu et al. 1984, Qian et al. 1993, Marinho et al. 1995, Picornell & Nazarian 1998, Cabarkapa et al. 1999, Mancuso et al. 2002). Recently, Vassallo et al. (2007) investigated the influence of suction history on G_0 of a compacted clayey silt. All these studies were limited to the measurements of a single average G_0. However, soils are generally anisotropic. As far as the authors are aware, systematic investigations of anisotropic G_0 of unsaturated soils have rarely been reported in the literature. Using multi-directional bender elements, the researchers at HKUST have studied the influence of suction and suction history on anisotropic G_0 of CDT.

4.3.1 Suction effects on anisotropic G_0 of recompacted CDT specimens
Figure 24 shows the relationships between the shear modulus and matric suction at different net mean stresses. In the figure, it can be seen that measured shear moduli, $G_{0(vh)}$, $G_{0(hh)}$ and $G_{0(hv)}$, increase with an increase in matric suction in a non-linear fashion. Similar results have been reported by Cabarkapa et al. (1999) and Mancuso et al. (2000, 2002) in their measurements of single G_0 values. At any given net mean

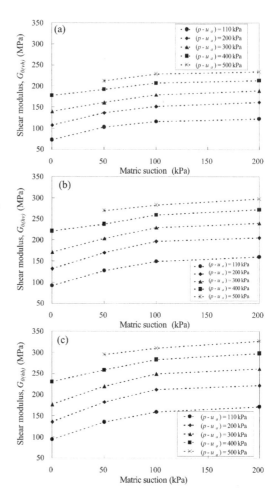

Figure 24. Variations in (a) $G_{0(vh)}$, (b) $G_{0(hv)}$, (c) $G_{0(hh)}$ with matric suction in recompacted CDT specimens from Fanling at different net mean stresses (Ng & Xu 2007).

stress, the shear moduli increase significantly as suction increases from 0 to 50 kPa. Beyond a suction of 50 kPa, the measured shear moduli continue to increase but at a reduced rate. These observations are similar to those reported by Mancuso et al. (2000, 2002). The estimated AEVs are 55 and 85 kPa at net mean stresses of 110 and 300 kPa, respectively, from measurements of SDSWCCs of this material (Ng & Xu 2007). Soil specimens subjected to a higher net mean stress tend to have higher AEVs. Since the net mean stresses applied in this series of tests are 110 kPa and higher, the AEVs of specimens should be equal to or larger than 55 kPa. The observed significant increase in shear moduli for suction up to 50 kPa (smaller than the AEV) may be because each specimen remains essentially saturated at any suction less than the AEV and so bulk water effects dominate the soil stiffness

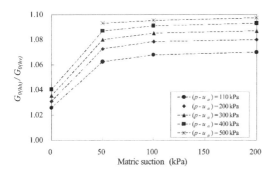

Figure 25. The influence of matric suction on the degree of stiffness anisotropy of recompacted CDT specimens from Fanling at different net mean stresses (Ng & Xu 2007).

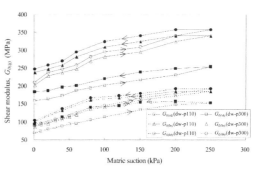

Figure 26. Variations in the shear moduli with matric suction during drying and wetting tests on recompacted CDT specimens from Fanling (Ng et al. 2007b).

responses (Mancuso et al. 2002). Any increase in suction is practically equivalent to an increase in the mean effective stress, resulting in a much stiffer soil. On the other hand, as suction increases beyond the AEV, each soil specimen starts to desaturate, resulting in the formation of an air-water meniscus (or contractile skin) at the contact points of soil particles. Although the meniscus water causes an increase in the normal forces that "hold" soil particles together and hence lead to higher shear moduli, this beneficial holding effect could not increase infinitely. This is because of progressive reduction in the meniscus radius when matric suction increases beyond 100 kPa (Mancuso et al. 2002, Ng & Yung 2007).

Figure 25 shows the influence of matric suction on the $G_{0(hh)}/G_{0(hv)}$. When the suction increases from 0 to 50 kPa, the measured $G_{0(hh)}/G_{0(hv)}$ increases by 3.6% to 5.1% at different net mean stresses. On the other hand, there is only a very small increase (less than 1%) in the measured $G_{0(hh)}/G_{0(hv)}$ when the suction increases from 50 to 200 kPa. The measured different results at different suction ranges seem to suggest that bulk water effects dominate soil behaviour when the suction is lower than the AEV of the soil (which is higher than 55 kPa) and meniscus water effects dominate soil behaviour when the suction is higher than the AEV of the soil. Although the magnitudes of increase in $G_{0(hh)}/G_{0(hv)}$ are not large, there is a clear trend that shows that $G_{0(hh)}/G_{0(hv)}$ increases with an increase in matric suction but at a reduced rate at a given net mean stress. The measured results seem to be consistent with recent theoretical and experimental work by Li (2003) and Mui (2005) who demonstrated that soil suction could induce anisotropic strains under isotropic stress conditions.

4.3.2 The influence of the suction history on anisotropic G_0 of recompacted CDT specimens

Figure 26 shows the variations in the measured $G_{0(vh)}$, $G_{0(hv)}$ and $G_{0(hh)}$ with matric suction during one drying and wetting cycle at net mean stresses of 110 (specimen dw-p110) and 300 kPa (specimen dw-p300). The variations in $G_{0(vh)}$, $G_{0(hv)}$ and $G_{0(hh)}$ with matric suction follow a similar trend. The shear moduli increase with increased matric suction along the drying path and decrease with decreased matric suction along the wetting path in a non-linear fashion. Similar to SWCC, there is hysteresis between the drying and wetting curves in the variations in shear moduli with matric suction. At a given suction, the shear moduli measured along the wetting path are consistently higher than those obtained during the drying process. This observation may be explained by using the constitutive model proposed by Alonso et al. (1990). According to the model, the elastic region of the unsaturated soil was enclosed by the loading-collapse (LC) and suction increase (SI) yield curves. At the early stages of the drying process, suction increased and elastic compression occurred within the elastic region. After the stress path reached the SI curve, the yield curve was pushed upward with increasing suction and plastic compression took place. Then, the SI curve moved to a new position (maximum applied suction) and the elastic region was enlarged after the drying process. Then, the test proceeded to the wetting phase. Since the stress path moved inside the elastic region for the wetting process, only elastic swelling resulted. Therefore, there was a net compression of the specimen after the drying and wetting cycle (Ng et al. 2007b). In other words, the soil became stiffer after the drying and wetting cycle. In addition to the hardening effect caused by the drying and wetting cycle as discussed earlier, the hydraulic hysteresis may also contribute to the different G_0 along the drying and wetting paths. Due to hydraulic hysteresis, the water content along the wetting path was lower. This might lead to more inter-particle contacts affected by water meniscus, resulting in a stronger capillary effect on the stabilising solid particles. Therefore, the soil was stiffer along the wetting path than along the drying path.

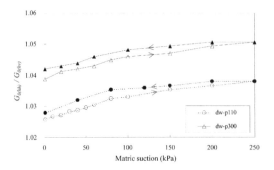

Figure 27. Variations in the degree of stiffness anisotropy with matric suction during drying and wetting tests on recompacted CDT specimens from Fanling (Ng et al. 2007b).

Figure 27 shows the variations in $G_{0(hh)}/G_{0(hv)}$ with the matric suction during the drying and wetting cycle. At different net mean stresses, the variations in $G_{0(hh)}/G_{0(hv)}$ with the matric suction follow a similar trend. When the matric suction increases, $G_{0(hh)}/G_{0(hv)}$ increases but at a gradually reduced rate. Though the magnitude of the increase in $G_{0(hh)}/G_{0(hv)}$ with the matric suction is very small, there is a clear trend showing that $G_{0(hh)}/G_{0(hv)}$ increases with the matric suction at a given net mean stress. Similar to SWCC, there is hysteresis in $G_{0(hh)}/G_{0(hv)}$ between the drying and wetting curves, but the size of the hysteresis loop is relatively small. The changing rates of $G_{0(hh)}/G_{0(hv)}$ and the sizes of the hysteresis loops at two different net mean stresses are almost the same. These observations illustrate that $G_{0(hh)}/G_{0(hv)}$ is independent of the net stress under isotropic stress conditions. This is consistent with the theoretical derivations by Ng & Yung (2007).

4.3.3 Comparison between recompacted and natural CDT specimens

Yung (2004) carried out two controlled suction isotropic compression tests for natural CDT specimens to investigate the influence of suction on small strain characteristics. Figure 28 shows the measured $G_{0(hh)}$ and $G_{0(hv)}$ at constant matric suctions of 0 and 100 kPa for both natural (i-iso-s0 and i-iso-sl00) and recompacted (r-iso-s0 and r-iso-sl00) specimens. It can be seen that the shear moduli increase with the net mean stress in a non-linear fashion at different suctions for both natural and recompacted specimens and the rate of increase appears to decrease with the net mean stress. At a given net mean stress, the moduli increase with suction for both natural and recompacted specimens. The shear moduli of natural specimens are generally higher than those of recompacted specimens. This is mainly because natural specimens experience high overburden stress in the field and hence possess stiffer soil structures than do recompacted specimens (Yung 2004).

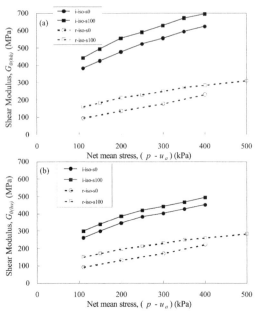

Figure 28. Comparisons of shear moduli of recompacted and natural CDT specimens from Fanling (dada from Yung 2004).

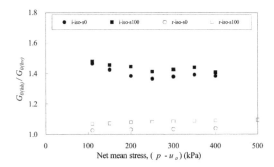

Figure 29. Comparisons of stiffness anisotropy of recompacted and natural CDT specimens from Fanling (data from Yung 2004).

As shown in Figure 28, the shear moduli of recompacted specimens increase more significantly than do those of natural specimens as the suction increases from 0 to 100 kPa. This is probably because of the denser structure of natural specimens and hence smaller average pore size distribution.

Figure 29 shows the measured $G_{0(hh)}/G_{0(hv)}$ of both natural and recompacted specimens of CDT under isotropic net mean stresses at various values of matric suction. At a given net mean stress, the measured $G_{0(hh)}/G_{0(hv)}$ appears to increase very slightly with the matric suction. Over the range of the applied net mean stress, the degree of stiffness anisotropy is essentially constant in both types of specimen. This is consistent

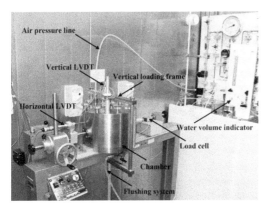

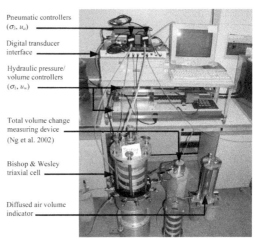

Figure 30. A modified direct shear box for testing unsaturated soils (Zhan 2003).

Figure 31. A computer-controlled triaxial stress path apparatus for testing saturated and unsaturated soils (Ng & Chen 2006).

with the theoretical deviations by Ng & Yung (2007) for unsaturated soils. The recompacted specimens consistently show a lower degree of stiffness anisotropy (1.0–1.1) than do the natural specimens (1.4–1.5). The anisotropic soil fabric of natural CDT was destroyed in the recompacted specimens, resulting in similar shear moduli in the horizontal and vertical planes (Ng & Yung 2007).

5 SHEAR STRENGTH

The shear strength of unsaturated soils can be determined from in-situ and laboratory shear tests. Two types of apparatus, i.e., a direct shear box and a triaxial apparatus, have been developed at HKUST to study the shear strength characteristics of unsaturated soils.

5.1 *Shear strength characteristics of CDV and CDG*

5.1.1 *Apparatuses*
Figure 30 shows the direct shear box developed at HKUST (Zhan 2003, Ng & Zhou 2005). The apparatus was modified from the one designed by Gan et al. (1988). Suction is applied to a soil specimen based on the axis-translation technique (Hilf 1956). A ceramic disc with a high AEV (500 kPa) is installed in the lower portion of the shear box. A water chamber beneath the ceramic disc is designed to serve as a water compartment as well as a channel for flushing diffused air. A desired matric suction is applied to the soil specimen by maintaining a constant u_a in the air pressure chamber and a constant u_w in the water chamber below the ceramic disc. The u_a and u_w in the soil are then allowed to come to equilibrium with these applied pressures. The matric suction in the soil is then equal to the applied u_a since the applied u_w is maintained at the atmospheric pressure. The direct

shear apparatus is equipped with five measurement monitoring devices. There are two LVDTs for monitoring horizontal and vertical displacements, a load cell for measuring the shear force, a pressure transducer for monitoring the u_w, and a newly designed autologging water volume indicator for monitoring water volume changes in the soil specimen. The accuracy of the water volume indicator is on the order of $2\,mm^3$, which is approximately equivalent to 0.003% water content change for a $50.8 \times 50.8 \times 21.4\,mm^3$ specimen. All the electrical instruments are connected to a data logger for data acquisition. Details of the shear box apparatus are given by Zhan (2003).

Figure 31 shows the computer-controlled stress-path triaxial system for unsaturated soils developed at HKUST. It is equipped with two water pressure controllers and two pneumatic controllers. The matric suction is applied to a specimen through one water pressure controller and one air pressure controller using the axis translation principle proposed by Hilf (1956). The u_a is applied/measured at the base of the specimen through a ceramic disc with an AEV of $500\,kPa$. The u_w is applied at the top of the specimen through a sintered copper filter. Axial displacement is measured externally with an LVDT. Similar to the apparatus illustrated in Figure 5, a new double-cell measuring device (Ng et al. 2002) is used to measure total volume changes of unsaturated specimens. To flush diffused air and measure the diffused air volume, a diffused air volume indicator (DAVI) is attached. This system is controlled by a closed-loop feedback scheme, which is capable of performing both stress- and strain-controlled tests in the triaxial stress space. Details of this triaxial system are given by Zhan (2003) and Chen (2007). A similar triaxial system based on

131

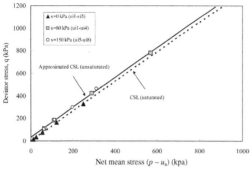

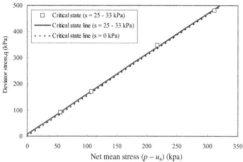

Figure 32. The shear strength envelopes of recompacted (a) CDV from Victoria Peak (Ng & Chiu 2001) and (b) CDG from CKL (Ng & Chiu 2003) specimens at different suctions.

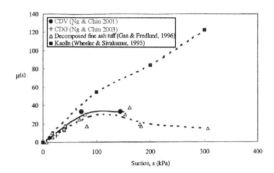

Figure 33. The relationship between the intercept $\mu(s)$ and suction.

(Gan & Fredlund 1996). Because ϕ' at the critical state is a strength parameter related to the frictional characteristic of the interparticle contacts, which is an intrinsic property of the soil, it should be independent of the stress state variables including suction.

Figure 33 compares the intercepts of the shear strength envelopes, $\mu(s)$, determined from Figure 32 at different suctions for CDV and CDG. In addition, measured results from a decomposed fine ash tuff (Gan & Fredlund 1996) and a kaolin (Wheeler & Sivakumar 1995) are included for comparisons. The CDV and the decomposed fine ash tuff exhibit a nonlinear increase in the intercept with suction. There is relatively little increase in the intercept when the suction exceeds about 80 kPa in the CDV and the decomposed fine ash tuff. On the contrary, there is a reduction in the intercept for the decomposed fine ash tuff as the suction increases beyond around 150 kPa. The contribution of suction to the shear strength is the smallest in CDG but the largest in Kaolin. This means that the suction makes a larger contribution to shear strength as the soil particle sizes decrease.

another suction control technique, the osmotic technique, was also developed at HKUST. Ng et al. (2007a) compared the triaxial test results on an expansive clay using both axis-translation and osmotic techniques to control suction in the tests. Consistent results were obtained.

5.1.2 Suction effects on the shear strength characteristics of recompacted CDV and CDG specimens

Ng & Chiu (2001, 2003) studied suction effects on the shear strength of recompacted specimens of CDV and CDG at various stress paths in a triaxial apparatus, respectively. Figure 32 shows the measured shear strength of recompacted CDV and CDG specimens at different suctions. As shown in Figure 32a, the shear strength envelope for unsaturated CDV specimens is approximately parallel to the one for saturated specimens. Similar observations are obtained from CDG specimens (see Fig. 32b). This may suggest that the gradient of the shear strength envelope (i.e. the angle of internal friction, ϕ') is a constant within the range of applied suctions for both CDV and CDG. Hence, the suction does not appear to affect the angle of internal friction for both CDV and CDG. Similar observations were obtained from other weathered soils including a Korean decomposed granite (Lee & Coop 1995) and a Hong Kong decomposed fine ash tuff and granite

5.1.3 Suction effects on the dilatancy of recompacted CDG specimens

Ng & Zhou (2005) conducted a series of direct shear tests to investigate the suction effects on the dilatancy of recompacted CDG specimens. Figures 34a and 34b show the relationships of the stress ratio ($\tau/(\sigma_v - u_a)$) and dilatancy versus horizontal displacement (δx), respectively. In the figure, dilatancy is defined as the ratio of incremental vertical displacement (δy) to incremental horizontal displacement (δx). A negative sign (or negative dilatancy) indicates expansive behaviour. As shown in Figure 34a, at zero suction and suctions of 10 kPa and 50 kPa, the stress ratio-displacement curves indicate strain hardening behaviour. With an increase in the suction, strain softening behaviour was observed at suctions of 200 kPa and 400 kPa. Generally, measured peak and ultimate stress ratios increased with suction, except for the ultimate stress ratio measured at 200 kPa. These results are

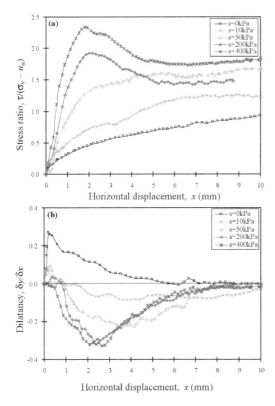

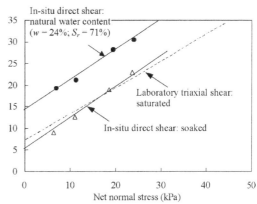

Figure 35. Shear strength envelopes obtained from in-situ and laboratory direct shear tests on CDG (modified from Brand et al. 1983).

Figure 34. Variations of the (a) stress ratio and (b) dilatancy of recompacted CDG specimens from BH subjected to shear at different controlled suctions (Ng & Zhou 2005).

consistent with triaxial results on CDV (Ng & Chiu 2001) and CDG (Ng & Chiu 2003).

Figure 34b shows the effects of suction on the dilatancy of CDG in the direct shear box tests. Under saturated conditions, the soil specimen showed contractive behaviour (i.e. positive dilatancy). On the other hand, under unsaturated conditions, all soil specimens displayed contractive behaviour initially but then dilative behaviour as the horizontal displacement continued to increase. The measured maximum negative dilatancy was enhanced by an increase in suction but at a reduced rate. This measured trend was consistent with test results on a compacted silt reported by Cui & Delage (1996). The increase in maximum negative dilatancy was likely attributed to a closer particle packing (i.e. a smaller void ratio) under a higher suction. At the end of each test, the dilatancy of all specimens approached zero, indicating the attainment of the critical state.

5.2 In-situ shear tests on CDG

The in-situ shear strength of soils can be determined from various methods including field vane shear and direct shear box tests. Brand et al. (1983) reported two series of in-situ direct shear tests on CDG, one at natural water content and the other after the material had been soaked.

Figure 35 shows the shear strength envelopes obtained from in-situ direct shear tests on CDG at the natural water content and after soaking. The natural gravimetric water content (S_r) was 24% and the corresponding degree of saturation was 71%. The shearing rate was slow (0.075 mm/min). It can be seen that the strength envelope for soaked CDG is parallel to that for CDG at the natural water content. This suggests that the internal friction angle is independent of the degree of saturation and thus of the suction level. This observation is consistent with laboratory tests on recompacted CDG specimens (see Fig. 32). Regarding the intercept (i.e. cohesion), there is a marked decrease upon soaking. The decrease in the cohesion upon soaking is caused by suction loss. This demonstrates the contribution of suction on the shear strength in the field is consistent with the laboratory test results from recompacted CDG specimens (see Fig. 32).

Figure 35 also includes results obtained from laboratory triaxial tests on saturated specimens from the same location. The shear strength envelope for the soaked CDG shows a slightly smaller friction angle and larger cohesion. These differences may be caused by strength anisotropy since the direct shear and triaxial shear tests have different principal stress directions.

6 SUMMARY AND CONCLUSIONS

This paper presents and discusses some state-of-the-art findings on the soil-water characteristics, the small strain stiffness and the shear strength of unsaturated decomposed geomaterials obtained from laboratory tests on recompacted and natural specimens and in-situ tests. Key conclusions are summarised in the following paragraphs.

6.1 Soil-water characteristics

For recompacted CDV and CDG specimens, there is a distinct difference in the SWCCs between the first and the second drying and wetting cycles but no major difference is identified between the second and the subsequent cycles. In addition to the significant reduction in the size of the hysteresis loop of SWCCs from the first to the second cycle, the difference between the initial and final volumetric water contents within each cycle is also greatly reduced in the second cycle as compared with that in the first cycle.

For both CDV and CDG under 1D stress conditions, the higher the vertical stress applied, the lower desorption rate and the smaller size of the hysteresis loop of the SDSWCCs. For CDG under the same applied mean net stress but different deviator stresses, the higher the stress ratio $(q/(p-u_a))$ imposed, the lower the desorption rate, the smaller size of the hysteresis loop and the higher the residual water content of the SDSWCCs. It is clear that both stress and suction effects on soil-water characteristics (i.e. SDSWCC) should be considered and measured. Obviously a SDSWCC is theoretically correct and more relevant than an unstressed SWCC for engineering applications.

By comparing recompacted and natural specimens of geomaterials, it is revealed that the SDSWCCs of a natural specimen (both CDV and CDG) are largely different from those of a recompacted specimen even although the two specimens had similar initial water content and density and were tested under the same stress level and conditions. A natural decomposed specimen has a smaller hysteresis loop likely due to its non-uniform pore size distribution as compared with that of a recompacted specimen. In CDV, the natural specimen has a slightly lower AEV and higher desorption and adsorption rates than those of the recompacted specimen. In CDG, both natural and recompacted specimens have similar AEVs and desorption and adsorption rates.

It appears that in-situ measured SDSWCCs of CDG show negligible hysteresis during wetting-drying cycles. The field observations do not seem to be consistent with the vast amount of laboratory tests on natural soil specimens. Further field study on CDG is necessary. On the other hand, in-situ measured SDSWCCs of CDT are stress-dependent and hysteretic, and they are consistent with the results from laboratory tests on natural soil specimens.

6.2 Small strain stiffness G_0

For saturated block and Mazier CDT specimens, the measured shear moduli in three different polarization planes, $G_{0(vh)}$, $G_{0(hv)}$ and $G_{0(hh)}$, of block specimens are all higher than the corresponding ones from the Mazier specimens due to different degrees of sample disturbance. The inherent stiffness anisotropies of saturated natural CDT specimens obtained from isotropic compression tests are 1.48 and 1.36 for block and Mazier specimens, respectively. The measured largest $G_{0(hh)}$ from the natural CDT reflects a strong layering structure in the horizontal plane, possibly resulting from various geologic processes. The lower degree of anisotropy in the Mazier specimens may be due to severe sampling disturbance. The measured values of the degree of stiffness anisotropy, which is expressed in terms of $G_{0(hh)}/G_{0(hv)}$, at the anisotropic stress state $(\eta = q/p' = 1.0)$ in both block and Mazier specimens are consistently lower than those measured at the isotropic states. This may be explained theoretically to be caused by the stress-induced anisotropy (in the opposite sense) along the anisotropic stress path applied to the specimen (i.e. reduced the initial degree of inherent stiffness anisotropy by anisotropic stresses).

By comparing the shear modulus-shear strain relationship of saturated CDG determined in triaxial and in-situ tests, it is concluded that despite some scatter in the measured stiffness values at shear strains larger than 0.01%, reasonable consistency can be found between the normalized secant shear stiffness-shear strain relationships obtained using SBPM and that derived from the triaxial tests. Assuming that shear moduli deduced from the suspension S-wave tests correspond to shear strains in the order of 0.001% or smaller, the deduced in situ shear moduli are also reasonably consistent with the corresponding values measured in the laboratory.

Based on the study of the influence of suction and suction history on anisotropic G_0 of unsaturated recompacted CDT specimens, it can be concluded that the measured shear moduli in three different polarization planes, $G_{0(vh)}$, $G_{0(hv)}$ and $G_{0(hh)}$, increase with an increase in matric suction at a reduced rate. Similar to the SWCC, there is hysteresis in the measured shear moduli during the drying and wetting cycles. After the first drying and wetting cycle, the specimen became stiffer due to the irrecoverable plastic compression and/or the irreversible decrease in the water content. The measured $G_{0(hh)}/G_{0(hv)}$ increases with an increase in the matric suction but at a gradually reduced rate. This means that soil suction could induce anisotropic stiffness even under isotropic stress conditions. The degree of stiffness anisotropy also shows hysteretic behaviour during a drying and wetting cycle, but the size of the hysteresis loop is relatively small.

The measured $G_{0(vh)}$, $G_{0(hv)}$ and $G_{0(hh)}$ of natural CDT specimens are generally higher than the corresponding ones of recompacted specimens at the same suction and net mean stress. Suction generally has a smaller effect on the shear moduli in natural specimens than on those of recompacted specimens. The degree of stiffness anisotropy of natural specimens is larger than that of recompacted specimens. This is mainly due to the anisotropic soil fabric of natural CDT being destroyed during the recompaction process in the laboratory, resulting in a smaller degree of stiffness anisotropy.

6.3 *Shear strength*

Triaxial tests on recompacted CDV and CDG specimens show that the internal friction angel, ϕ', is independent of the suction and the measured shear strength increases nonlinearly with the suction. The suction makes a smaller contribution to the shear strength of CDG than to that of CDV since the former has a coarser particle size distribution.

Direct shear tests on recompacted CDG specimens reveals that the maximum negative dilatancy is enhanced (i.e. stronger dilative behaviour) by an increase in suction but at a reduced rate.

Field direct shear tests on CDG show that the measured internal friction angles, ϕ', are similar for CDG with natural water content and after soaking, but there is a marked decrease in the cohesion intercept upon soaking. The field measurements are consistent with the laboratory results.

ACKNOWLEDGMENTS

This work presented was supported by research grants HKUST6053/97E and HKUST6046/98E provided by the Research Grants Council of the Hong Kong Special Administrative Region (HKSAR), DAG00/001.EG36 from HKUST and OAP06/07.EG01 from Ove Arup and Partners (HK) Ltd (OAP). The authors would like to acknowledge the support provided by the Geotechnical Engineering Office (GEO) of the Civil Engineering and Development Department to the instrumentation work and field monitoring carried out at Tung Chung. The efforts to make this field work possible by Ir H.N. Wong and Ir Dr H.W. Sun of GEO and Ir Dr Jack Pappin and Ir Stuart Millis of OAP are acknowledged. Moreover, the authors would like to thank Ms Xu Jie and Ms Tse Yin Man for assisting in checking and preparing some figures for the paper.

REFERENCES

Alonso, E.E., Gens, A. & Josa, A. 1990. A Constitutive model for partially saturated soils. *Géotechnique* 40(2): 405–430.

Basile, A., Ciollaro, G. & Coppola, A. 2003. Hysteresis in soil water characteristics as a key to interpreting comparisons of laboratory and field measured hydraulic properties. *Water Resources Research* 39(12): 1355. doi:10.1029/2003WR002432.

Bell, F.G. 1992. *Engineering properties of soils and rocks.* Butterworth-Heinemann Ltd., Woburn, Ma.

Brand, E.W., Phillipson, H.B., Borrie, G.W. & Clover, A.W. 1983. In-situ direct shear tests on Hong Kong residual soils. *Proc., Int. Symp. on Soil and Rock Investigations by In-Situ Tesing,* Paris: Vol. 2, 13–17.

Burland, J.B. 1990. On the compressibility and shear strength of natural clays. *Géotechnique* 40(3): 329–378.

Cabarkapa, Z., Cuccovillo, T. & Gunn, M. 1999. Some aspects of the pre-failure behaviour of unsaturated soil. *Proc. II Int. Symp. On Prefailure Deformation Characteristics of Geomaterials,* Torino: Vol. 1, 159–165.

Chen, R. 2007. *Experimental study and constitutive modelling of stress-dependent coupled hydraulic hysteresis and mechanical behaviour of an unsaturated soil.* Ph.D. Thesis, The Hong Kong University of Science and Technology, Hong Kong.

Clarke, B.G. 1995. *Pressuremeters in geotechnical design.* Blackie Academic and Professional.

Clayton, C.R.I. & Khatrush, S.A. 1986. A new device for measuring local axial strains on triaxial specimens. *Géotechnique* 36(4): 593–597.

Cui, Y.J. & Delage, P. 1996. Yielding and plastic behaviour of an unsaturated compacted silt. *Géotechnique* 46(2): 291–311.

Fredlund, D.G. 1996. *The emergence of unsaturated soil mechanics.* The fourth Spencer J. Buchanan Lecture, College Station, Texas, A & M University Press.

Fredlund, D.G. & Rahardjo, H. 1993. *Soil mechanics for unsaturated soils.* John Wiley and Sons Inc., New York .

Fredlund, D.G., Rahardjo, H., Leong, E.C. & Ng, C.W.W. 2001. Suggestions and recommendations for the interpretation of soil-water characteristic curves. *Proc. 14th Southeast Asian Geotechnical Conference,* Hong Kong: Vol.1, 503–508.

Gan, J.K.M., Fredlund, D.G. & Rahardjo, H. 1988. Determination of the shear strength parameters of unsaturated soil using the direct shear test. *Canadian Geotechnical Journal* 25(8): 500–510.

Gan, J.K.M. & Fredlund, D.G. 1996. Shear strength characteristics of two saprolitic soils. *Canadian Geotechnical Journal* 33: 595–609.

GCO. 1987. *Guide to Site Investigation (GEOGUIDE 2).* Geotechnical Control Office, Civil Engineering Services Department of Hong Kong.

GCO. 1988. *Guide to rock and soil descriptions (GEOGUIDE 3).* Geotechnical Control Office, Civil Engineering Services Department of Hong Kong.

Hilf, J.W. 1956. *An investigation of pore water pressure in compacted cohesive soils.* Technical Memo 654. Denver: Bureau of Reclamation.

Hillel, D. 1998. *Environmental soil physics.* Academic Press, San Diego, Calif.

Ho, K.M.Y., Ng, C.W.W., Ho, K.K.S. & Tang, W.H. 2006. State-dependent soil-water characteristic curves (SDSWCCs) of weathered soils. *Proc. 4th Int. Conf. on Unsaturated Soils,* Arizona, USA: 1302–1313.

Ho, K.M.Y., Tse, J.M.K. & Ng, C.W.W. 2007. Influence of drying and wetting history and particle size on state-dependent soil-water characteristic curves (SDSWCCs).

Proc. 3rd Asian Conf. on Unsaturated Soils, Nanjing, China: 213–218.

Jefferson, I., Tye, C. & Northmore, K.J. 2001. Behaviour of silt: the engineering characteristics of loess in the UK. *Proc. Symp. On Problematic Soils*, Nottingham, UK: 37–52.

Lai, C.H. 2004. *Experimental study of stress-dependent soil-water characteristics and their applications on numerical analysis of slope stability*. M.Phil. Thesis, The Hong Kong University of Science and Technology, Hong Kong.

Lee, I.K. & Coop, M. R. 1995. The intrinsic behaviour of a decomposed granite soil. *Géotechnique* 45(1): 117–130.

Li, A.G., Tham, L.G., Yue, Z.Q., Lee, C.F. & Law, K.T. 2005. Comparison of field and laboratory soil-water characteristic curves. *Journal of Geotechnical and Geoenvironmental Engineering* 131(9): 1176–1180.

Li, X.S. 2003. Effective stress in unsaturated soil: a microstructural analysis. *Géotechnique* 53(2): 273–277.

Liu, T. 1988. *Loess in China*. Springer-Verlag, Berlin.

Mancuso, C., Vassallo, R. & d'Onofrio, A. 2000. Soil behaviour in suction controlled cyclic and dynamic torsional shear tests. *Proc. 1st Asian Regional Conference on Unsaturated Soils*, Singapore: 539–544.

Mancuso, C., Vassallo, R. & d'Onofrio, A. 2002. Small strain behavior of a silty sand in controlled-suction resonant column – torsional shear tests. *Canadian Geotechnical Journal* 39: 22–31.

Marinho, E.A.M., Chandler, R.J. & Crilly, M.S. 1995. Stiffness measurements on an unsaturated high plasticity clay using bender elements. *Proc. 1st International Conference on Unsaturated Soils*, Paris: Vol. 2: 535–539.

Mui, T.S. 2005. *The shearing effect of suction*. M.Phil. Thesis, The Hong Kong University of Science and Technology, Hong Kong.

Ng, C.W.W. & Chen, R. 2006. Advanced suction control techniques for testing unsaturated soils (In Chinese). *Chinese Journal of Geotechnical Engineering* 28(2): 123–128.

Ng, C.W.W. & Chiu, C.F. 2001. Behavior of a loosely compacted unsaturated volcanic soil. *Journal of Geotechnical and Geoenvironmental Engineering* 127(12): 1027–1036.

Ng, C.W.W. & Chiu, C.F. 2003. Laboratory study of loose saturated and unsaturated decomposed granitic soil. *Journal of Geotechnical and Geoenvironmental Engineering* 129(6): 550–559.

Ng, C.W.W., Chui, C.F. & Shen, C.K. 1998. Effects of wetting history on the volumetric deformations of an unsaturated loose fill. *Proc. 13th Southeast Asian Geotechnical Conference*, Taipei, Taiwan: Vol. 1, 141–146.

Ng, C.W.W., Cui, Y., Chen, R. & Delage, P. 2007a. The axis-translation and osmotic techniques in shear testing of unsaturated soils: a comparison. *Soils and Foundations* 47(4): 675–684.

Ng, C.W.W. & Leung, E.H.Y. 2006. Invited paper: Small-strain stiffness of granitic and volcanic saprolites in Hong Kong. *International Workshop on Natural Soil 2006*, Dec. Singapore, Vol.4: 2507–2538.

Ng, C.W.W. & Leung, E.H.Y. 2007. Determination of shear wave velocities and shear moduli of completely decomposed tuff. *Journal of Geotechnical and Geoenvironmental Engineering* 133(6): 630–640.

Ng, C.W.W., Leung, E.H.Y. & C.K. Lau. 2004. Inherent anisotropic stiffness of weathered geomaterial and its influence on ground deformations around deep excavations. *Canadian Geotechnical Journal* 41: 12–24.

Ng, C.W.W. & Menzies, B. 2007. *Unsaturated Soil Mechanics and Engineering*. Taylor & Francis, UK & USA.

Ng, C.W.W. & Pang. Y.W. 2000a. Influence of stress state on soil-water characteristics and slope stability. *Journal of Geotechnical and Geoenvironmental Engineering* 126(2): 157–166.

Ng, C.W.W. & Pang, Y.W. 2000b. Experimental investigations of the soil-water characteristics of a volcanic soil. *Canadian Geotechnical Journal* 37(6): 1252–1264.

Ng, C.W.W., Pun, W.K. & Pang, R.P.L. 2000. Small strain stiffness of natural granitic saprolite in Hong Kong. *Journal of Geotechnical and Geoenvironmental Engineering* 126(9): 819–833.

Ng, C.W.W., Wang, B. & Gong, B.W. 2001. A new apparatus for studying stress effects on soil-water characteristics of unsaturated Soils. *Proc. 15th ICSMGE*, August, Istanbul, Turkey. Vol. 1, 611–614.

Ng, C.W.W. & Wang, Y. 2001. Field and laboratory measurements of small strain stiffness of decomposed granites. *Soils and Foundations* 41(3): 57–71.

Ng, C.W.W. & Xu, J. 2007. Anisotropic small strain shear moduli of unsaturated completely decomposed tuff. Proc. 3rd Asian Conference on Unsaturated Soil, Nanjing, China: 47–65.

Ng, C.W.W., Xu, J. & Yung, S.Y. 2007b. Effects of wetting-drying and stress ratio on anisotropic small strain stiffness of an unsaturated soil. To be submitted to *Canadian Geotechnical Journal*.

Ng, C.W.W. & Yung, S.Y. 2007. Determination of the anisotropic shear stiffness of an unsaturated decomposed soil. *Géotechnique*: in press.

Ng, C.W.W., Zhan, L.T. & Cui, Y.J. 2002. A new simple system for measuring volume changes in unsaturated soils. *Canadian Geotechnical Journal* 39(2): 757–764.

Ng, C.W.W. & Zhou, R.Z.B. 2005. Effects of soil suction on dilatancy of an unsaturated soil. *Proc. 16th ICSMGE*, Osaka, Japan: Vol. 2, 559–562.

Picornell, M. & Nazarian, S. 1998. Effects of soil suction on the low-strain shear modulus of soils. *Proc. 2nd International Conference on Unsaturated Soils*, Beijing, China: Vol. 2, 102–107.

Qian, X., Gray, D.H. & Woods, R.D. 1993. Voids and granulometry: effects on shear modulus of unsaturated sands. *Journal of Geotechnical Engineering ASCE* 119(2): 295–314.

Tse, M.K. 2007. *Influence of stress states on soil-water characteristics, conjunctive surface-subsurface flow modeling and stability analysis*. M.Phil. Thesis, The Hong Kong University of Science and Technology, Hong Kong.

Tse, Y.M. 2008. *Laboratory and field studies of drying-wetting effects on slope stability*. M.Phil. Thesis, The Hong Kong University of Science and Technology, Hong Kong.

Vanapalli, S.K., Fredlund, D.G. & Pufahl, D.E. 1999. The influence of soil structure and stress history on the soil-water characteristics of a compacted till. *Géotechnique* 49(2): 143–159.

Vassallo, R. Mancuso, C. & Vinale, F. 2007. Effects of net stress and suction history on the small strain stiffness of a compacted clayey silt. *Canadian Geotechnical Journal* 44: 447–462.

Viana da Fonseca, A., Matos Fermandes, M. & Silva Cardoso, A. 1997. Interpretation of a footing load test on a saprolitic soil from granite. *Géotechnique* 47(3): 633–651.

Wang, Y. & Ng, C.W.W. 2005. Effects of Stress Paths on the Small-strain Stiffness of Completely Decomposed Granite. *Canadian Geotechnical Journal.* 42 (4), 1200–1211.

Watson, K.K. 1966. An instantaneous profile method for determining the hydraulic conductivity of unsaturated porous materials. *Water Resources Research* 2: 709–715.

Wheeler, S.J. & Sivakumar, V. 1995. An elasto–plastic critical state framework for unsaturated soil. *Geotechnique* 45(1): 35–53.

Wong, Y.L., Lam, E.S.S., Zhao, J.X. & Chau, K.T. 1998. Assessing Seismic Response of Soft Soil Sites in Hong Kong Using Microtremor Records. *The HKIE Transaction* 5(3): 70–78.

Wu, S., Gray, D.H. & Richart, F.E. 1984. Capillary effects on dynamic modulus of sands and silts. *Journal of Geotechnical and Geoenvironmental Engineering ASCE* 110(9): 1188–1203.

Yung, S.Y. 2004. *Determination of shear wave velocity and anisotropic shear modulus of an unsaturated soil.* M.Phil. Thesis, The Hong Kong University of Science and Technology, Hong Kong.

Zhan, L.T. 2003. *Field and laboratory study of an unsaturated expansive soil associated with rain-induced slope instability.* Ph.D. Thesis, The Hong Kong University of Science and Technology, Hong Kong.

Geotechnical and Geophysical Site Characterization – Huang & Mayne (eds)
© 2008 Taylor & Francis Group, London, ISBN 978-0-415-46936-4

Sampling and sample quality of soft clays

H. Tanaka

Hokkaido University, Sapporo, Japan

ABSTRACT: Using various types of samplers, sample quality was examined. This paper discusses the key points for getting high quality sample. Assessments for sample quality are also presented and examined. It is found that the loss of residual effective stress (also called suction) does not necessarily distract the soil structure. Taking the example of the site investigation for construction of the Kansai Airport as a case history, sample quality for great depth is also assessed. Values of assessment indices such as volumetric strain (ε_{vo}) or change in void ratio ($\Delta e/e_o$) do not increase with the depth. Instead, they remain almost constant with the depth. The yield consolidation pressure at this site is somewhat scattered, and OCR lies in the range of 1.0 to 2.0. From the site investigation using CPT, which was newly developed for great depth penetration, it is revealed that the occurrence of such variation is not due to laboratory testing techniques including sampling quality. Instead, they are due to the heterogeneity of the targeted soil layers.

1 INTRODUCTION

Sample quality has been recognized by geotechnical engineers as one of the most important issues for obtaining reliable geotechnical parameters from laboratory testing (for example, Tanaka, 2000; Hight and Leroueil, 2003). A lot of trials have been conducted to make the soils sample free from the soil disturbance which is caused in the sampling processes or during the preparation of specimen for laboratory tests. A number of sampling methods including samplers for retrieving high quality soil samples have been developed and some of them are being used in practical investigations. Assessments of sample quality or methods for correcting values derived from poor quality samples have been proposed by many researchers. However, these sampling or evaluating sample quality methods are strongly dependent on local differences, which are caused by difference in geotechnical properties and skills of the driller. For example, in the northern part such as Scandinavia or North America, low plastic clays having been influenced by glacier are widely distributed. Geotechnical properties of these soils may be different from those in temperate or tropical areas. It should also be kept in mind that unlike materials used in the civil engineering such as steel or concrete, natural soils are not manufactured in factories under strict quality control. In addition, there are few opportunities for drillers to exchange their experiences such as in the conferences, etc. It is nearly impossible to make manuals describing detailed procedures of drilling and sampling. For these reasons, sampling including drilling has been originally and independently developed in each region.

The geotechnical group at Port and Airport Research Institute (PARI) (formally called PHRI), to which the author used to belong, carried out sampling at various sites of Japan as well as overseas. Based on experiences from these investigations, this paper will describe key factors for retrieving high quality sample and sample quality assessment methods valid for various soils. In this paper, soft clayey soils are mainly focused.

2 COMPARISON STUDIES USING DIFFERENT SAMPLERS

Comparison of sample quality was made at the Ariake site whose geotechnical properties have been extensively investigated and published by many researchers, for example, Hanzawa et al. (1990); Tanaka (2000); Ohtsubo et al. (2007). A typical feature of Ariake clay in comparison to other Japanese clays is that the natural water content of these clays exceeds liquid limit. Henceforth, their sensitivity is very large.

2.1 Samplers used in comparison study

In this study, six different samplers were used. Their main features are indicated in Table 1. Both Sherbrooke and Laval samplers were developed at the same period and place: i.e., Quebec, Canada (Lefebvre and Poulin, 1979; La Rochelle et al., 1981, respectively).

Table 1. Main features of samplers used by the comparison study at the Arika site.

Sampler	Inside diameter (mm)	Sampler length (mm)	Thickness (mm)	Area ratio (%)	Piston
JPN	75	1000	1.5	7.5	yes
LAVAL	208	660	4.0	7.3	no
Shelby	72	610	1.65	8.6	no
NGI54	54	768	13	54.4	yes
ELE100	101	500	1.7	6.4	yes
Sherbrooke	350*	250*	–	–	no

*Dimension of the soil sample.

They are earning a reputation for collecting high quality sample. Diameter of the sample obtained from both samplers exceeds 20 cm which make them expensive in comparison with other conventional tube samplers. Therefore, their use is restricted only in research works or special projects. JPN, ELE100 and NGI54 samplers are a stationary piston type sampler but the geometry of their sampling tube is slightly different as shown in the table. ELE100 sampler has the largest inside diameter whereas NGI54 sampler has the smallest. NGI54 sampler has the thickest wall because it is a composite sampler consisting of outer and inner tubes (Andresen and Kolstad, 1979). Sampling using this sampler is usually carried out by so-called "displacement method", where the sampler is directly penetrated at the desired sampling depth without borehole. In this study, however, a borehole was driven even for the NGI54 sampler as well as for other samplers. Shelby tube is the most common sampler used in the world. It does not have a piston and is made up of mild steel. Sampling was carried out by the same Japanese driller to avoid difference for the skill of the driller in the sample quality. More information on sampler and sampling are given by Tanaka et al. (1996) and Tanaka (2000).

2.2 Unconfined compression strength

Sample quality was assessed by the unconfined compression test. Figures 1 and 2 show unconfined compression strength (q_u). Significant differences in q_u for the samples retrieved by the different samplers are clearly seen in both the figures. Sherbrook sampler shows the highest value, followed by Laval and JPN samplers. The reason for inferior performance of NGI54 may be attributed to its thicker wall of the sampling tube. Sample quality collected by ELE sampler is the lowest which is similar to that by Shelby tube. It may be easily accepted that the performance of Shelby tube is not good because there is no piston. However, the ELE100 sampler is a piston sampler

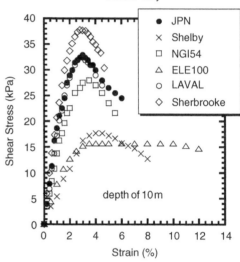

Figure 1. Comparison of stress strain relation from unconfined compression test at Ariake Site (after Tanaka, 2000).

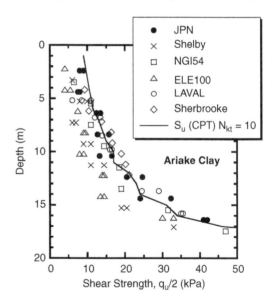

Figure 2. q_u values of samples retrieved from various types of samplers at Ariake site (after Tanaka, 2000).

which is same as JPN and NGI54 samplers. Furthermore, taking account of diameter as well as thickness of the sampling tube, the ELE100 sampler could give high quality sample equivalent or superior to the JPN sampler.

Reasons for good performance of the JPN sampler were discussed by several researchers and following

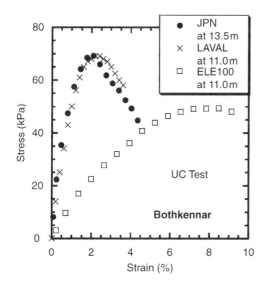

Figure 3. Comparison of stress strain relation from unconfined compression test at Bothkennar site (after Tanaka, 2000).

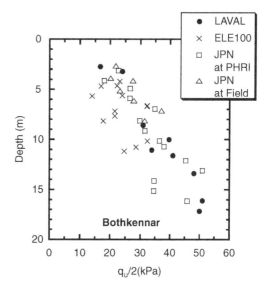

Figure 4. q_u values of samples retreived from three types of samplers. Tests for samples collected by JPN sampler were extruded at the sampling site and the lobarotry of PARI (after Tanaka, 2000).

comments were given: 1) The JPN sampler was developed to be suitable for Japanese soils. The high quality collected by the JPN sampler is a matter of course; 2) All the samples were collected by the Japanese driller, who is familiar with the JPN sampler and had never seen or touched other overseas samplers before the comparison of the study. It might be considered that the inferiority of sample quality collected by non-Japanese samplers might be due to these handicaps. To cope with above comments, sampling with the JPN sampler was carried out outside Japan, and test results of Bothkennar, UK and Drammen, Norway are presented in this paper. All samplings were carried out under supervision of the author.

Figures 3 and 4 show the comparison of stress ~ strain behavior and q_u values for the samples retrieved by JPN and ELE100 samplers. Sampling with both the samplers was carried out by British drillers who usually use the ELE100 sampler as routine work. Samples retrieved by Laval sampler were provided by the British group (Hight, et al. 1992). Test results at the Bothkennar site are the same as those at the Ariake site: the q_u values of the samples obtained from the Laval and the JPN samplers are nearly in the same order and those retrieved by the ELE100 sampler are definitely smaller than those from JPN an Laval samplers. Another example is shown in Fig. 5, which was carried out at the Drammen (Lierstranda) site using JPN and Sherbrooke samplers (sample for Sherbrooke sampler was provided by NGI). Difference in the q_u value at Drammen site is much larger than that at the Ariake site, and hence the superiority of Sherbrook sampler

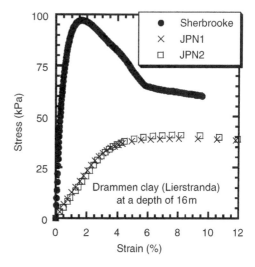

Figure 5. Comaprison of stress strain relation at Drammen site (after Tanaka, 2000).

was admitted. The most distinguished geotechnical properties at the Drammen site compared with Bothkennar or Ariake sites, is that the plasticity index (I_p) of Dramme site is very small (I_p for Ariake 60 to 70; for Bothkennar 40; for Drammen 15 to 20). From this investigation, it may be concluded that except for low I_p clay, tube samplings are capable of retrieving high quality sample equivalent to the large size diameter sampler such as Sherbrooke and Laval samplers. The reason for poor performance of the ELE100 sampler will be discussed later again.

2.3 Effects of transportation

Transportation of samples is considered to be an important factor affecting the sample quality as well as sampling. This factor becomes prominent in overseas site investigations. All samples outside Japan were transported by the private transportation service, keeping the soil sample in the sampling tubes which were further wrapped with rubber sponge. At the site of Bothkennar, some samples were extruded from the sampling tube and unconfined compression test was performed on site. As shown in Fig. 4, no distinguished difference in q_u values obtained on site and at laboratory of PARI was found. Similar tests were conducted for Drammen and Louiseville clay also (a Champlein clay, Quebec, Canada), but no difference was recognized. From these investigations, it may be concluded that the effect of sample transportation is not important, provided that proper counter measures were taken.

2.4 Effects of drillers

Sample quality is also strongly dependent on the skill of driller. This aspect complicates situations for introducing the design method based on reliability to geotechnical engineering. In addition to natural heterogeneity, human factors cause further variation in geotechnical properties. Figure 6 shows an example of variation in q_u values caused by the difference in the skills of the drillers. This example is referred to a technical report submitted by a commercial company which conducts site investigations for port construction works in Japan. Geotechnical properties at this site are quite homogeneous and sampling was carried out at every 20 m interval. Water depth at this site is about 20 m. Undrained shear strength was evaluated by unconfined compression test, which is a typical evaluation method in Japan. The sampling method is strictly regulated by the government specification and followed by technical standard Japanese Geotechnical Society (JGS), i.e., using the JPN sampler. This sampling was carried out by several drilling groups (A to G groups. See Fig. 6). The value of q_u considerably varies with the different groups, and the D group always provides the lowest strength at every depth. In the report, these values were not omitted, but a design line was drawn by the regression analysis for all measured strengths.

2.5 Oedometer test

Comparison of e-log p curves measured by Constant Rate of Strain (CRS) oedometer test is shown in Fig. 7 for Bothkennar clay. Similar to the q_u value, it can

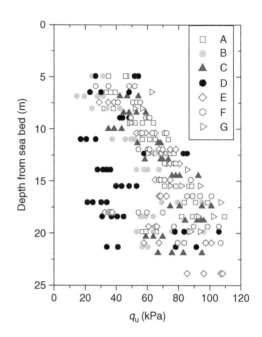

Figure 6. Variation in q_u values due to the difference in the skills of drillers.

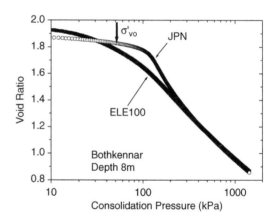

Figure 7. Comparison of e-log p curves at the Bothkennar site (after Tanaka, 2000).

be seen that the e-log p relation is also prominently affected by sampler (sample quality). For a sample retrieved by the JPN sampler, a clear bending point, which may correspond to the yield consolidation pressure (p_c) can be seen. On the other hand, the ELE100 sampler provides a moderate e-log p curves possibly showing lower yield point. From this comparison study, it can be said that if the sample quality is poor, p_c is underestimated and the change in void ratio (Δe) at the in situ effective stress (σ'_{vo}) becomes large.

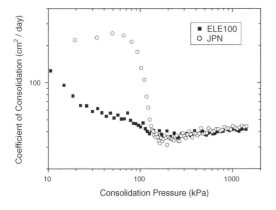

Figure 8. Comparison of c_v for ELE100 and JPN samplers at the Bothknnar site.

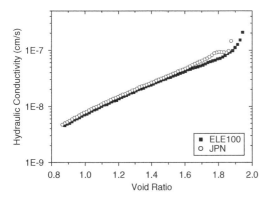

Figure 9. Comparison of e-log k relation for different sample quality at the Bothkennar site.

A coefficient of consolidation (c_v) is also influenced by the sample quality, especially, when the consolidation pressure is at the overconsolidated state. As shown in Fig. 8, c_v values at OC state are nearly ten times smaller than that for poor quality samples retrieved from the ELE sampler. However, the e-log$k$ (k: hydraulic conductivity) curves for different quality samples is nearly identical (see Fig. 9), indicating almost no effect of sampler. This indicates that the reason for the lower value of c_v at the OC state is not due to reducing k values. Instead, the high compressibility (m_v) for poor quality sample makes c_v smaller.

Unlike the influence of the sample quality on consolidation properties for Bothkennar clay, test results from Ariake clay show completely different behavior. Comparison of the e-log$p$ curve for JPN, ELE100 and Sherbrooke samplers is shown in Fig. 10. The e-log$p$ curves obtained for these samplers are nearly identical, despite of the fact that the q_u values for the samples are so much different as shown in Figs 1 and 2. Similar test results are seen for Drammen clay, where e-log$p$

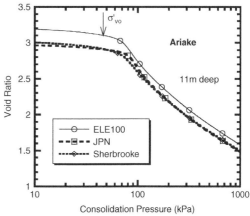

Figure 10. Influence of sample quality on e-log$p$ curves at the Ariake site (after Tanaka).

curve for JPN and Sherbrooke sampler is nearly the same (see Tanaka, 2000).

2.6 Triaxial test

To investigate the influence of sample quality, triaxial test so called the recompression method was also carried out, where a specimen was consolidated under the in situ effective overburden pressure (σ'_{vo}) and $K_0\sigma'_{vo}$ in vertical and horizontal directions, respectively. Here, K_0 is the coefficient of the earth pressure at rest and assumed to be 0.5. Similar to the oedometer test, the influence of sample quality on the behavior measured by the triaxial test are different for different soils. For the Ariake site, as shown in Figs 11 and 12, any systematic differences due to sample quality cannot be observed. Slight differences with different samplers may be due to the heterogeneity of soils.

On the other hand, for Bothkennar site, even though the specimen was consolidated under in situ stress conditions, the stress strain relation is different for different sample quality (see Figs 13 and 14). The same observation was made by Lunne et al. (2006) for Norwegian clays, where samples were retrieved by Sherbrooke and NGI54 samplers. These interesting differences from triaxial recompression and oedometer test will be discussed later.

3 INDICES FOR ASSESSMENT OF SAMPLE QUALITY

3.1 Conventional indices

As already shown in Figs 1 through 6, if the sample quality is poor, following features are observed: (1) low q_u strength; (2) low tangential stress strain relation,

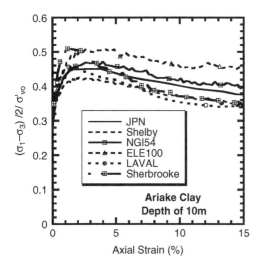

Figure 11. Stress strain relation from recompression triaxial test for Ariake clay (after Tanaka).

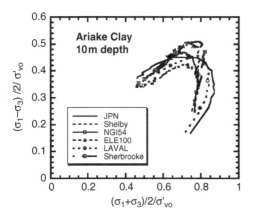

Figure 12. Stress path from recompression triaxial test for Ariake clay (after Tanaka, 2000).

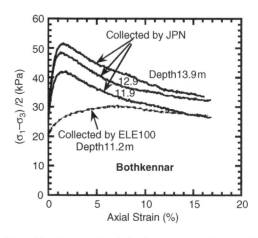

Figure 13. Stress strain relation from recompression triaxial test for Bothkennar clay (after Tanaka, 2000).

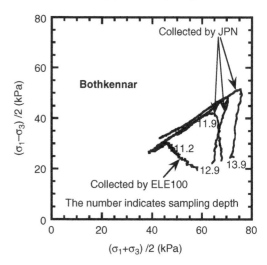

Figure 14. Stress path from recompression triaxial test for Bothkennar clay (after Tanaka, 2000).

i.e., small E or E_{50}; (3) large strain at failure (ε_f). If there exists a reference value of q_u or E, then it can be judged whether the measured q_u or E is smaller or the same as that of reference value from good quality samples. Without such a reference, it is very hard or nearly impossible to judge the sample quality only from the value of q_u and E. In this sense, the order of ε_f may be a better index to assess the sample quality. However, ε_f varies with different soils. For example, for Ariake clay in Fig. 1, the order of ε_f is around 3% while for Bothkennar clay in Fig. 3, ε_f is 2%. It is difficult to draw criteria with ε_f to determine sample quality, which is valid for all kinds of clays.

As already shown in Fig. 7, if sample disturbance is appeared in the e-log$p$ curves, then non-linearity at the

NC state, the C_c value at pressures just after p_c and the p_c value itself are affected by the sample quality: i.e., for poor quality sample, non-linearity is disappeared and C_c as well as p_c becomes small. Also, c_v coefficient at the OC state is reduced. In the extreme case, c_v at the OC state becomes as low as that at the NC state.

However, as shown in Ariake clay in Fig. 10, no difference can be recognized in the e-log$p$ relation for samples with different quality. Some other extreme examples are shown in Figs 15 and 16. Both clays behave overconsolidated manner, probably due to the erosion caused by river (for Yubari clay, Tanaka et al., 2006a; for Louiseville, Leroueil, et al., 2003). These samples were retrieved by the JPN sampler. For Yubari clay, the gradient of e-log$p$ curve at the OC state

144

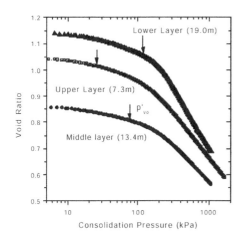

Figure 15. The e-log p curves for Yubari clay (Hokkaido, Japan) (after Tanaka et al., 2006a).

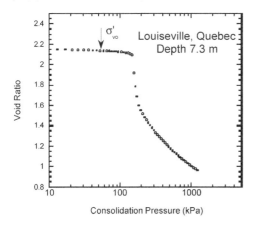

Figure 16. A typical e-log p curve for Louiseville clay (Quebec, Canada).

is so large that the bending point at the p_c value is not clear and the volume change at σ'_{vo} is very large. These characters may be typical features showing that the sample is disturbed, according to the conventional assessments. Contrary to Yubari clay, Louisville clay shows complete different e-log p curves. Argument point here is whether this difference is truly due to the sample quality or not: i.e., the sample quality of Louisville is superior to that of Yubari clay. However, the sample quality of Yubari clay is not poor, referring to the value of the measured suction, which will be mentioned later.

3.2 Volume change at recompression stresses

Since old days, the NGI group has proposed criteria for sample quality from a volume change at σ'_{vo} (Andresen and Kolstad, 1979; Lacasse and Berre, 1988; Lunne, et al., 1997). The criteria of Andresen and Kolstad is

Table 2. Sample quality assessment using volumetric strain (after Andersen and Kolstad, 1979).

$\varepsilon_{vo}(\%)$	Quality of Sample
<1	very good
1–2	good
2–4	fair
4–8	poor
>8	very poor

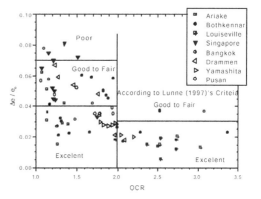

Figure 17. Assessment of sample quality according to Lunne et al's (1997) criteria for various clays that were retrieved by JPN sampler (after Tanaka et al., 2002).

based on the volumetric strain (ε_{vo}) generated at σ'_{vo}. Their criteria are shown in Table 2. Lacasse and Berre's criteria are also based on ε_{vo}, but they are dependent on OCR. The idea of dependency of OCR on ε_{vo} is that when σ'_{vo} is far from the p_c value, ε_{vo} becomes small. Lunne et al's criteria is based on change in void ratio (Δe), which is correlated to ε_{vo} ($\varepsilon_{vo} = \Delta e/(1 + e_o)$), and normalized by e_o. In this sense, their idea is same as that of Andresen and Kolstad, but the influence is to be divided by e_o or $(1 + e_o)$.

Quality of samples retrieved by the JPN sampler is judged by Lunne et al's criteria and shown in Fig. 17. As shown in the figure, some samples are classified into "poor", e.g., some Singapore clay. For samples with relatively large OCR greater than 2.0, most of them are Louiseville clay and grouped into "Excellent". If the figure includes Yubari clay (in the figure, Yubari clay is excluded), its quality will be classified into "Good to Fair" or "Poor".

4 SUCTION FOR EVALUATION OF SAMPLE QUALITY

4.1 Suction in sampled soil

Assessments for the sample quality described above are a destructive method: i.e., once the samples are

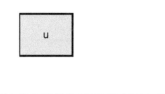

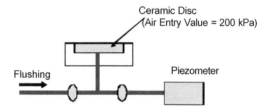

Figure 19. Sketch of apparatus for suction measurement (after Tanaka and Tanaka, 2006b).

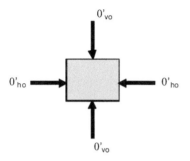

Figure 18. Stress conditions of the in situ and after sampling.

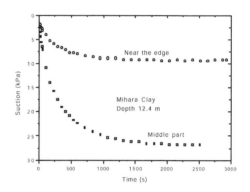

Figure 20. Example of relation between time and suction.

assessed for their sample quality, assessed samples do not exist and no more tests can be performed for the same sample. As non-destructive methods for assessing sample quality, measurement of suction and shear wave velocity using bender element has been recently paid attention (for example, Tanaka and Nishida, 2008).

A soil element may be subjected to σ'_{vo} and σ'_{ho} in vertical and horizontal directions respectively, at the in situ, as shown in Fig. 18. When the soil is sampled and is exposed to the atmosphere, the total stresses subjected to the sample become zero. However, some amounts of negative pore water pressure remains in the sample in the form of suction or sometimes called residual effective stress (p'_r). Unlike σ'_{vo} or σ'_{ho}, p'_r is an isotropic stress, in another word, it is a scalar number.

In the ideal conditions, i.e., without mechanical disturbance, the value of p'_r should be equal to the mean stresses of σ'_{vo} and σ'_{ho}, i.e., $p'_m = (\sigma'_{vo} + 2\sigma'_{ho})/3$. If the suction is equal to p'_m, this condition is called "perfect sampling" (Ladd and Lambe, 1963). In reality, however, p'_r is much smaller than p'_m due to disturbance occurred during sampling or during the preparation of a specimen for testing. When the suction is measured before mechanical testing, the sample quality can be assessed, compared to the in situ p'_m.

4.2 Measurement of suction

Measurement of the suction usually uses a ceramic disc whose air entry value is large enough to prevent the entrance of air bubbles into the measurement system (Fredlund and Rahardjo, 1993). Figure 19 shows

the apparatus that the author used in the study. The air entry value of the ceramic disc is 200 kPa and it is saturated with de-aired and distilled water using vacuum pump. Thenafter the specimen was placed on the saturated ceramic disc and it was left to reach a constant value. When the suction becomes nearly 100 kPa, cavitation may occur in the ceramic disc. To prevent the cavitation, some amounts of confining pressure should be applied to the specimen. However, since tested soil in this study is below the ground water table and the sampling depth is at most 40 m, the specimen during measurement of the suction was left in the atmosphere. To prevent the specimen from drying, it was surrounded by thin plastic film.

Relationship between time and suction is shown in Fig. 20. In this figure, suction measured for two specimens with different sample quality (one is extruded from the central part and another is extruded from near the edge of the sampling tube) is shown. In this figure, the constant suction for both specimens is attained after about 20 min. However, some specimens required much longer time to reach a constant suction value. It is not still identified which factor governs the duration to reach to a constant suction value (Tanaka and Tanaka, 2006b).

4.3 Change in suction with time

Some papers reports that suction value is reduced with time (for example, Hight, et al., 1992). In contrary to

146

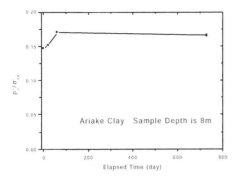

Figure 21. Change in suction with time for Ariake Clay (after Tanaka and Tanaka, 2006b).

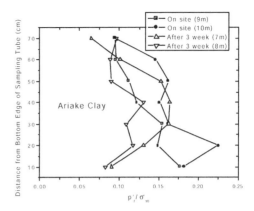

Figure 22. Distribution of suction in a sampling tube (after Tanaka and Tanaka, 2006b).

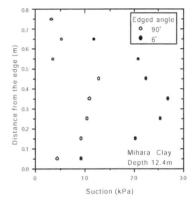

Figure 23. Comparison of distribution of suction for samples retrieved by samplers with different cutting edges.

this, different test result was observed for Ariake clay as shown in Fig. 21. To avoid variation in suction value due to heterogeneity in soil properties, a specimen with 10 cm in height was cut into 4 pieces, keeping the diameter as same with that of inside diameter of JPN sampler which is 7.5 cm. The cutting of the specimen and the measurement of suction for one piece were done on the site immediately after sampling. Other three pieces were wrapped by thin plastic and covered by paraffin with pine resin in the same way as in the usual investigation. These samples were stored in the room with constant temperature (20°C), and the measurement was carried out almost within 2 years after the sampling. Change in p'_r with time is shown in Fig. 21. It may be concluded from the test result that the p'_r value is constant for considerably longer time period in this study, provided that the sample is properly sealed and well stored.

This test result is also supported by Fig. 24. At the Bothkennar site, some samples were extruded immediately after sampling at the site and the suction was measured. Other samples were transported as described earlier and suction was measured at the laboratory of PARI. Although some scattering in the measured p'_r at the site and at the laboratory are seen, no distinct difference between different samples cannot be observed.

4.4 Effect of location in the sampling tube

Figures 22 and 23 show the distribution of p'_r in the sample tube. It can be seen from these figures that p'_r considerably varies with locations in the sampling tube. Stress conditions of the sampled soil in the sampling tube are very complicated, but they are important so as to interpret the variation in p'_r within the sampling tube. When soil is sampled and extracted from the ground, the total vertical stress on the soil sample would ideally become zero. Due to the existence of friction between the inside wall of the sampling tube and the soil sample, however, some fraction of

the vertical stress remains. Also, some amount of the horizontal stress is acting on the sample, because the soil sample is confined by the sampling tube. These stresses may not be uniformly distributed and they vary considerably with the location of the sample in the sampling tube. Some excess pore water pressure exists in the soil sample, due to the volume change caused by pushing the sampling tube into the ground, as well as the friction acting on the inside wall of the sampling tube. This pore water pressure also may vary with the location of the sampling tube. However, such variation is redistributed and equilibrated to a certain value after a certain elapsed time. As in usual investigation practices, where samples are extruded after several days from sampling, it is likely that the excess pore water pressure is redistributed and attains a uniform value throughout the sample in the sampling tube. When the sample is extruded from the sampling tube at the laboratory, the sample suffers from shearing and additional excess pore water (negative or positive) may be generated in the sample. When the specimen is completely exposed to the atmosphere, total stresses become zero and the resultant negative pore water pressure (p'_r) is finally attained in the specimen.

The p'_r values in Fig. 22 were measured at two different stages: on the site and after 3 weeks from the sampling, when the excess pore water pressure is considered to be uniformly distributed. Former samples were extruded on the site immediately after extracting the sampler from the ground. The soil samples were cut into 10 cm long and wrapped in the same manner as mentioned in Fig. 21.

In Fig. 22, distribution of p'_r extruded at two different times is shown. Nearly same value was observed for both samples, except near the cutting edge. This result demonstrates that the distribution of p'_r is not mainly determined during the process of the sampling, but that is created during the extrusion process of the sample. For the interpretation of the p'_r distribution shown in Fig. 22, the extrusion method employed in this investigation should be noted. Two methods for the extrusion of specimen from the sampling tube are allowed by the Japanese Geotechnical Society (JGS) as a standard for the preparation of undisturbed soil sample (JGS012-2000): one is the extrusion from the cutting edge (in which the direction of the soil sample is opposite to that in sampling) and in contrary, another method is opposite to the former method (direction is in the same way as during the processes of both sampling and extrusion). In this investigation, the sampling tube was vertically installed with the cutting edge on the top, and the sample was extruded from the cutting edge, i.e., the former method.

Not only during the process of sampling, but also during the extrusion, the soil in the sampling tube may suffer from shearing action that occurs between the inside wall and sampled soil, which reduces p'_r values. Farther from the edge of the sampler, the intensity of the shearing becomes more prominent because of longer travel distance during both sampling and extrusion. Therefore, the p'_r distribution in Fig. 22 is quite understandable. Also we have to consider the influence of disturbance occurred by drilling. The reason of the largest p'_r value at the cutting edge of the sampler is most likely to be due to the creation of vacuum during the extraction from the sampling tube. The large negative pore water pressure may have been redistributed and disappeared, when the sample was extruded after three weeks from the sampling.

Figure 23 shows another example of the p'_r distribution, where two types of sampling tubes were used: one with the angle of the cutting edge is 6° (Standard) and another with 90°, which intentionally creates disturbance in the sample. Sample extrusion was carried out after about 1 week from the sampling, which is a general way in practice. It can be seen here that the p'_r

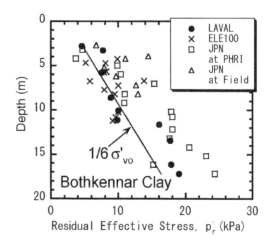

Figure 24. Suction for different samplers for Bothkennar clay (after Tanaka, 2000).

distribution is also same as that in Fig. 22, but absolute value of p'_r for 6° and 90° is clearly different: i.e., p'_r for 90° sampler is less than that for 6° sampler. This result provides a reason why the performance of the ELE100 sampler is worse than that of the JPN sampler. The edge angle of the ELE sampler is 30° while that of the JPN sampler is 6°. It also should be borne in mind that the length of the ELE100 sampler used in this investigation is 50 cm. From the graph, it is observed that the sample quality at both top and bottom of the sampling tube is not so good. As a result, high quality sample cannot remain in the 50 cm length sampler. The importance of cutting angle of the sampling tube has been stated by Clayton et al. (1998), by numerical method. Hight and Leroueil (2003) also suggested using a modified ELE sampler with 5° cutting edge to obtain high quality sample at the Bothkennar site.

4.5 Loss of suction and destruction of structure

Figures 24 and 25 show the measured values of p'_r for samples recovered from various samplers described previously in the comparison study. At both sites of Bothkennar and Ariake, tendency of p'_r for different samplers is same as that of q_u. The p'_r value for Sherbrooke, Laval and JPN samplers is remarkably greater than that for the ELE100 sampler or Shelby tube. From this result as well as with the evidences previously presented, the strength test results obtained from the recompression triaxial test will be able to interpret.

For Bothkennar clay, sample quality influences the q_u value, the e-$\log p$ curves and the recompression triaxial test. Whereas for Ariake clay, effect of sample quality appears only in the unconfined compression test, no effect is seen in other tests. From the above, it is observed that the p'_r value is strongly affected by

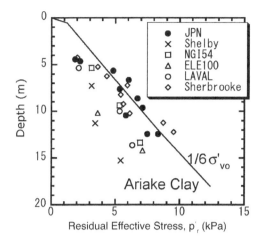

Figure 25. Suction for different samplers for Ariake clay (after Tanaka, 2000).

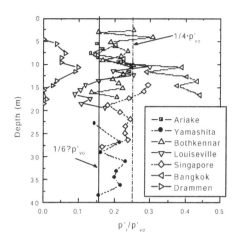

Figure 26. Suction normalized by the in-situ vertical effective pressure for various clays. All clays were retrieved by the same sampling method, i.e., using JPN sampler (after Tanaka and Tanaka, 2006b).

sample quality for both clays. From these results, it may be concluded that loss of p_r' and destruction of the structure (rearrangement of soil particles) should be considered separately. In another word, reduction of p_r' does not always bring the destruction of structure. By comparing the shape of the e-log p curve, it can be judged whether the structure is destructed or not. For example, Ariake clay retrieved by the ELE100 sampler lost a large amount of p_r' but its structure is not destructed. The q_u value is measured under unconfined stress condition so that the specimen with small p_r' presents small q_u value. If the specimen loses only p_r' and keeps the structure unchanged, then the stress strain relation including the strength is recovered to the original one (before sampling), once the sample is consolidated at the in situ effective stresses. Indeed, for Ariake clay, no remarkable difference can be detected in the recompression triaxial test (see Figs 13 and 14). If the structure is destructed by improper sampling, then the strength cannot be recovered even if the specimen is consolidated under the in situ stresses. Bothkennar clay is a typical example showing this type of soil.

4.6 Factors governing suction

4.6.1 Suction for various soils
Let us take a look at Figs 24 and 25, again. In these figures, the line indicating $(1/6)\sigma_{vo}'$ is drawn, and it is found that for Ariake clay, $(1/6)\,\sigma_{vo}'$ corresponds to the p_r' value for high quality samples. However, for Bothkennar $(1/6)\,\sigma_{vo}'$ indicates the minimum value of p_r'. The p_r' value for high quality sample seems much greater than $(1/6)\sigma_{vo}'$. The p_r'/σ_{vo}' ratio for various soils is plotted against depth in Fig. 26. It is found that p_r'/σ_{vo}' ratio is not only different for different

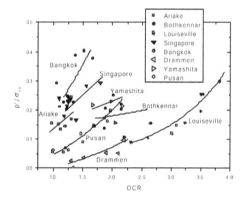

Figure 27(a). Relationship between the p_r'/σ_{vo}' ratio and OCR for various clays with small OCR (after Tanaka and Tanaka, 2006b).

soils, but also p_r'/σ_{vo}' is not constant to depth for some soils, which increases or decreases with the increase in depth. In the following discussions including Fig. 26, data are restricted only for samples retrieved by the JPN sampler and they represent the central part of the sampling tube.

4.6.2 Relation between suction and OCR
When p_r'/σ_{vo}' ratio is plotted against OCR as shown in Fig. 27(a), a clear tendency of increment of p_r'/σ_{vo}' with the increase in OCR is seen. In Fig. 27(b), p_r'/σ_{vo}' ratios of Yubari and Singapore (T) clays are added and much more prominent increment in p_r'/σ_{vo}' ratio with OCR is observed. For Singapore (T) clay, some points of p_r'/σ_{vo}' exceed 1.0. Two reasons for existence of the relation between p_r'/σ_{vo}' and OCR may be pointed out: 1) p_r'

149

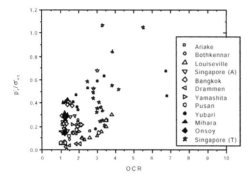

Figure 27(b). Relationship between the p'_r/σ'_{vo} ratio and OCR for various clays including large OCR.

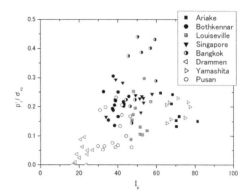

Figure 29. Relationship between the p'_r/σ'_{vo} ratio and the plasticity index (after Tanaka and Tanaka, 2006b).

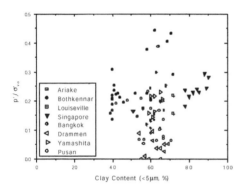

Figure 28. Relationship between the p'_r/σ'_{vo} ratio and the clay content (after Tanaka and Tanaka, 2006b).

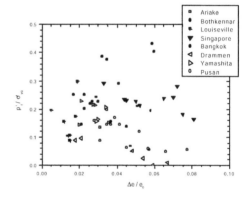

Figure 30. Relationship between the p'_r/σ'_{vo} and the $\Delta e/e_o$ index proposed by Lunne et al. after Tanaka and Tanaka, 2006b).

should be related to the mean in situ effective stress (p'_m) rather than σ'_{vo}. For clays with large OCR, p'_m becomes larger at the same σ'_{vo} as K_o increases with OCR. 2) As Hight (1992) has pointed out, when the sampling tube is pushed into the heavily overconsolidated clay, the negative pore water pressure may be generated by shearing stresses because of the positive dilatancy.

4.6.3 Other factors

It is understood that the value of p'_r is strongly controlled by OCR. But as shown in Figs 27 (a) and (b), even at the same OCR, p'_r/σ'_{vo} ratio considerably varies. It is anticipated that there are some other factors which govern the p'_r value. Data shown in Fig. 27 (a) are replotted against the clay content as shown in Fig. 28 and against I_p as shown in Fig. 29. When the clay content decreases, the hydraulic conductivity (k) may increase. The capacity of holding negative pore water pressure may be affected by k, so that p'_r/σ'_{vo} should decrease with clay content. However, no remarkable relation between them is found. Indeed, although Bangkok clay contains the same clay content of Drammen clay and nearly the same order of

OCR, their p'_r/σ'_{vo} ratios are differed by more than ten times. Instead of clay content, p'_r/σ'_{vo} can be related with I_p, though there is large scatter in their relation. At the present stage, it seems too early to determine the criteria for sample quality in terms of p'_r/σ'_{vo}.

It is anticipated that there is a relation between p'_r/σ'_{vo} ratio and $\Delta e/e_o$ or ε_{vo}, because when the p'_r is smaller compared with σ'_{vo}, large volume change due to reduction of the effective stress may be generated. This relation is shown in Fig. 30 where the points for each clay are plotted. In Fig. 31, points are grouped by the OCR range. In both figures, no systematic relation can be recognized. These facts lead to the following important interpretations. The approach using $\Delta e/e_o$ or ε_{vo} at σ'_{vo} is useful for clay whose e-log p curve has a sharp bending point like Louisville clay (see Fig. 16, and sometimes called structured clay). For such clay, if the sampling is done improperly, then e-log p curve gets deformed as Bothkennar clay (see Fig. 7).

On the other hand, as seen from Fig. 15, Yubari clay presents totally different behavior from those

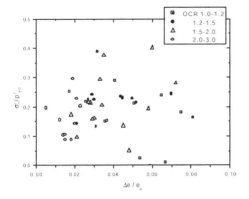

Figure 31. Replotting of Fig. 30 in reference to OCR group (after Tanaka and Tanaka, 2006b).

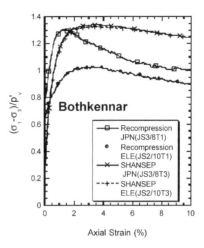

Figure 32. Comparison of Recompression and SHANSEP methods for Bothkennar clay (after Tanaka et al., 2003a).

mentioned above. Although the ratio of p_r'/σ_{vo}' for Yubari clay ranges from 0.3 to 0.6, considerable volume change is observed in these stress ranges, because of the large slope of e-log p at OC states. Is such a different behavior of Yubari clay brought by unsuitable sampling method? If the sampling were carried out properly, does the e-log$p$ curve become the same as Louiseville clay (see Fig. 16)? Although a lot of unsolved problems are still left on these materials, it is true that a large amount of Δe or ε_{vo} is generated in spite of high suction.

5 APPLICATION OF SHANSEP METHOD

Many methods have been proposed for correcting values derived from unfavorable quality sample (for example, Shogaki, 1996; Mitachi et al., 2001; Lunne et al., 2006). Among them, SHANSEP method may be the most well known method (Ladd and Foott, 1974). It is also true that there are many criticisms to this method. In the SHANSEP method, there are two major assumptions: 1) yield consolidation pressure (p_c) can be precisely obtained even from poor quality sample. 2) The roles of OCR in the in situ are the same as those of OCR created at laboratory. The p_c value of natural soil deposits is more or less greater than the in situ effective overburden pressure (σ_{vo}'), which is thought to be created by long term consolidation. In the SHANSEP method, such pseudo-overconsolidation can be artificially created by mechanical overconsolidation at laboratory: i.e., after a specimen is consolidated under normally consolidated state, then the specimen is swelled to be the same OCR of the in situ sample. It is criticized by many researches that such an artificial OCR and OCR created by long term consolidation, in another word, ageing effect are completely different so that

strength obtained by SAHNSEP method underestimates the real strength (for example, Leroueil and Vaughan, 1990). In spite of these criticisms, however, it is true that this method is widely used in practice.

The applicability of SHANSEP method will be examined for various soils of the world, using the triaxial and the Mikasa's direct shear apparatus (Takada, 1993; Hanzawa et al., 2007). In this study, the strength from the recompression test for a sample retrieved by the JPN sampler is considered as the reference strength. For more detailed information on testing procedures and testing materials, see Tanaka, et al., (2003a). Test results for Bothkennar clay is shown in Fig. 32, where the stress stain curve was measured by triaxial compression test. In this test, samples with two different qualities were tested: retrieved by the JPN and the ELE100 samplers. As already mentioned, for the ELE100 samples, the peak strength is smaller and strain at failure is greater than those for the JPN samples. The SHANSEP test was performed for ELE100 and JPN samples. In the SHNASEP method, a specimen is consolidated at pressure greater than the p_c value and hence as shown in figure, the effect of sample quality is disappeared: the peak strength as well as stress strain relationship are almost same for both JPN and ELE100 samples. And also, it is very interesting to note that the peak strength is nearly the same as that from recompression test for JPN sample, although the strain at failure from SHANSEP method is somewhat larger than that from the recompression test.

Comparison of SHANSEP and recompression methods was made for various clays and their comarison is shown in Fig. 33, where the vertical axis represents the ratio of strengths obtained from SHANSEP and recompression methods. In contrary to the anticipation, the strength ratio for various soils lies within

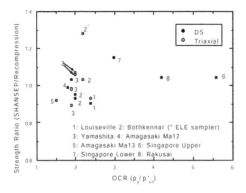

Figure 33. Comparison of Recompression and SHANSEP methods for various clays (after Tanaka et al., 2003a).

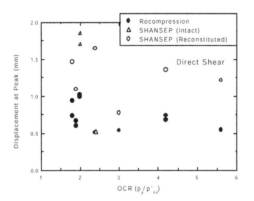

Figure 34. Comparison of displacement at failure for Recompression and SHANSEP methods measured by direct shear test (after Tanaka et al., 2003a).

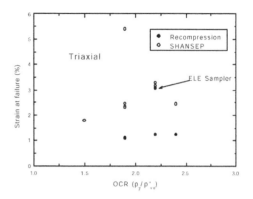

Figure 35. Comparison of strain at failure for Recompression and SHANSEP methods measured by triaxial test (Tanaka et al., 2003a).

which is considerably faster than that of the conventional incremental loading (IL) test. Inevitably, if the p_c value is measured by IL test, the strength ratio in Fig. 33 might become smaller, because the p_c value for CRS test is greater than that for IL, for example, its ratio is about 1.2 to 1.3, as shown in Table 3 (Leroueil and Jamiolkowski, 1991). Another issue is the assumption in SHANSEP method that the precise p_c value is able to be measured even from poor quality sample. However, the e-log$p$ curve for some clays or samples are affected by the sample quality as shown in Fig. 7.

6 A CASE HISTORY – EVALUATION OF SAMPLE QUALITY FROM GREAT DEPTHS

Since available lands are restricted in spite of large population, Japan has utilized coastal areas from the old days. A typical example is the construction of airport on the man-made island: i.e., the Kansai and the Chubu international airports. The Kansai airport was constructed 5 km off the main land in the Osaka bay, where the sea water depth is about 20 m. Since the end of Tertiary period, the Kansai area has been sinking due to the plate tectonics activity. Hence, the soft sediment of Quaternary period is thickly deposited in this area. A typical example of soil profile at the construction site of the Kansai Airport is shown in Fig. 36. It is known that in Quaternary period, the earth suffered from the cold climate nearly every 100,000 years. On those days, the sea water table lowered because of the formation of huge glaciers in the north and south poles. On the other hand, in warm climate such as the present, the sea water level rose and soft marine sediments were deposited. In the Kansai area, the number of these marine layers is counted as 14 in total. They are numbered in chronological order from 0 to 13 and

the relatively small range between 0.85 and 1.15. In case of Bothkennar clay using ELE100 sample, this ratio is as large as about 1.3, because of poor quality of the sample. It can be seen in the figure that SHANSEP method does not always underestimate the strength (the ratio is not always less than 1.0).

However, stress strain curve obtained from the SHANSEP method is clearly different from that obtained from the recompression test. In Fig. 34, displacement at the peak strength measured by the direct shear test (d_f) is compared, while in Fig. 35, strain at failure in the triaxial test (ε_f) is shown. The d_f or ε_f obtained from the direct shear and triaxial compression tests, respectively, for the SHANSEP method is clearly greater than that for the recompression method. From these test results, it can be concluded that for samples with unsuitable quality the SHANSEP method provides reliable strength unless the stress strain relation is concerned. However, two important issues should be pointed out. One is that in this study, the p_c value is measured by CRS oedometer test with $3.3 \times 10^{-6}\,\text{s}^{-1}$,

Table 3. Yield consolidation pressure measured by Constant Rate of Strain (CRS) and conventional 24 hours incremental loading oedometer (conv) (after Leroueil and Jamiolkowski, 1991).

Location	$\sigma'p$CRS/$\sigma'p$ conv	Reference
Champlain Sea, Quebec	1.28	Lerouiel et al., 1983
Finland	1.16	Kokisoja et al., 1989
Fucino, Italy	1.2	Burghignoli
Ariake and Kuwana, Japan	1.3–1.4	Hanzawa, 1991
Yokohama, Japan	1.25	Okumura & Suzuki, 1991
Osaka, Japan[b]	1.3–1.5	Hanzawa et al., 1991

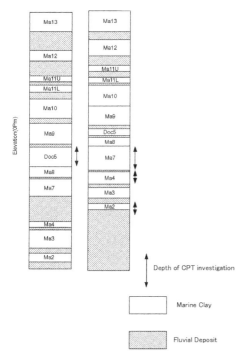

Figure 36. Soil profiles at the Kansai International Airport and the locations where CPT was carried out (after Tanaka et al., 2003b).

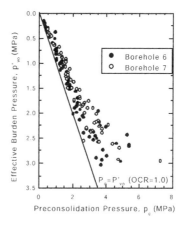

Figure 37. Yield consolidation pressure measured by Constant Rate of Strain (CRS) test for Osaka bay clays (after Tanaka et al., 2002).

time-settlement relation precisely, sampling was carried out until 400 m depth. Hereafter, sample quality at this investigation will be described.

6.1 P_c values measured

Figure 37 shows the distribution of the p_c values plotted against the σ'_{vo}. Figure 38 shows the distribution of OCR at the same points of the p_c values shown in Fig. 37. The p_c values are somewhat greater than σ'_{vo}, not due to mechanical overconsolidation, but probably due to effects of long term consolidation such as ageing or cementation effects (Tanaka, et al., 2004). As can be seen, the p_c values and OCR are somewhat scattered. The problem is whether such a scatter is caused by heterogeneity of soil properties, or by techniques of laboratory testing including sampling. Because of great depth, the ordinary sampling technique could not be used. For improvement of efficiency in sampling work, a wire line method was employed for sampling and drilling (Kanda, et al., 1991). Up to about 150 m, sample was recovered by hydraulic piston sampler (which is also called Osterberg sampler),

named such as Ma 13. "Ma" means marine layer. 13 is the number of the layers and represents the youngest layer, e.g., Ma13 is the marine layer deposited in the Holocene era.

In the construction of the Kansai airport, the Ma13 was improved by sand drain method so that the settlement in the Ma13 layer was completed before the airport was on operation. However, marine layers below the Ma12, which were deposited during the Pleistocene era, are under consolidation and considerable settlement are being observed. To predict the

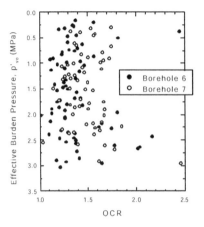

Figure 38. OCR measured by CRS test for Osaka bay clays (after Tanaka et al., 2002)

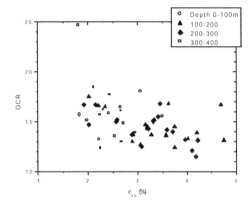

Figure 40. Relationship between ε_{vo} and OCR for Osaka bay clays (after Tanaka et al., 2002).

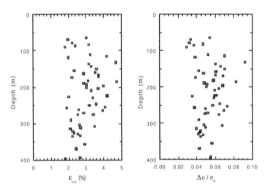

Figure 39. Volumetric strain (ε_{vo}) and change in void ratio ($\Delta e/e_0$) at σ'_{vo} measured by CRS for Osaka bay clays (after Tanaka et al., 2002).

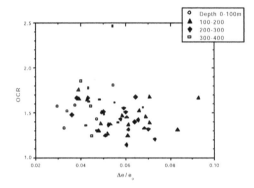

Figure 41. Relationship between $\Delta e/e_0$ and OCR for Osaka bay clays (after Tanaka et al., 2002).

but its sampling tube is somewhere different from the JPN sampler because of great stiffness: inside diameter and thickness are 81.1 mm and 4.0 mm, respectively. Below 150 m deep, a rotary core sampler was used with the same sampling tube as the Hydraulic sampler. Samplers and drilling are descried in more detail by Tanaka et al. (2002).

6.2 Sample evaluation in terms of ε_{vo} and $\Delta e/e_0$

Volumetric strain (ε_{vo}) and change in void ratio ($\Delta e/e_0$) at σ'_{vo} measured by CRS test are shown in Fig. 39. In spite of expectation, neither ε_{vo} nor $\Delta e/e_0$ increases with the depth. Instead, those values are almost constant with depth. It should be also kept in mind that ε_{vo} or $\Delta e/e_0$ is not affected by different sampling method, i.e., hydraulic or rotary core sampling. According to the criteria of Andresen and Kolstad (1979), most of the samples are classified into "good" and "fair". Similarly, according to Lunne et al.'s (1997) criteria,

these samples lie in "Good to Fair" zone shown in Fig. 17.

If the variation in the p_c value is caused by sample disturbance and if its quality can be accessed by ε_{vo} or $\Delta e/e_0$ indices, then there should exist a relationship between OCR and ε_{vo} or $\Delta e/e_0$. Some correction methods for the p_c value using these indices are proposed by several researchers (for example, Shogaki, 1996). As shown in Figs 40 and 41, OCR slightly decreases with ε_{vo} or $\Delta e/e_0$. This fact indicates that underestimation of OCR or the p_c value may be partly due to sample disturbance. However, taking account of large variation in OCR at the same ε_{vo} or $\Delta e/e_0$, the decrease in OCR due to ε_{vo} or $\Delta e/e_0$ can be neglected. Other examples in Figs 42 and 43 show difficulty in explanation of small p_c value due to sample quality. Figure 42 shows a relationship between p_c and depth, obtained for Borehole-9. Comparing the p_c values at depths of 202 m and 204 m, as much as 0.3 MPa difference in the p_c is found. Their e-log p curves are shown in Fig. 43. Both show very nice curves, suggesting that they represent high quality samples. Only from the e-log p curve,

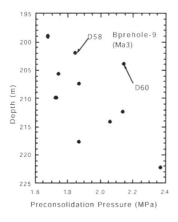

Figure 42. Detailed distribution of yield consolidation pressure at Osaka bay (after Tanaka et al., 2002).

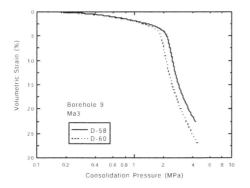

Figure 43. Comparison of e-log p relationship for D-58 and D-60 shown in Fig. 42 (after Tanaka et al., 2002).

it is difficult to consider that the low p_c value for D-60 sample is due to the poor quality sample, like Bothkennar clay shown in Fig. 7.

6.3 Investigation using CPT

In geotechnical engineering, it is a hard task to judge whether the variation in soil properties is caused by human's factor or naturally varied properties. This point makes it difficult to introduce design method based on the reliability to the geotechnical engineering. Unlike laboratory or in situ testing, cone penetration test is one of a few tests which are nearly free from human dependency. Because unlike the standard penetration test (SPT), no borehole is required to drill and a cone probe is simply penetrated at a constant speed, measuring resistances during the penetration (sometime in addition to tip resistance, sleeve friction and pore water pressures are also measured). At the construction site of the Kansai Airport, a penetrometer for great depth was developed (Tanaka et al., 2003b).

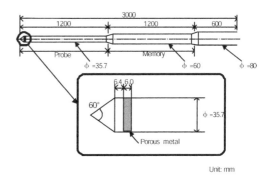

Figure 44. Newly developed cone penetrometer for great depth for Osaka bay clays (after Tanaka et al., 2003b).

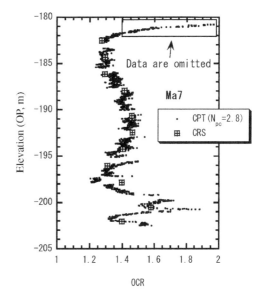

Figure 45(a). Comparison between OCR measured by laboratory test and estimated by great depth cone penetrometer for Ma7 clay layer (after Tanaka et al., 2003b).

Its schematic diagram is shown in Fig. 44. The diameter and the angle of the conical tip of the cone are same as those for the ordinary CPT: i.e., 35.7 mm and 60°, respectively. Pore water pressure during penetration was measured at the shoulder behind the conical tip. However, skin friction was not measured. The maximum capacity of tip resistance and pore water pressure measurements are 30 MPa and 20 MPa, respectively. Due to the great depth and the presence of several sandy layers as shown in Fig. 36, a borehole was driven up to the targeted depth prior to the penetration of CPT. To avoid complicated procedures for the insertion of electrical cables into CPT's rods (total length is over 100 m), the probe has a memory as shown in Fig. 44 where the measured values are stored.

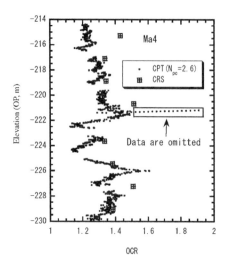

Figure 45(b). Comparison between OCR measured by laboratory test and estimated by great depth cone penetrometer for Ma4 clay layer (after Tanaka et al., 2003b).

Test results are shown in Figs 45(a) and (b), where OCR is obtained by the following equations.

$$p_c = (q_t - \sigma_{vo})/N_{pc} \qquad (1)$$

$$OCR = p_c/\sigma'_{vo} = (q_t - \sigma_{vo})/(N_{pc}p'_{vo}) \qquad (2)$$

where, q_t and σ_{vo} are the tip resistance of the CPT considering the ratio of effective sectional area and the in situ total overburden pressure, respectively. N_{pc} is the cone factor for the yield consolidation pressure and it is determined that these values fit well with the p_c values measured by CRS oedometer test for sampled soil. CRS test was carried out at a rate of $3.3 \times 10^{-6} s^{-1}$ (0.02%/min). Back-calculated N_{pc} factor is slightly different from those investigated for each layer as shown in figure 45 (N_{pc} is 2.8 for Ma7, and 2.6 for Ma4). Likewise, the cone factor for the undrained shear strength, N_{pc} may be also influenced by various soil properties. Nevertheless, OCR measured by CRS and estimated by CPT are very close to each other. And it is also interesting to know that OCR, in another word, tip resistance considerably varies with depth. This test result suggests that the investigated layer does not have homogeneous geotechnical properties. Instead, they show considerable heterogeneity.

7 CONCLUSIONS

Based on the author's experiences, sampling of soft clays are described, including the influence of sample quality on the values measured by laboratory test. Important findings are as follows:

(1) Even with the tube samplers, high quality can be retrieved, provided that the proper considerations

are taken, for example, stationary piston, cutting edge angle, and thickness of the wall. For low plastic clay, however, large diameter sampler may be superior.

(2) Loss of suction (residual effective stress) does not necessarily bring destruction of soil structure (rearrangement of soil particles). When the structure is destructed, the e-$\log p$ relation is affected.

(3) Assessments for sample quality should be established for the introduction of reliability analysis. However, still there is no assessment method valid for all kind of soils. Amounts of changes in $\Delta e/e_o$ or ε_{vo} at the in situ effective stress are dependent on soil properties. The suction remained in the specimen is also strongly affected by soil properties, especially OCR.

(4) SHANSEP provides reasonable strengths for various soils. However, it is doubtful to obtain the precise yield consolidation pressure (p_c) from the poor quality sample.

(5) Sample quality assessments were carried out at the construction of the Kansai Airport. Samples were recovered from Pleistocene clay layers which are as deep as 400 m. Scattering of the p_c value was found which is not due to the difference in the sample quality. Instead they are due to the heterogeneity of the investigated soil layers.

REFERENCES

Andresen, A. & Kolstad, P. 1979. The NGI 54 mm sampler for undisturbed sampling of clays and representative sampling of coarser materials. *Proceedings of International Symposium of Soil Sampling*, Singapore, 13–21.

Clayton, C. R., Siddique, A. & Hopper, R. J. 1998. Effects of sampler design on tube sampling disturbance –numerical and analytical investigations, *Geotechnique* 46(6): 847–867.

Fredlund, D. G. & Rahardjo, H. 1993. Soil mechanics for unsaturated soils, John Wiley & Sons, Inc.

Hanzawa, H., Fukaya, T. & Suzuki, K. 1990. Evaluation of engineering properties for an Ariake clay. *Soils and Foundations* 30(4): 11–24.

Hanzawa, H., Nutt, N., Lunne, T., Tang, Y. X. & Long, M. 2007. A comparative study between the NGI direct simple shear apparatus and the Mikasa direct shear apparatus, *Soils and Foundations* 47(1):47–58.

Hight, S. W. 1992. A review of sampling effects in clays and sands. *Proc. Int. Conf. Offshore Foundations and Site Investigation*, Society for Underwater Technology, Klewer, 115–146.

Hight, D. W., Boese, R., Butcher, A. P., Clayton, C. R. I. & Smith, P. R. 1992. Disturbance of the Bothkennar clay prior to laboratory testing, *Geotechnique* 42:199–217.

Hight, D. W. & Leroueil, S. 2003. Characterization of soils for engineering purposes, *Characterization and engineering properties of natural soils* (1): 255–362.

Kanda, K., Suzuki, S. & Yamagata, N. 1991. Offshore soil investigation at the Kansai international Airport. *Proc. of the inter. Conf. on Geotechnical Engineering for Coastal Development, -theory and practice on soft ground- Geo-Coast '91*(1): 33–38, Port and Harbour Research Institute.

Lacasse, S. & Berre, T. 1988. Triaxial Testing Methods for soils. Advanced Triaxial Testing of Soil and Rock. *ASTM STP 977:* 264–289.

Ladd, C. C. & Foott, R. 1974. New design procedure for stability of soft clays. *ASCE* 100(GT7): 763–786.

La Rochelle, P., Sarrailh, J., Tavenas, F., Roy, M. & Leroueil, S. 1981. Cause of sampling disturbance and design of a new sampler for sensitive soils. *Canadian Geotechnical Journal* 18(1): 52–66.

Ladd, C. C. & Lambe, T. W. 1963. The strength of "undisturbed" clay determined from laboratory tests, Laboratory shear testing of soils, *ASTM*, ATP361:342–371.

Lefebvre, G. & Poulin, C. 1979. A new method of sampling in sensitive clay. *Canadian Geotechnical Journal* 16(1): 226–233.

Leroueil, S. & Jamiolkowski, M. 1991. Exploration of soft soil and determination of design parameters, general report, *Proc. of the inter. Conf. on Geotechnical Engineering for Coastal Development, -theory and practice on soft ground- Geo-Coast '91*(2): 969–998, Port and Harbour Research Institute.

Leroueil, S. & Vaughan, P. R. 1990. The general and congruent effects of structure in natural soils and weak rocks, *Geotechnique* 40:467–488.

Leroueil, S., Hamouche, K., Tavenas, F., Boudali, M., Locat, J., Virely, D. & Roy, N. 2003. Geotechnical characterization and properties of a sensitive clay from Quebec, In T.S. Tan et al. (eds), *Characterization and Engineering Properties of Natural Soils* (1):363–394. Taylor & Francis.

Lunne, T., Berre, T. & Strandvik, S. 1997. Sample disturbance effects in soft low plastic Norwegian clay. *Proceedings of the Conference on Recent Developments in Soil and Pavement Mechanics,* Rio de Janeiro:81–102.

Lunne, T., Berre, K. H., Strandvik, S. & Sjursen, M. 2006. Effects of sample disturbance and consolidation procedures on measured shear strength of soft marine clay Norwegian clays, *Canadian Geotechnical Journal* 43:726–750.

Mitachi, T., Kudoh, Y. & Tsushima, M. 2001. Estimation of in situ undrained strength of soft soil deposits by use of unconfined compression test with suction measurement, *Soils and Foundations* 41(5): 61–71.

Ohtsubo, M., Higashi, M., Kanayama, M. & Takayama, M. 2007. Depositional geochemistry and geotechnical

properties of marine clays in the Ariake bay area, Japan; In T.S. Tan et al. (eds), *Characterization and Engineering Properties of Natural Soils* (3):1893–1938. Taylor & Francis.

Shogaki, T. 1996. A method for correcting consolidation parameters for sample disturbance using volumetric strain. *Soils and Foundations* 36(3):123–131.

Takada, N. 1993. Mikasa's direct shear apparatus, test procedures and results, *Geotechnical Testing Journal, GTJODJ.* 16(3): 314–322.

Tanaka, H. 2000. Sample quality of cohesive soils: lessons from three sites, Ariake, Bothkennar and Drammen, *Soils and Foundations* 40(4): 57–74.

Tanaka, H., Oka, F. & Yashima, A. 1996. Sampling of soft soils in Japan. *Marine Georesources and Geotechnology* 14:283–295.

Tanaka, H., Ritoh, F. & Omukai, N. 2002. Sample quality retrieved from great depth and its influence on consolidation properties. *Canadian Geotechnical. Journal* 39 (6):1288–1301.

Tanaka, H., Shiwakoti, D. R., & Tanaka, M. 2003a. Applicability of SHANSEP method to six different natural clays, using triaxial and direct shear tests. *Soils and Foundations* 43(3): 43–56.

Tanaka, H., Tanaka, M., Suzuki, S. & Sakagami, T. 2003b. Development of a new cone penetrometer and its application to great depths of Pleistocene clays. *Soils and Foundations* 43(6):51–61.

Tanaka, H., Kang, M. S. & Watabe, Y. 2004. Ageing effects on consolidation properties–based on the site investigation of Osaka Pleistocene clays-. *Soils and Foundations* 44(6):39–51.

Tanaka, H., Nishida, K., Hayashi, H., Fukazawa, T., Nakamura, A., Yoshimura, M., Koizumi, K. & Nakajima, A. 2006a. Characterization of Yubari peat and clay by in situ and laboratory testing. *Proc. of Technical reports* 46: 239–246, Hokkaido branch of Japanese Geotechnical Society (in Japanese).

Tanaka, H. & Tanaka, M. 2006b. Main factors governing residual effective stress for cohesive soils sampled by tube sampling, *Soils and Foundations* 46(2):209–219.

Tanaka, H. & Nishida, K. 2008. Sample quality assessment of soft clay by suction and shear wave velocity measurements, *Proceedings ISC-3 on site characterization* (submitted).

157

Geotechnical and Geophysical Site Characterization – Huang & Mayne (eds)
© 2008 Taylor & Francis Group, London, ISBN 978-0-415-46936-4

An overview on site investigation for Hsuehshan Tunnel project

Tseng, Dar-Jen
Taiwan Area National Expressway Engineering Bureau, MOTC

Hou, Ping-Cheng
Geotechnical Engineering Department, Sinotech Engineering Consultants, LTD

ABSTRACT: Among the five tunnels on National Taipei-Ilan expressway, the 12.9 km Hsuehshan tunnel, partly bored by TBM and partly excavated by D&B, is the longest one which cuts through the northern section of Hsuehshan Range. To understand the tectonic environment, much effort have been done to explore and characterize the geological condition along the tunnel alignment including satellite image and air photo interpretation, field mapping, in-situ rock test, rock mechanics test, hydrogeological study and on-going exploration during construction. The explorations indicated that the tunnel is lithologically cutting through the Oligocene argillite and quartz sandstone in the eastern part of its alignment and the Miocene sedimentary rocks in the western section. Structurally, seven major faults and two regional fold structures were identified along the tunnel, aside from numerous small-scale shear zones. Of the seven major faults, six are congested within the eastern 3 km, resulting in extremely poor ground conditions, especially in the Szeleng Sandstone formation–a highly jointed rock unit showing extremely high compressive strength. This paper briefly overviews the exploration and site characterization for the tunnel project.

1 INTRODUCTION

The Lanyang area is separated from the rest of northern Taiwan by the precipitous Central Mountain Range and the Hsuehshan Mountain Range. Although the Lanyang area belongs to the general area of northern Taiwan, lacking of efficient means of transportation isolated the area from the rest of northern Taiwan. As a result, while the Taipei city area was enjoying prosperity brought on by the "Taiwan Economic Miracle", the 'locked-out' Lanyang area lagged behind in economical development. Then, it was Realized that constructing a modern expressway linking the Lanyang area with the rest of northern Taiwan as well as the entire island of Taiwan would be the necessary means to unlock the Lanyang area and boost its economy. The government began planning on an expressway connecting the Lanyang area to the Taipei City area. Planning of an expressway between the Lanyang area and Taipei City commenced in 1982. The expressway starts from the Nankang District in Taipei City and ends at the township of Toucheng in Ilan County. The total length of the expressway is approximately 31 km. The expressway penetrates the Hsuehshan Range and connects the Taipei Basin with the Lanyang Plain geographically. A long tunnel was designed as means of cutting through the precipitous mountains of the Hsuehshan Range. This tunnel, later

named Hsuehshan Tunnel, became the critical-path to the entire Taipei-Ilan Expressway Project.

The Hsuehshan Tunnel is a twin-tube two-direction highway tunnel, 12.9 km in length. It cuts through the northern part of the Hsuehshan Range. Besides the two main tubes, a pilot tunnel has also been designed. The pilot tunnel is located below the line linking the center of two main tubes. The expected functions of the pilot tunnel, in construction stage, include collecting detailed geological information, providing access for main tubes construction, performing necessary ground treatment and draining the ground before excavation of main tubes. In the meantime, after the tunnels are completed, the pilot tunnel is reserved as a service tunnel. Along the alignment of the tunnel there are three sets of vertical ventilation shafts, their depths are: 501 m, 249 m and 459 m respectively. The diameter of the main tunnel tubes is 11.8 m; that of the pilot tunnel 4.8 m. The original design used three TBMs for both pilot tunnel and the main tunnel excavation. The construction of the pilot tunnel and the main tunnels started in 1991 and 1993 respectively. Since the beginning, due to the rather adverse geological conditions and the presence of good quantity of groundwater, tunnel constructions progressed in great difficulty, and the schedule lagged behind. Under persistent endeavour from a team effort, construction of the tunnel was eventually completed.

Situated on the plate collision suture zone, Taiwan is marked with extreme topographic relieves and highly variable geological conditions. A long tunnel running under high rock cover can only be expected to have nothing but a vague geological picture of the area penetrated by the planned tunnel alignment, even though good quality geologic investigation had been performed. Nevertheless, geological investigations performed under various scales in accordance to requirements for the various study stages yielded results and information that proved to be of crucial importance in the construction of the Hsuehshan Tunnel, and the validity of this geological information was verified when the actual geology was finally revealed on the excavated walls of the tunnel. This paper presents an account on the results and methodologies for the investigations and site characterization performed for the Hsuehshan tunnel project in various stages.

2 THE HSUEHSHAN RANGE

The Hsuehshan Range is famous for its highest peak Hsuehshan with elevation of 3886 m. The Hsuehshan Tunnel, cutting through the Hsuehshan Range geological subprovince which is situated on the western wing of the Central Mountain Range, is one of the most complicated geological areas of the island. Rock formations in the Hsuehshan Range comprise folded Tertiary sedimentary sequence, and belong to the fold-and-thrust belt in the active orogenic belt. Most geological structures are regional folds and thrust faults whose strikes are generally in NEE-SWW direction. Normal faults, however, were also formed locally as result of later tectonic activities. Following the major phase of orogenic movement, back-arc basin spreading occurred in the sea outside of Ilan. This on-going spreading of the back-arc basin resulted in normal faulting in areas in the immediate vicinity of Ilan. The Hsuehshan Tunnel is located right in the centre of these geologic complexities.

The earliest available geological information on the northern half of the Hsuehshan Range was contained in the 1:50,000-scale "Geologic Map Sheet and Explanatory Text of the Hsintien Area" by the Japanese geologist Ichikawa, 1932. Names of the major geological features appearing on this map sheet are still in use at present. These names include: the Szeleng Sandstone, the Kankou Formation, the Tsuku Sandstone, the Tatungshan Formation, the Yingtzulai Syncline, the Taotiaotzu Syncline and the Shihtsao Fault. Based on this foundation, later contributions to the geology of the area included topics on structural geology, historical geology, stratigraphy and sedimentology from the Chinese Petroleum Corporation and academic studies. When the Central Geological Survey, Ministry of Economic Affairs, published the 1:50,000 "Geologic

map sheet of the Toucheng Area" in 1989, the geologic framework of the project area had been quite well established. In geological investigation for the Hsuehshan Tunnel Project, the geological model and framework of the project area followed these publicly accepted results. Detailed information on geological structures, distribution of rock formations, the rock mechanical properties as well as hydrogeological behaviours were all recorded and presented during each study stage for use as reference in engineering design and construction operation.

3 STUDY PROGRAM

The township of Pinglin is situated mid-way of the route alignment of Taipei-Ilan Expressway; it also marks an area of simple straightforward geographic settings to its west, and an area of open plain on the Ilan Plain to its east. In route selection, the alignment from Nankang to Pinglin had always been comparatively simple since there is only one logical choice for the candidate route, and the alignment selected during the route selection stage became the final route alignment. From Pinglin to Ilan, however, the open Ilan Plain offers many alternatives, and there were as many routes as there were issues to consider. Hence for the sake of selecting the best possible route for road engineering design purposes, geological investigations were performed during various study stages of the Hsuehshan Tunnel project. The first generalized geological investigation on the Hsuehshan Tunnel area was completed in 1982. From that date on, a number of geological investigations for preliminary study, feasibility study, route evaluation and selection study, basic design study as well as detail design study had been completed. Main contents of these investigations include:

1. Remote sensing and air photo interpretation was performed with the purpose of understanding the regional geological settings of the project area.
2. Field geological reconnaissance survey was carried out to verify findings and results of the remote sensing and air photo interpretation, and to gain in-depth, detailed understanding on the geological conditions along the tunnel alignment.
3. Seismic reflection survey, with 11,350 m in total length, were performed on the structural conditions, distribution of rock formations and their sonic velocity characteristics within 50 m deep along the tunnel alignment.
4. Subsurface exploration through boreholes was designed to unreveal the distribution of rock formations underground, their lithology and rock mass quality. For example, boreholes as deep as 490 m were drilled at the locations of vertical shafts to

reach the tunnel alignment elevation, and in-situ stress measurements were also conducted.

5. Exploration through trenching was conducted for locations of major faults and to collect geologic and geotechnical data and information.
6. Testing adits were excavated in the south portal area for understanding the rock mass behavior during tunnel excavation.
7. In-situ tests such as hydrofracturing were performed at the sites for understanding the in-situ stress conditions and rock mass mechanical behaviors.
8. Laboratory tests were conducted to collect and understand relevant rock mechanical parameters.

Geological investigations of various scales were conducted during the various stages. Through results of these investigations the basic geological models along the tunnel alignment were established. Also based on the results, cross section and rock mass quality classifications were established. A brief description on these geological investigations is presented in the followings:

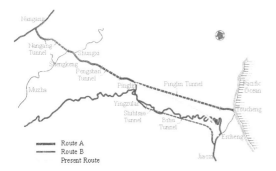

Figure 1.   Candidate routes from Nankang-Toucheng in year of 1984.

3.1   *Preliminary study stage (1982–1984)*

A preliminary study on construction of a better quality highway between Taipei and Ilan was carried out by former Taiwan Provincial Bureau of Highways in 1982. A report, entitled "Feasibility Study on Constructing a Tunnel-Highway between Nankang and Toucheng", was published. In this report two routes were proposed (see Figure 1 for details).

To appraise whether the proposed routes were feasible, Taiwan Provincial Bureau of Highways entrusted consulting firms to conduct a geological evaluation of these two routes in 1984. A report entitled "Geological Evaluation Report on a Tunnel Highway between Nankang and Toucheng" was completed. This report presented evaluation on the geological conditions along the proposed routes and proposed initial engineering layouts. Geological investigation works conducted during this stage are presented in Table 1.

At time of preparation of the report, design standards for ordinary provincial highway were adopted. The alignment of the proposed route managed to avoid the Chinying and the Shanghsin Faults by passing through the northern side of the fractured Szeleng Sandstone formation. The eastern portal of the roposed long tunnel was located about 3 km north to the eastern portal of Hseuhshan Tunnel. The proposed route was shelved when the highway was later on up-graded to an expressway standard.

3.2   *Feasibility study stage (1987–1988)*

The Institute of Transportation, Ministry of Transportation and Communications, conducted feasibility study on a highway between Taipei and Ilan. In 1987, a international consulting firm was entrusted to conduct the feasibility study. However, no additional geological investigation was performed. And, the existing geological information were further reviewed. During

Table 1.   Summary of geological exploration along Hsuehshan tunnel.

Stage item		Preliminary/ Feasibility study (1984–1988)	Route selection stage (1989–1990)	Basic design stage (1990–1991)	Detail design state (1992–1994)	Construction (1991–2004)	Subtotal
Field Mapping		√	√	√	√	√	
Remote Sensing and Airphoto Interpretation		–	√	√	–	–	
Boreholes	hole	16	15	30	15	7	91
	m	1144.5	1036.6	2246.0	859.0	1320.0	6606.1
Refraction	line	4	9	1	–	2	16
Seismic	m	1150.0	12190.0	13110.0	–	2000.0	28450.0
RIP	line	–	–	–	–	1	1
	m	–				1500.0	1500.0
Trench	set	–	–	7	–	–	7
	m³	–	–	2099.3	–	–	2099.3
Adit	set	–	–	1	–	–	1
	m	–	–	150.0	–	–	150.0

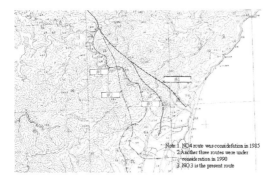

Figure 2.   Candidate routes of Hsuehshan tunnel.

the feasibility study three route alignments were proposed (Figure 2). Among these three proposed routes, route No. 1 was taken as widening of some sections of the existing Provincial Highway No. 9, adding some short tunnels of variable lengths. The ending point for No. 1 route was slightly due south. It passed through the tourist spa area to the mountainside of the town of Chiaochi to connect up with Taiwan Highway No. 9 at Ilan again. Meanwhile, both routes No. 2 and No. 3 consisted with a long tunnel of 12.9 km. The ending point for route No. 2 was in the vicinity of the Wufeng Falls, and the route ran south to connect with Taiwan Highway No. 2. The ending point for route No. 3 was at Toucheng and connected with Taiwan Highway No. 9. All three routes were planned with standards for expressway. Evaluation performed at this stage indicated that route No. 2 would be the one to recommended. However, the difference between routes No. 2 and No. 3 was very insignificant.

3.3   Route selection stage (1989–1990)

In 1989, the Taiwan Area Expressway Engineering Bureau (TANEEB) became the supervisory authority over planning and construction of the Taipei-Ilan Expressway. In this stage, De Leuw Cather was entrusted to conduct route evaluation study and Sinotech Engineering Ltd. to conduct additional geological investigation on various routes of the Hsuehshan Tunnel. Items of geological investigation included surface geological mapping, subsurface exploration through boreholes, geophysical surveys and remote sensing and air photo interpretation. Results of these investigations had furnished information in route selection. Table 1 presents contents of these investigations.

Results of field geological investigations and information from drilling indicated that the geological conditions of the western portal area are good. Rock mass quality for 3/4 length of the western part of the tunnel alignment falls between fair and good; rock mass quality for the remaining 1/4 length of the tunnel

on the eastern part is poor. Among these, area of around 1 km in the vicinity of the eastern portal is fractured, loose rock mass. Judging from the regional geological and structural geological models, this fractured, loose rock mass persists in the mountainside area between Chiaochi and Toucheng, none of the routes can avoid traversing this poor rock mass. At this stage, adding an exploratory pilot tunnel between the two main tunnel was first been discussed. It was expected that information collected from this exploratory pilot tunnel would serve as reference in design and construction of the main tunnels.

Considering to connect the eastern portal of the long tunnel at Toucheng township, preliminary candidate routes No. 3, No. 3A, and No. 3B were proposed. These routes were evaluated based on engineering contents, costs, topography, geologic structures, number of faults crossed, distance to landslide area, thermal spring and geothermal area, traffic safety, and glare at portal etc. Route 3A was then selected and became the route under construction. Tunnelling with TBM was carefully assessed in this stage in response to the ground conditions along the tunnel alignment.

3.4   Basic design stage (1990–1991)

During basic design stage, remote sensing analysis and air photo interpretation was performed, this was aimed at providing topics and issues for field investigation to establish a geological model for the project area. Also, refraction seismic survey was conducted along the entire length of the selected tunnel alignment. For geological investigation, field investigation was supplemented with borehole exploration. The geological conditions at the portals, along the tunnel alignments as well as the locations and the characteristics of faults were also investigated in detail. Exploration through trenching was conducted to pinpoint locations of faults. Through boreholes in-situ stresses and downhole deformation were measured. Core samples were tested in the laboratory for rock mechanical strengths as well as other relevant tests to get rock mechanical parameters required in designs.

Investigation results indicated that groundwater would be abundant in the project area. Tunnel construction may encounter difficulty of groundwater seepage especially when penetrates through fault gouges and fractured Szeleng Sandstone. On the eastern portion, tunnel alignment passes through gently dipping rock formations of the Szeleng Sandstone. Gas emission from the carbonaceous argillites and possible failure due to shear gouges in the shear zones should be watched. It was recommended that since the geological conditions at eastern portion of the tunnel is rather inferior, detailed record on geological features during excavation should be made so as to provide reference in detailed design. At certain sections, rock cover is as thick as 750 m, special attention should be rendered to rock supports at these locations.

3.5 Detailed design stage (1992–1993)

During detailed design stage, a review of the geological information from the previous stages was made. And, supplementary geological investigation was conducted where required. The supplementary geological investigation was mainly on realizing the characteristics of the fault zones. In addition, subsurface exploration through boreholes were performed at foundations of important structures and bridge abutments, coalmine pits, known landslides, indistinct fault locations, portals of tunnels, cut-slopes and waste disposal sites. During basic design stage drilling records showed that the site for vertical ventilation shaft No. 3 was not suitable, a new borehole (PH-29-1) was made slightly east of the original site. The new borehole record showed that although the rock formations were locally poor, however, at tunnel level the rock mass was rather good as a vertical shaft location.

3.6 Supplementary investigation during construction (1991–2003)

3.6.1 Investigation during pilot tunnel construction

As safety measures, during construction of the pilot tunnel various advance geological probing investigations were persistently performed, these included HSP and TSP seismic surveys. Initially, horizontal seismic profiling (HSP) was conducted and 5 profiling with total surveyed length of 1065 m were completed. Later, tunnel seismic profiling (TSP) with higher resolution was adopted, and 27 surveys totaling 4,115 m were completed in the pilot tunnel. In the main tunnel 9 TSP surveys totaling 1037 m were conducted. Horizontal boreholes of reverse circulation coring were done for 229 m and wire line coring were also done for 1330 m. Furthermore, the length of non-coring probe-drilling in pilot tunnel excavation was 5661 m.

3.6.2 Investigation for groundwater influx evaluation

The excavation of the pilot tunnel and main tunnels were severely obstructed with huge groundwater inflow in the early stage of construction. A research program was then setup to evaluate groundwater influx in 1998. Works for this investigation included in-situ field reconnaissance survey, collection of relevant information, flow discharge monitoring, and supplementary deep borehole for hydrogeological investigation (1 deep hole to 300 m). Groundwater table at several locations were monitored and analyzed. Meanwhile, radioactive isotopic dating and analysis

(C14 and H3) was carried out. Eventually, a conceptual groundwater model was established. In turn, analysis on hydrogeological conditions during tunnel construction was analyzed. Furthermore, possible effect of tunnel groundwater influx on nearby reservoir was also evaluated.

The tunnel alignment was longitudinally divided into three areas; while on the vertical direction it was divided into shallow-seated and deep-seated groundwater, a conceptual term "drawer box" was proposed as hydrogeologic model. The above concept was manifested from groundwater monitoring result, isotopic dating of groundwater and dynamic analysis on groundwater influx. In addition, groundwater influx simulation and hydraulic parameter analysis revealed that permeability coefficient for the fractured Szeleng Sandstone were seemingly steady (ca. 3.0×10^{-4} m/s), and when a fracture zone over several meters in width is encountered in the future, the groundwater influx could be expected to exceed 100 l/s.

3.6.3 Supplementary geological investigation recommended in 6th consulting board meeting

The 6th Consulting Board Meeting, held in 1999, recommended that the characteristics of the Shihtsao Fault and the characteristics of the boundary between the hard Szeleng Sandstone and the soft Kankou Formation argillite should be verified through surface geological investigation, seismic refraction survey, ground resistivity imagery profiling and deep borehole drilling.

Two seismic line totaling 2,000 m were surveyed, and 1,710 m of resistivity imagery profiling were performed. The results indicated that the fault fractured zone for the Shihtsao Fault varied from 40 to 70 m with intercalated clay gouge 0.5 to 1.3 m in thickness. Tunnel excavation would encounter fractured rock mass estimated to be 230 m wide. It was also estimated that south of the fault zone was a concentrated fold zone. There were minor faults at the fold axis and minor fault fractures. The boundary between the Szeleng and Kankou Formations inclined at 22°, tunnel excavation would come across longer than expected Szeleng Sandstone rocks, the hard quartzite would extend for an additional 500–529 m toward the west. At the boundary there existed a fracture zone about 20–30 m wide.

4 SITE CHARACTERIZATION OVERVIEW

Through the major investigation works carried out in every stage described above, the project site were geologically characterized and are briefly summaried as follows:

4.1 *Remote sensing and photo interpretation study*

Remote sensing were performed in the early stage of the basic design stage, using the images from Landsat Thematic Mapper, panchromatic SPOT image and SLAR (Side Looking Airborne Radar) and aerial photos. The remote sensing study was aimed to assist identifying the adverse geological conditions along the tunnel alignment based on lineaments, drainage system, landslides etc. Major conclusions from the remote sensing study indicated that several prominent lineaments were identified in vicinity of the tunnel alignment. These lineaments are, however, not consistent with any of the known fault zones. Field checks indicated that the lineament at the southeast of the eastern portal of the tunnel, for example, seemed to run along the ridge and parallel to the trend of the Okinawa Trench and does not comply with the general structure trend. Also no any known faults seem to be in line with the identified lineaments and the most prominent lineaments are not consistent with any fault traces.

4.2 *Field mapping*

Field mapping is one of the most crucial works in the whole investigation program. Field mapping began in September 1989 and completed in February 1990. The field based studies included stratigraphic study and structural study and map of 1 to 5,000 scale was used as base map during field work.

4.2.1 *Stratigraphic study*
The mapping units established for the field investigation in the project area are Szeleng Sandstone, Kankou formation, Tsuku formation, Tatungshan Formation, Makang Formation and Fangchiao Formaion. Among them, the former 4 formation names were used following Janpanese geologist Prof. Ichino (1932) and the latter 2 were first used by Mr. Tang and Yang in 1976. It was revealed during the investigation that the lithology and thickness of each formations mentioned above are equivalent to those mapped in the nearby Feitsui Reservoir area. Major differences are:

1. The key fossil zones used to identify the Tsuku Formation-Turritalla and Amussiopecten were not found along the tunnel alignment.
2. In the area east of the tunnel alignment, the grain size of Tsuku sandstone become much finer so that it is hard to tell the difference between Tsuku and Kangkou Formation.
3. The Kangkou Formation in the Reservoir area contains an alternation of sandstone and siltstone for about 180 m in thickness. The alternation was not found in tunnel area.

4. In the southeastern area of the tunnel alignment, the Tsuku Formation and Kangkou Formation are massive with less bedding planes, which made the altitude measurement in field investigation much difficult.

During the field mapping, the following lithological characters were used to differentiate the rock formations:

1. The quartz sandstone, dark gray fine grained sandstone partly with tuffaceous material, and carbonaceous shale are the major characteristic of Szeleng Sandstone.
2. The dark gray siltstone and argillite are characteristic of Oligocene and Eocene formation in the northern Hsuehshan Range.
3. The gray, very fine grained sandstone with muddy flakes are characteristic of Tsuku Sandston which can be used to differentiate the Kankou Formation and Tatungshan Formation.
4. The light gray, fine to medium grained, massive sandstone are characteristic of Tatungshan, Makang and Fanchiao Formations.
5. Grayish coarse grained sandstone and coal bed are unique to the Fanchiao Formation.

During the mapping process, all the outcrops were carefully observed, examined, photographed, located and recorded for further analysis.

4.2.2 *Structure analysis*
Major structures including bedding, folding, fault and shear, and joint system were analyzed to show their spatial variation and understand their tectonic significance.

4.2.2.1 Bedding attitude
There are totally 743 bedding attitude data recorded in the 1 to 5,000 scale base map. Some of these data are shown in the 1 to 25000 scale map. All these data were analyzed with lower hemisphere stereographical projection method as shown in Fig. 3. It could be seen from the analysis that the poles to the bedding plane are distributed roughly along a great circle, the normal to which represents the striking of the folding axis with dipping direction of N 54° E and plunging 70°. The standard deviation is 26°.

4.2.2.2 Folding
Rock strata subject to tectonic stress may be deformed in different ways including syncline, anticline folding and fault. During the field mapping, the attitudes of rock strata were carefully mapped to understand if the rock strata are deformed or displaced. All the deformation data were mapped and marked in the base map. Some of the small scale faults were observed in the axial portion of folding system. Figure 4 is the major folding structures in the project area.

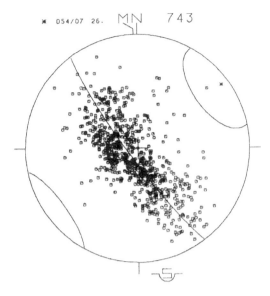

Figure 3. Distribution of bedding poles on a stereographic projection of project area.

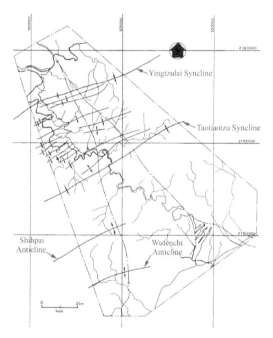

Figure 4. Locations of major foldings of project area.

4.2.2.3 Fault and shear

In the field mapping, faults with stratigraphic displacement exceeding 150 to 200 m, which can be shown in 1 to 25,000 scale map, are considered as major fault, and those with small stratigraphic displacement and having difficulty adequately shown in

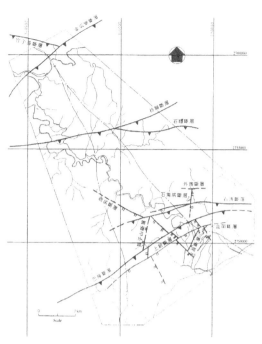

Figure 5. Locations of major fault structures of project area.

the map are called secondary faults. Those with only shear phenomenon or gouge but did not show significant displacement are called shear. From the mapping, there are 13 major faults in the mapped area and 7 of them passing through the Hsuehshan tunnel alignment (Fig. 5). The secondary faults and shears were found in 55 locations, some of which were observed directly in the outcrops or some were inferred when preparing the final draft map.

4.2.2.4 Joint

There are 51 locations out of the entire mapped area were chosen to map in detail all the joint systems. The survey points are totally 2728. All these data are analyzed statistically.

4.2.3 *Zoning of structure homogeneous area*

In the study of the project area, the concept of structural homogeneous area was adopted to characterize the condition of the geological condition of the tunnel area. The structural homogeneous areas (SHA) are defined as limited area with similar variation of bedding attitude. Areas of one limb of major fold, for example, having unique structure characters can be classified as a SHA of first order. Some of small folds may be found in the SHA and can be considered as secondary order. In this study only SHA of first order were considered. In view of the quantities of the structural data collected from the field, 12 SHA of first order were differentiated in study area as shown in Fig. 6.

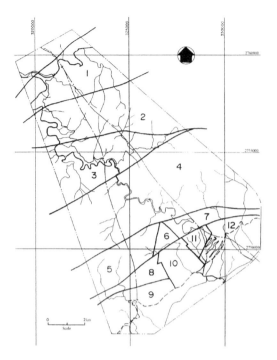

Figure 6. Zoning based structure homogeneity of project area.

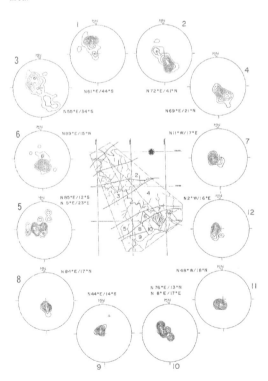

Figure 7. Bedding attitude analysis of project area.

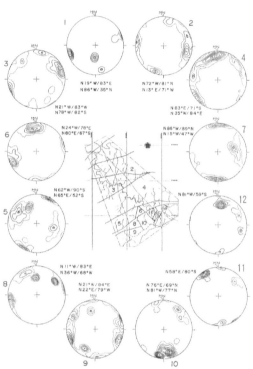

Figure 8. Joint orientation analysis of project area.

Bedding and joint were statistically analyzed to show the spatial variation and to understand their impact on the tunnel engineering. Figure 7 is the spatial variation of bedding attitude and Fig. 8 is the spatial variation of joint system.

4.3 Deep borehole drilling

There are three ventilation shafts along the Hsuehshan Tunnel, overburden at these ventilation shafts were as high as 270–460 m. Among them, the shaft No. 3 is then the deepest and needed to be explored by deep boring. However, the depth of exploratory borehole seldom exceeded 200 m for previous site investigation in Taiwan.

Deep borehole PH-29 for vertical shaft No. 3 encountered poor geological conditions, the underlying rock mass was intensely fractured, hard and abrasive. One local drilling company failed to complete the tough task. This might already indicate the future difficulty in vertical shaft excavation. A new borehole PH29-1, 490 m deep, was then re-sited. The contract was awarded to an international contractor. The borehole was drilled with a 24-hour non-stop schedule. It was finished in 102 days and the depth reached 496 m. The net boring rate amounted to 25–30 m per day. Furthermore, this operation achieved a recovery rate of 90%. The reasons for such good

166

Figure 9. Stylolite in the vertical drill core of Szeleng sandatone.

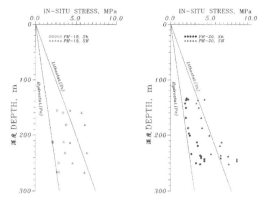

Figure 10. Correlation between in-situ stress magnitude and depth.

performance were mainly due to the contractor's excellent professionalism and management.

4.4 Hydrofracturing tests for in-situ stresses

Available geologic information indicated that the major geologic structure of Taiwan was formed through a mid-Tertiary orogenic movement. The Philippine sea plate moved in a SE-NW direction causing a series of thrust faults and fold structures. The axis of these thrusts and folds are parallel to the main axis of Taiwan. The northern part of Taiwan including the location of the present project is located within the compressive zone. The major stress axis should coincide with the compressional direction and ran parallel to the strike of the tunnel alignment. This stress axis gradually rotated consequent to opening and spreading of the Okinawa Trench, and normal faults were formed. Some stylolite shown in the vertical drill cores of Szeleng sandstone was observed during deep hole drilling, which implied that parts of the rocks in the northern Hsuehshan ranges was once subjected to higher vertical stress (Figure 9). The strike of this spreading was NW-SE.

Hence it could be concluded that the stress conditions along the tunnel alignment are highly complicated. Furthermore, overburden along the tunnel alignment varied in thickness from place to place, thus parts of the tunnel alignment might encounter very high in-situ stresses. To evaluate the regional tectonic stresses in the vicinity of the tunnel alignment, hydraulic fracturing test for in-situ stresses measurements were performed in deep boreholes PH-19 and PH-20, both of which were bored to tunnel invert elevation. In borehole PH-19 tests were at depths 155.9 m–258.9 m. In this test, hydraulic fracturing were conducted 12 times, fracture patterns duplicated 8

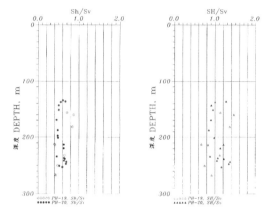

Figure 11. Correlation between stress ratio and depth.

times. In borehole PH-20 tests were at depths 134.1–246.9 m. In this test, hydraulic fracturing were conducted 20 times, fracture patterns duplicated 12 times.

The results of measurements indicated that the vertical stress (S_V) for overburden fell between the maximum horizontal stress (S_H) and minimum stress (S_h) (Figure 10), the maximum stress ratio ($K_H = S_H/S_V$) was 0.8–1.3, the minimum stress ratio ($K_h = S_h/S_V$) was 0.5–1.7 (Figure 11); the direction of maximum stress was NE-SW, approximately normal to the strike of the tunnel alignment. The measured results agreed well with current researches on geotectonics.

4.5 Geomechanical characteristics

The rock formations along the tunnel alignment comprised sandstone, quartz sandstone, siltstone and argillite. The rock mechanical tests included general rock physical properties, uniaxial compressive strength tests, triaxial compressive strength tests, static elasticity tests for rock cores, dynamic elasticity tests

Table 2. Index properties of rock materials.

Rock materials	Stratum	Unit weight $r_t(t/m^3)$	Water content $W_n(\%)$	Special gravity	Void ration (10^{-2})	Absorption Abs(%)
Argillaceous	FC	2.70	1.4	2.75	3.2	2.21
Rocks	MK	2.68	1.3	2.7	3.0	1.9
(ST)	TTS	2.68	1.2–1.7	2.72–2.75	3.0–4.1	2.0–2.9
	KK	2.70	1.3	2.74–2.75	3.0–3.4	1.8–1.91
	SL	2.55–2.69	1.6–3.1	2.68–2.73	4.0–8.4	1.7–6.74
Arenaceous	FC	2.64	1.1	2.71	3.0	1.85
Rocks	MK	2.68	–			
(SS)	TTS	2.68	1.4	2.73	3.7	2.84
	SL	2.57–2.68	0.6–3.7	2.71–2.75	1.0–11.2	1.10–6.17
Quartzitic Rocks(QTZ)	SL	2.55–2.67	0.3–0.7	2.64–2.68	1.0–5.6	0.5–2.20

Table 3. Average uniaxial compressive strength of rock.

Rock materials	Stratum	Monotonic uniaxial compressive strength Uc kg/cm²	Cyclic uniaxial compressive strength U_c cyclic kg/cm²	Estimated from point load test $U_c = 22Is$ (50)kg/cm²
Argillaceous	FC	464	551	342
Rocks	MK	750	550	618
(ST)	TTS	210–410	477–778	393–634
	KK	340–463	538–561	385–394
	SL	137–650	93–660	715–748
Arenaceous	FC	1556	1617	1226
Rocks	TTS	452–620	766–855	818
(SS)	KK	800		
	SL	53	101	
Quartzitic Rocks(QTZ)	SL	986–1670	1264–1775	1287–2394

for rocks, point load tests, Talber abrasive tests, petrographic analysis, X-ray analysis tests, general physical property tests for clay seams, Schmidt hardness tests, swelling tests, slaking tests and creep tests. These test results were used in synthetic study to derive the geomechanical properties along the tunnel alignment that included general properties of rocks, strengths of rocks, and rock deformability.

The characteristics of rock formations occurring along the tunnel alignment are as follows: Relevant to siltstone-type of rocks quartzitic rocks and sandstone-type of rocks are generalized in having higher specific gravity, lower porosity and lower water absorption (Table 2). Attention was especially paid to the average quartz content of 82% in the Szeleng sandstone. Some samples of the quartz sandstones showed that the quartz content can be high up to 99%. The total hardness for the Szeleng sandstone, among others, is ranging between 99 and 157, which is statistically close to the total hardness value for similar rocks. All

rock materials are mostly medium to high durable materials with high resistance to corrosion but little expansion when water is encountered. Long term cumulative creep deformation is generally smaller than stress failure under load.

Statistical uniaxial compressive strengths for rocks is listed in Table 3. Quartzite in the project area reached strength of 2400 kg/cm². Other well-cemented sandstones such as sandstones of the Fangchiao Formation may also be as high as 1600 kg/cm². For siltstones, the strength averaged 400–600 kg/cm².

Synthesis of laboratory results from elasticity tests and in-situ down-hole deformation tests gave the mean elasticity moduli for rocks along the tunnel alignment as shown in Table 4. According to material classification by Deere and Miller (1996), materials along the tunnel alignment belonged to low to high strength with medium to low modular ratios. Some argillite showed high modular ratios with low strength characteristics. The quartzite and sandstone from the Fangchiao

Table 4. Average modulus of elasticity of rock.

Rock materials	Stratum	Static elastic modulus $E_s * 10^4 kg/cm^2$	Static elastic modulus $E_s * 10^4 kg/cm^2$	E_s/E_d
Argillaceous	FC	17.6	27.8	0.63
Rocks	MK	14.4	–	–
(ST)	TTS	6.2–11.2	23.9	0.36
	KK	8.5–32.7	34.6	0.6
	SL	4.2	12.95	0.32
Arenaceous	FC	34.5	29.9	1.15
Rocks	TTS	8.2–21.0	20.3	0.72
(SS)	SL	1.5	9.37	0.16
Quartzitic Rocks(QTZ)	SL	22–30.8	–	

Formation generally showed low to medium modular ratios.

4.6 Establishing geological model

The investigation results were subjected to evaluation and estimation for establishing a geological model for the area along the tunnel alignment. Plan geologic map and profile cross sections of tunnel alignment were prepared (Figures 12, 13). The Hsuehshan Tunnel passed through a series of slightly metamorphosed folded and faulted Tertiary sedimentary rocks. Regional folds thrust faults, strike slip faults and normal faults constituted complicated geologic structures. The geologic characteristics are summarized as follows:

In an old to young ascending order of Hsuehshan Tunnel rock formations are: the Eocene Szeleng Sandstone (SL), the Oligocene Kankou Formation (KK), Tsuku Sandstone (TSK), Tatungshan Formation (TTS), and the Miocene Makang (MK) and Fangchiao Formations (FC), all belonging to the Hsuehshan Range geologic subprovince. These rock formations had been subjected to vigorous orogenic movements, and were deformed by folding and cut by faults. These rock formations thus are not continuous and the thickness of the formations are not complete. The Szeleng Sandstone is a coarse to fine-grained quartzitic sandstone with a uniaxial compressive strength over 3,000 kg/cm². It distributes 4 km length on the eastern half of the tunnel. For the remainder of the rock formations their uniaxial compressive strengths were mostly between 500 and 800 kg/cm².

Major geologic structures along the tunnel alignment included two regional fold structures and 7 local fold structures. As for faults, in an east to west direction, they are: the Chinying, Shanghsin, Palin, Tachinmian, South Branch of the Shihpai Fault, Northern Branch of the Shihpai Fault and the Shihtsao Fault. All geologic structures had been verified through

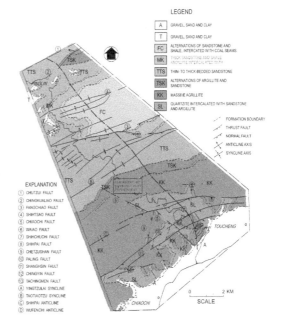

Figure 12. Geological plan map of Hsuehshan tunnel area.

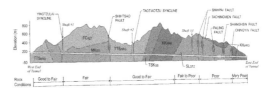

Figure 13. Geological profile along the Hsuehshan tunnel.

borehole drilling during design stage, and their spatial relationship with the tunnel had been clarified using available data. Except the Shihtsao Fault, being a thrust fault, all of the faults are normal faults and occur

169

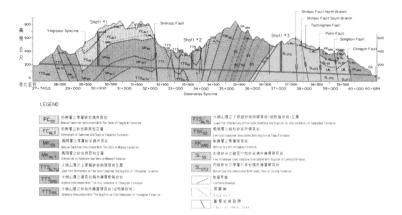

Figure 14.   The actual geological profile along Hsuehshan tunnel.

within a 4 km area on the eastern half of the tunnel alignment.

Basing on the lineament patterns, density of faults and rock mass quality from boreholes, it was estimated that for the 3/4 length of the western half of the tunnel alignment the geological conditions are similar: the rock mass quality fall between fair and good with a westward trend of improving; for the remaining 1/4 length on the eastern half the rock mass quality is poor. Especially within 1 km on the eastern side of the tunnel alignment, the rock mass is loose and fractured with very poor rock mass quality.

The pilot tunnel was designed for purpose of draining the groundwater during construction of the main tunnel. The possible total groundwater seepage from the pilot tunnel had been roughly estimated an accumulated of 3.5 cms. In fact, the thickness of overburden of about 7.5 km in length along the tunnel alignment exceeds 300 m. It was very difficult to reasonably estimate the groundwater inflow in the main tunnels. In the 3 km of the eastern portion of the tunnel alignment, the groundwater inflow varied greatly from section to section. And the groundwater inflow came out mainly from broken rock formation around faults and shear zones. The instantaneous influx could not be estimated due to presence of too many control factors.

4.7   Actual geological conditions compared

The pilot tunnel was designed for purpose of geological exploration. Detailed information concerning lithology of rock formations along the tunnel alignment, the location, extent and characteristics of geological structures and shear fracture zones were unravelled through the completion of pilot tunnel. The actual geology was recorded in detail as cross section profiles (Figure 14). The actual geological profile was found to be in close consistence with the geological framework obtained in the design stages. Lithologies and Geotechnical characteristics of rock formations along the tunnel alignment were also consistent with those estimated properties used in the planning and design stages. Locations of major faults coincided well also (Table 5). The prediction of poor rock mass quality for the eastern portal section is also relatively correct.

4.8   Geologic logging during tunnel construction

The pilot tunnel was designed for geologic information collecting. Data collected from the pilot tunnel was analyzed and processed, then constituted the main basis for deriving the geologic conditions of the main tunnel. These geologic informations were used in construction of the main tunnel. They included the location, occurrence, attitude and extent of incompetent zones likely to be encountered in main tunnel construction. These were then plotted and formed geologic plan maps and cross section profiles of the main tunnel. The primary aim of the pilot tunnel geologic investigation was features of significant scale. Joints, beddings, small shears or fractures, and minor quantity of groundwater seepage were omitted since these were common occurrences in any TBM excavated tunnel. Large-scale or extensive shear or fracture zones, faults and structural weak zones that might affect construction operation were investigated. Their location, extent and attitude were identified for future reference in construction operation.

The TBM selected to excavate the Hsuehshan Tunnel was a double-shield type. The choice was based on its speedy excavation and safe working environment. Nevertheless, this was highly inconvenient from the point of view of geological investigation. A double shield TBM only allowed geologic observation from a limited space at the wall and at a very short time

170

Table 5. Comparison of the measured and predicted fault system along pilot tunnel.

Major structure		Location	Fault width (m)	Width of disturbance zone (m)	Attitude
Shihtsao Fault	Predicted	33 + 250	10		N90E/80S
	Measured	33 + 260	20	40	N74E/80S
Shihpai Fault-north	Predicted	37 + 750	20–30		N75E/80S
branch	Measured	37 + 756	16	28	N80W/80S
Shihpai Fault-south	Predicted	37 + 900	10–20		N47E/80S
branch	Measured	38 + 150	8	14	N25E/77S
Paling Fault	Predicted	38 + 650	30		N40-70E/80S
	Measured	38 + 680	6	20	N85E/78S
Shanghsin Fault	Predicted	39 + 250	10		N60E/80S
	Measured	39 + 316	6	5	N50E/50S
Chingyin Fault	Predicted	39 + 700	20		N30E/70S
	Measured	39 + 816	7	11	N20E/70S
Extent of Szeleng	Predicted	36 + 400 ~ 39 + 650			
sandstone	Measured	36 + 145 ~ 39 + 816			

before supportive segment installation, hence it was not possible to record in detail the features revealed by TBM excavation. Although the collected data was less than desired, it contributed in certain limited manner towards TBM tunnel construction. The final actual excavation revealed that the pilot tunnel had achieved a preview on the geological conditions, and when large quantity of groundwater was encountered, the pilot tunnel allowed in time draining and lowering of the groundwater.

4.9 Groundwater dating and analysis

During tunnel construction groundwater from influx was analyzed that was also dated using radioactive isotopes to clarify the source of this large quantity of groundwater. Radioactive dating using hydrogen isotopes deuterium and tritium would reveal the age of the groundwater. This age of the groundwater was the time from the date the groundwater entered into the saturate zone in the groundwater column. It would also reflect the rate of groundwater circulation and its relationship with meteoric water and surface water. Isotopes C14 and H3 were used in dating and deriving the age of the groundwater that would in turn reveal the origin of the groundwater. The older the age of the groundwater the poorer the groundwater circulation was. Dating through H3 would reveal whether young, recent water had been mixed into the groundwater. If tritium was detected in the groundwater, it would indicate that the groundwater was a mixture of older groundwater and younger groundwater.

Twenty groundwater samples collected from various points of groundwater influx at the eastern end of the tunnel were used in the analysis. Except four samples showing analytical values smaller than 1TU, the concentration of tritium for the remainder of the samples was between 1–3.2TU, a value very close to

tritium concentration in rainwater in northern Taiwan for recent years. This analytical result showed that the groundwater possessed atomic explosion signature, and thus it contained recharge water after 1953. The H3 signal TU > 1 on groundwater at the eastern portal and a groundwater age of <6000 years revealed that surface water had seeped into the tunnel by ways of fracture zones. At the west portal, groundwater influx did not show signal of recent surface water influx.

5 ENGINEERING CONSIDERATIONS

Geologic investigations revealed that geological condition of the Hsuehshan Tunnel was very complicated. Engineering difficulties included encountering six major faults and a bulk of fractured, hard quartzite and large quantity of groundwater. Most of the tunnel alignment is under overburden more than 300 m. It was not possible to explore fully geologic conditions at tunnel elevation through deep boreholes. To solve this problem, construction of a pilot tunnel before the main tunnel is preferable.

The major drawback for a pilot tunnel is the increased engineering cost. On the other hand, through information provided by the pilot tunnel certain pre-treatments could be taken and thus reduce engineering risks that would ultimately result in actual saving in engineering costs. In Hsuehshan Tunnel project the final decision was to excavate a 5-m diameter pilot tunnel between the two tubes of the main tunnel with an elevation slightly lower than that of the main tunnel.

In fact, the pilot tunnel provided valuable detailed geological information for construction of main tunnels. Meanwhile, it was also used, for several times as supplementary access to the main tunnels during construction. This really helped to improve construction progress.

In Hsuehshan Tunnel project, TBM was adopted as the major excavation method, according to the original construction plan. However, due to the extremely poor ground condition at the eastern part of the tunnel, eventually, mixed excavation method was used to complete tunnel excavation. For example, a section of 1.2 km of pilot tunnel was cut by D&B method to overcome the difficult ground conditions in the Szeleng Sandstone formation.

6 CONCLUSION

6.1 A tremendous time and effort have been put into characterize the site condition of the Hsuehshan tunnel. Exploration works done to understand the ground condition included satellite image and photo interpretation, field mapping, geophysical sounding, borehole drilling, in-situ and lab rock mechanics tests, hydrogeological analysis and on-going exploration during construction. The exploration indicated that the rock formations in the Hsuehshan Range comprise folded Tertiary sedimentary sequence, and belong in the fold-and-thrust belt in the active orogenic belt. Most geological structures, in the forms of regional folds and thrust faults, strike generally in an NEE-SWW direction. Normal faults, however, were also formed locally as result of later tectonic activities. Following the major phase of orogenic movement, back-arc basin spreading occurred in the sea outside of Ilan. This on-going spreading of the back-arc basin resulted in normal faulting in areas in the immediate vicinity of Ilan. The Hsuehshan Tunnel is located right in the centre of these geologic complexities. Rock formations penetrated by the tunnel are: in an east to west direction, the Szeleng Sandstone, the Kankou, Tatungshan, Makang, and the Fangchiao Formations. About one quarter from the eastern end of the tunnel is comprised mainly of argillite and quartz sandstone. In this section, due to the dense occurrence of faults, the rock mass are intensely fractured. In the northern three quarters of the tunnel, the main lithologies comprise sandstone, shale, and alternation of sandstone and shale, and the rock mass are massive with good rock quality.

6.2 Due to topographic inaccessibility and exploration technique constrain, the geological exploration for deep-seated long tunnel can only be limited to certain area and with the data thus obtained to develop the rational model along tunnel alignment. In an area with geological condition so complicated as in Taiwan, it is extremely difficult to have the measured geological condition exactly the same as the predicted. For the Hsuehshan tunnel, the comparison between the actual ground condition and the predicted indicated that there is high degree of consistency in terms of lithology, tectonics, rockmass quality and area of potential difficult geology. This implies that the engineering geological assessment proposed in planning and design stages for Hsuehshan tunnel is acceptable in terms of alleviating the possibility of encountering geological surprise during tunneling.

6.3 Although the degree of consistency of the geological model is proved to be acceptable based on comparison with the actually excavated ground condition, there are still some blind spots, which are hardly detected from the surface beforehand. The blind spots, i.e. confined water pockets, that exist between the major fault zones might only be detected by probing during construction. Hence, probing, either with long horizontal drilling or geological pilot tunnel, is a must for a long and deep seated tunnel project in an area with such a complicated tectonic environment as Taiwan.

REFERENCES

Sinotech Engineering Consultant Inc., 1990, Final Report of Geological Exploration of Pinglin Tunnel, Route Selection Stage for Nankang-Ilan Freeway Project.

Chang, W.C. 1991, Site Investigation Technique of Taipei-Ilan Expressway, Modern Construction.

Taiwan Provincial Highway Bureau, 1984, Geological Assessment and Rock Mechanics Test Report of Tunnels between Nankang- Toucheng Highway.

TANEEB, 1989, Geological Exploration Report of Pinglin Tunnel, Nankang-Ilan Expressway Project, Route Selection Stage.

TANEEB, 1991, Geological Exploration Report of Pinglin Tunnel, Taipei-Ilan Expressway Project, Basic Design Stage.

TANEEB, 1993, Final Report on Geological Exploration of Pinglin Tunnel of Taipei-Ilan Expressway Project, Detail Design Stage.

TANEEB, 1997, Water Inflow Study of Pinglin Tunnel of Taipei-Ilan Expressway Project, Construction Stage.

TANEEB, 2000, Supplementary Geological Exploration of Pinglin Tunnel of Taipei-Ilan Expressway Project, Construction Stage.

TANEEB, 2004, Final Report on Geology of Hsuehshan Pilot Tunnel of Taipei-Ilan Expressway Project, Construction Stage.

Chi-Tso Chang, Ping-cheng Hou, Chih-Shae Liu, Ting-Huai Hsiao, 2005, Geological and Geotechnical Overview of the Hsuehshan Tunnel.

Statistical analysis of geotechnical data

M. Uzielli
Georisk Engineering, Florence, Italy

ABSTRACT: This paper attempts to provide an overview of the main domains of application of statistical-based techniques to geotechnical data. Examples and references to literature contributions are provided.

1 INTRODUCTION

1.1 *Uncertainty, variability and determinism*

The geotechnical engineer processes testing data to obtain parameters for characterization and design. In practice, *information is never sufficient in quantity, nor entirely precise and accurate*. Geomaterials, moreover, are *naturally complex and variable at all scales*, ranging from the microstructure to regional scale. The competent engineer must account for this lack of uniformity and information while parameterizing and modeling the physical world. The *level of explicitness* with which this occurs depends upon the selected approach. In *deterministic approaches*, variability is not addressed explicitly as in *uncertainty-based approaches*.

In the technical literature – and geotechnical engineering is no exception – the terms *variability* and *uncertainty* are often employed interchangeably. Strictly speaking, this is not correct. *Variability* is an observable manifestation of heterogeneity of one or more physical parameters and/or processes. *Uncertainty* pertains to the modeler's state of knowledge and strategy, and reflects the decision to recognize and address the observed variability in a qualitative or quantitative manner.

Deterministic methods lie at the basis of virtually every technological science, and geotechnical engineering is no exception. However, the importance of explicitly modeling and assessing the variability of geotechnical parameters (i.e. quantifying, processing and reporting the associated uncertainty) is increasingly recognized in geotechnical design and characterization. Most evolutionary design codes operate in an uncertainty-based perspective, requiring explicit quantification not only of most suitable values (usually termed 'characteristic' or 'nominal'), but also of the level of uncertainty and confidence in the selection of such values.

The progressive shift towards an uncertainty-based perspective may be motivated by the fact that this may be, on the whole more *convenient* in terms of *safety*, *performance* and *costs*. The explicit parameterization of uncertainty allows to provide more *complete* and *realistic* information regarding the level of risk associated with design. Addressing uncertainty does not *per se* increase the level of safety, but allows the engineer to rationally calibrate his decisions on a desired or required *reliability* or *performance level* of a geotechnical system. Being able to select the performance level and reduce undesired conservatism, in turn, is generally beneficial in the economic sense.

Among the main trade-offs for the positive aspects of uncertainty-based approaches, which hinder a more rapid diffusion among geotechnical practitioners, are the necessity to rely on specialized mathematical techniques and, not infrequently, large computational expense. While ever-increasing computational power is constantly reducing the relevance of the latter, a correct implementation of uncertainty-based techniques requires at least some degree of comprehension on the side of the engineer. The results of uncertainty-based analyses can be used confidently for engineering purposes only if preceded, accompanied and followed by geotechnical expertise and judgment.

1.2 *Rationale and scope of the paper*

Among mathematical disciplines which allow consistent modeling, processing, evaluation and assessment of uncertainty, *statistical theory* (usually employed jointly and iteratively with probability theory) provides a well developed, widely understood and accepted framework. Statistical theory encompasses a broad range of topics. A notable advantage of statistics over other uncertainty-addressing techniques such (e.g. fuzzy logic) is – at present – the vast bulk of statistical software packages which are available.

The main goal of the paper is to provide a wide – though necessarily synthetic – overview of the main domains of applications of statistical analysis of geotechnical data for practical implementation in uncertainty-based characterization and design. This is attempted herein through:

a) Description of the main factors of variability in geotechnical testing data, and definition of the sources of geotechnical uncertainty;
b) Description of the domains of application of statistical analyses for geotechnical purposes;
c) Presentation of selected examples from the geotechnical literature.

Statistical techniques are not addressed from a purely theoretical perspective. Let alone the specialized literature, a number of textbooks are available for a more comprehensive insight into the formal aspects of statistical science in the context of geotechnical engineering: Ang & Tang (1975); Ayyub & McCuen (2003); Baecher & Christian (2003) among others.

It is generally deemed preferable not to 'lose sight of the forest for the trees', i.e. to focus on simpler, more easily applicable methods rather than better-performing techniques which are hardly utilizable in geotechnical practice. The main reason is that there are large, at present unavoidable uncertainties in geotechnical parameters and calculation models which can generally be expected to exceed the potentially existing bias and error resulting from a less refined analysis. In any case, it is attempted to emphasize that *one should never apply even the most simple techniques uncritically*.

This paper focuses on statistical approaches based on the *frequentist* perspective. The rapidly increasing use of Bayesian statistics (which are not addressed here) in geotechnical engineering attests for the advantage of exploiting the dual nature of probability in an integrated, rigorous manner. Interested readers are referred to Ayyub & McCuen (2003), Baecher & Christian (2003) and Christian (2004) for a comparative insight between frequentist and Bayesian statistics in the engineering disciplines.

Many of the procedures which have been developed by geotechnical researchers and practitioners for uncertainty-based analysis make use of statistics. However, they often rely on the synergic integration of statistical methods with other techniques. Hence, the attribute of 'statistical' should not be assessed too rigorously in this paper.

2 VARIABILITY AND UNCERTAINTY IN GEOTECHNICAL TESTING DATA

The variability in geotechnical data can be ascribed to both the soils and the investigators. Soils are natural materials, which are formed and continuously modified by complex processes, as discussed in detail by Hight & Leroueil (2003). The variety and complexity of such processes result in physical heterogeneity and, consequently, in the variability of quantitative parameters. *Inherent soil variability* describes the variation of properties from one spatial location to another inside a soil mass. Inherent soil variability is parameterized by *aleatory uncertainty*, which can be observed at virtually any scale at which properties are measured, and which is not necessarily reduced by increasing the numerosity and quality of data. *Epistemic uncertainty* exists as a consequence of the investigator's invariably limited information and imperfect measurement and modeling capabilities.

2.1 Shortcomings of total variability analyses

Uncertainty-based analyses can, in principle, neglect the compound nature of geotechnical uncertainty, and address *total variability* instead. However, there are at least three extremely important reasons for which this approach may not be desirable. The first reason is that analyses performed in the perspective of total variability are strictly site-specific. The strong influence of factors such as the specific testing equipment and personnel, soil type, depositional history and in-situ state parameters on the results of statistical analyses makes it virtually impossible to replicate the overall conditions at any other site (or, possibly, even at the same site at different times or with different equipment). Exporting results of total variability analyses to other sites uncritically generally results in incorrect assessments of uncertainty, and should be avoided as far as possible.

The second main shortcoming of total variability analyses is the overly conservative assessment of uncertainty due to the fact that the important effect of spatial averaging of variability, which is parameterized by a reduction of variance of aleatory uncertainty and which will be described in the paper, cannot be addressed. Total variability analysis thus hinders one of the main goals of uncertainty-based analysis, namely the reduction of excess conservatism in geotechnical characterization and design, or at least a rational assessment of the level of conservatism itself.

Third, a separate assessment of the magnitude of each uncertainty component allows, if desired, specific actions aimed at the reduction of total uncertainty. For instance, if it is seen that the highest contributor to total uncertainty is estimation uncertainty, then it could be decided that more data should be collected. If, on the other hand, measurement uncertainty were deemed too significant, supplementary campaigns making use of more repeatable testing methods could be planned.

2.2 How useful are literature values?

Examples and references to literature values of relevant output parameters of statistical analysis are provided in the paper. As the explicit categorization and separation of total uncertainty into aleatory and epistemic components (and sub-components thereof) is not frequently encountered in the geotechnical literature, most literature values should be regarded as deriving from total variability analyses.

On the basis of what has been stated in Section 2.1, while it is conceptually appropriate to view such data as a plausible range of values, the effects of *endogenous* factors (i.e. pertaining to the compositional characteristics of soils) and *exogenous factors* (e.g. related to in-situ conditions, groundwater level and geological history) which may influence the magnitudes of a given geotechnical parameter, should be inspected in the light of geotechnical knowledge if it is of interest to export literature values to other sites.

Hence, literature values should not be exported to other sites without a critical analysis of such endogenous and exogenous factors.

2.3 *Geotechnical uncertainty models*

If it is of interest to bypass the undesirable effects of total variability analyses, a rational assessment of the level of uncertainty in a measured or derived geotechnical parameter can be made by use of an *uncertainty model*. Phoon & Kulhawy (1999b), for instance, proposed an additive model for the total coefficient of variation of a point design parameter obtained from a single measured property using a transformation model:

$$COV_{tot,D}^2 = \frac{\delta}{L}COV_{\omega}^2 + COV_m^2 + COV_{se}^2 + COV_M^2 \quad (1)$$

in which $COV_{tot,D}$ is the total coefficient of variation of the design property; δ/L is an approximation of the *variance reduction* due to *spatial averaging* (see Section 4; δ is the scale of fluctuation of the design property (Section 4.2); L is the spatial extension of interest for design; COV_{ω} is the coefficient of variation of inherent variability of the measured property (Section 4.2); COV_{se} is the coefficient of variation of statistical estimation uncertainty (Section 9.1); COV_m is the coefficient of variation of measurement uncertainty of the measured property (Section 9.2); and COV_M is the coefficient of variation of transformation uncertainty of the transformation model (Section 9.3). The coefficient of variation as a measure of dispersion and uncertainty is defined in Section 3.

In quantitative uncertainty models, the components of uncertainty are usually assumed to be statistically uncorrelated; hence, the absence of correlation terms in Eq. (1). This is only approximately true: for instance, the quantification of variability requires data

from measurements of soil properties of interest. Different geotechnical measurement methods, whether performed in the laboratory or in–situ, will generally induce different failure modes in a volume of soil. This usually results in different values of the same measured property. Measured data are consequently related to test-specific failure modes. The type and magnitude of variability cannot thus be regarded as being inherent properties of a soil volume, but are related to the type of measurement technique. Scatter due to small-scale but real variations in a measured soil property is at times mistakenly attributed to measurement error. Given the above, it is extremely difficult to separate the components of geotechnical uncertainty. The hypothesis of uncorrelated uncertainty sources, though approximate, is especially important as it justifies separate treatment of the variability components, thereby allowing the application of the most suitable techniques for each.

Sections 3 through 8 address a number of domains of application of statistical methods for geotechnical characterization and design. The procedures presented therein do not address epistemic uncertainty explicitly. Hence, one should keep in mind, whenever pertinent, that a full quantification of uncertainty requires the adoption of an uncertainty model such as the one in Eq. (1), and the quantitative parameterization of epistemic uncertainties, for instance as described in Section 9.

3 SECOND-MOMENT STATISTICAL ANALYSIS

If a data set pertaining to a specific parameter is addressed statistically, such parameter can be referred to as a *random variable*. The term *sample statistic* refers to any mathematical function of a data sample. An infinite number of sample statistics may be calculated from any given data set. For most engineering purposes, sample statistics are more useful than the comprehensive *frequency distribution* (as given by frequency histograms, for instance). For geotechnical data, given the typically limited size of samples, it is usually sufficient to refer to *second-moment analysis*. In second-moment approaches, the uncertainty in a random variable can be investigated through its first two moments, i.e. the mean (a *central tendency* parameter) and variance (a *dispersion* parameter). Higher-moment statistics such as skewness and kurtosis are not addressed.

The *sample mean*, i.e. the mean of a sample $\xi_1, \ldots, \xi_n$ of a random variable Ξ is given by

$$m_\xi = \frac{1}{n}\sum_{i=1}^{n}\xi_i \quad (2)$$

Table 1. Second-moment sample statistics of Szeged soils (adapted from Réthàti 1988).

	Soil type	w (%)	w_L (%)	w_P (%)	I_P (%)	I_C	e	S_r	γ (kN/m^3)
Mean	S1	31.1	44.5	24.2	20.6	0.62	0.866	0.917	18.8
	S2	22.3	32.3	19.4	13.0	0.82	0.674	0.871	19.7
	S3	24.5	32.2	20.7	11.5	0.63	0.697	0.902	19.7
	S4	28.3	54.0	24.2	29.9	0.86	0.821	0.901	19.3
	S5	28.6	52.8	25.4	27.4	0.82	0.823	0.905	19.2
COV	S1	0.30	0.39	0.34	0.57	0.49	0.26	0.10	0.07
	S2	0.18	0.13	0.12	0.37	0.37	0.13	0.12	0.04
	S3	0.15	0.13	0.11	0.36	0.47	0.13	0.11	0.04
	S4	0.17	0.21	0.14	0.34	0.20	0.11	0.10	0.03
	S5	0.17	0.27	0.15	0.44	0.32	0.12	0.10	0.03

The *sample variance* of a set of data is the square of the *sample standard deviation* of the set itself. The latter is given by

$$s_\xi = \sqrt{\frac{1}{n-1}\sum_{i=1}^{n}\left(\xi_i - m_\xi\right)^2} \qquad (3)$$

The *sample coefficient of variation* is obtained by dividing the sample standard deviation by the sample mean. It provides a concise measure of the relative dispersion of data around the central tendency estimator:

$$COV_\xi = \frac{s_\xi}{m_\xi} \qquad (4)$$

The coefficient of variation is frequently used in variability analyses. It is dimensionless and in most cases provides a more physically meaningful measure of dispersion relative to the mean. However, it should be noted that it is a ratio which is very sensitive, for instance in case of small mean values. Harr (1987) provided a 'rule of thumb' by which coefficients of variation below 0.10 are considered to be 'low', between 0.15 and 0.30 'moderate', and greater than 0.30, 'high'.

Though the coefficient of variation is defined as a statistic and, in principle, is to be calculated using Eq. (4), it is paramount to acknowledge the multi-faceted nature of probability: the 'frequentist' perspective relies on *objective* calculation of sample statistics as described previously; the 'degree of belief' perspective requires subjective quantification of input parameters on the basis of experience, belief or judgment. Hence, as shall be discussed in the following, second-moment statistics can be quantified using either perspective.

3.1 Second-moment analysis of independent random variables

Parameters such as water content, plasticity index and unit weight are in principle not univocally related to specific in-situ conditions. For instance, they are not affected by the in-situ stress state and, thus, do not always show a trend with depth. These can be defined as *independent random variables* for the purpose of statistical analysis.

As a best-practice literature example of second-moment analysis independent random variables, the results of an investigation on soil physical characteristics (see Réthàti 1988) for 5 soil types underlying the town of Szeged (S1: humous-organic silty clay; S2: infusion loess, above groundwater level; S3: infusion loess, below groundwater level; S4: clay; S5: clay-silt) using the results of approximately 11000 tests from 2600 samples are shown. Mean values and COVs of water content w, plasticity index I_P, consistency index I_C, void ratio e, degree of saturation S_r, unit weight γ are reported in Table 1. The subdivision of data by soil type and quality of samples, along with the considerable size of the data samples, ensures a greater significance of results.

3.2 Second-moment analysis of dependent random variables

The magnitude and variability of many geotechnical parameters is bound to other parameters. For instance, the undrained shear strength of a cohesive soil generally varies with depth (even if the soil is compositionally homogeneous) due to increasing overburden stress, overconsolidation effects and other in-situ factors. These parameters, when addressed statistically, can be referred to as *dependent random variables*. Such variables generally display *spatial trends* which are due to one or more independent random variables. Trends can be complex and difficult to characterize analytically *a priori* as they are generally a superposition of trends from various independent variables. Recognizing and addressing such trends quantitatively is very important in geotechnical uncertainty-based analysis, because this allows to assess how much of the total spatial variability can be attributed to the independent variables which generate the spatial trend, and how much to the inherent variability of the dependent variable about the trend. *Decomposition* of data from samples of dependent random variables into a deterministic trend and random variation is often achieved using *statistical regression*.

Least-squares regression is widely used to estimate the parameters to be fit to a set of data and to characterize the statistical properties of the estimates. The main outputs of regression analysis are the *regression parameters* which describe the trend analytically, and the parameters which assess the reliability of the output regression model. Such parameters include the *determination coefficient* (commonly denoted R^2 as the square of the *correlation coefficient*), which provides an indication of how much of the variance of the dependent variable can be described by the independent variables which are included in the regression model and how much by the inherent variability, and the *standard error* of the regression model, which equals the standard deviation of the model's errors of prediction, and provides a measure of the uncertainty in the regression model itself. In a best-practice perspective it is very important to report these parameters explicitly along with the regression model (see Figure 2).

Least-squares can be implemented in several versions: *ordinary least-squares* (OLS) relies on the hypothesis of *homoscedasticity* (i.e. constant variance of the residuals) and does not assign weights to data points; generalized least-squares (GLS) relaxes the hypothesis of homoscedasticity in favor of an independent estimate of the variance of the residuals, and allows for weighing of data points as a consequence of the variance model. Today, regression is commonly performed using dedicated software. Far too often, however, regression is approached mechanically and uncritically; the hypotheses underlying specific regression methods are ignored and neglected, leading to incorrect regression models and biased assessment of the uncertainty associated with regression itself. The reader is referred, for instance, to Ayyub & McCuen (2003) for theoretical aspects of regression procedures.

The hypotheses and implications of decomposition are discussed in detail in Uzielli et al. (2006a). It is important to note that there is no univocally 'correct' trend to be identified, but rather a 'most suitable' one. The choice of the trend function must be consistent with the requirements of the mathematical techniques adopted, and must, more importantly, rely on geotechnical expertise. An example of decomposition of undrained shear strength values from consolidated anisotropic undrained triaxial compression tests on Troll marine clays (Uzielli et al. 2006b) is shown in Figure 1.

Another important application of decomposition is in the assignment of 'characteristic' design

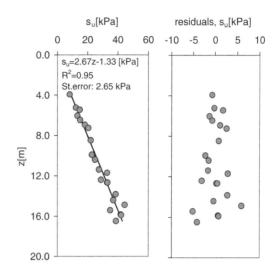

Figure 1. Decomposition of undrained shear strength data from CAUC testing in 2 homogeneous soil units of Troll marine clays: (a) linear trend; (b) residuals of detrending.

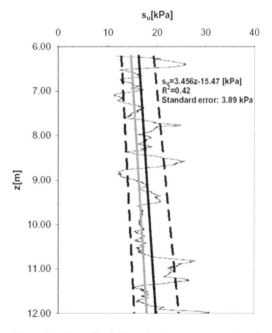

Figure 2. Generalized linear least-squares regression of undrained shear strength calculated from CPTU testing, and subjective design profile (data from Lacasse et al. 2007).

values for dependent random variables. For instance, characteristic values could be taken as the trend function minus a certain number of standard deviations (depending on the degree of desired conservatism). Such approach to the definition of characteristic

Table 2. Multiplicative coefficient for the estimation of the standard deviation of a normally distributed data set with known range (e.g. Snedecor & Cochran 1989).

n	N_n	n	N_n	n	N_n
		11	0.315	30	0.244
2	0.886	12	0.307	50	0.222
3	0.510	13	0.300	75	0.208
4	0.486	14	0.294	100	0.199
5	0.430	15	0.288	150	0.190
6	0.395	16	0.283	200	0.180
7	0.370	17	0.279		
8	0.351	18	0.275		
9	0.337	19	0.271		
10	0.325	20	0.268		

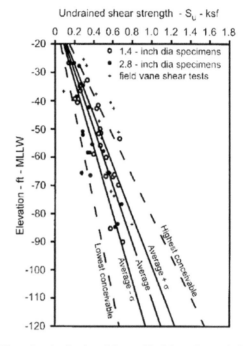

Figure 3. Application of the graphical three-sigma rule for the estimation of the standard deviation of undrained shear strength of San Francisco Bay mud (Duncan 2000).

values would reduce the degree of subjectivity in design. Lacasse et al. (2007) compared characteristic design values (previously assigned) of undrained shear strength with the trend obtained from statistical regression on the same data. Figure 2 illustrates an example of such analysis, in which the gray line represents the design values assigned subjectively and the black line is the result of linear generalized least squares regression. It was assessed that there are varying degrees of conservatism in subjectively assigned characteristic values.

3.3 Approximate estimation of sample second-moment statistics

It may be necessary to assign second-moment parameters in cases in which the complete data set is not available but other sample statistics are known. A number of techniques yielding approximate estimates of sample statistics have been proposed.

For data which can be expected (on the basis of previous knowledge) to be symmetric about its central value, the mean can be estimated as the average of the minimum and maximum values; hence, knowledge of the extreme values would be sufficient. If the range and sample size are known, and if the data can be expected to follow at least approximately a Gaussian (normal) distribution, the standard deviation can be estimated by Eq. (5) (e.g. Snedecor & Cochran 1989), in which the coefficient N_n depends on sample size as shown in Table 2:

$$s_\psi \approx N_n \left(\psi_{max} - \psi_{min} \right) \qquad (5)$$

3.3.1 Three-sigma rule

Dai & Wang (1992) stated that a plausible range of values of a property whose mean value and standard deviation are known can vary between the mean plus or minus three standard deviations. If it is of interest to assign a value to the standard deviation, the statement

can be inverted by asserting that the standard deviation can be taken as one sixth of a plausible range of values. The three-sigma rule does not require hypotheses about the distribution of the property of interest even though its origin relies on the normal distribution. In case of spatially ordered data, Duncan (2000) proposed the *graphical three-sigma rule method*, by which a spatially variable standard deviation can be estimated. To apply the method, it is sufficient to select (subjectively or on the basis of regression) a best-fit line through the data. Subsequently, the *minimum* and *maximum conceivable bounds lines* should be traced symmetrically to the average line. The *standard deviation lines* can then be identified as those ranging one-third of the distance between the best-fit line and the minimum and maximum lines. An application of the graphical three-sigma rule (Duncan 2000) is shown in Figure 3. While the three-sigma rule is simple to implement and allows exclusion of outliers, Baecher & Christian (2003) opined that the method may be significantly unconservative as persons will intuitively assign ranges which are excessively small, thus underestimating the standard deviation. Moreover, for the method to be applied with confidence, the expected distribution of the property should at least be symmetric around the mean, which is not verified in many cases.

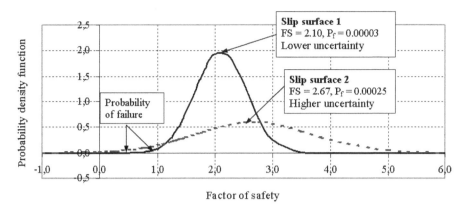

Figure 4. Comparative deterministic and probabilistic assessment of stability for 2 slip surfaces (Nadim & Lacasse 1999).

An exemplifying case which shows the importance of probabilistic approaches even at the simplest second-moment level is provided by Nadim & Lacasse (1999), who performed an undrained slope stability analysis on a slope consisting of 2 clay layers under static and seismic loading using Bishop's method. Analyses were performed both deterministically (i.e. evaluating the factor of safety) and probabilistically (i.e. estimating the probability of failure) using the First-Order Reliability Method (FORM). In the latter case, undrained strength, unit weight and model uncertainty (the latter consisting of a 'mean' model bias term and a COV of model dispersion) were modeled in the second-moment sense. Results showed that, due to the effect of spatial averaging phenomena, the critical surface with the lowest deterministic safety factor was not the critical surface with the highest probability of failure (Figure 4). This example attests for the importance of the additional information which can be provided by probabilistic approaches in comparison with methods which are solely deterministic.

4 SPATIAL VARIABILITY ANALYSIS

Second-moment statistics alone are unable to describe the spatial variation of soil properties, whether measured in the laboratory or in-situ. Two sets of measurements may have similar second-moment statistics and frequency distributions, but could display substantial differences in spatial distribution. Figure 5 provides a comparison of two-dimensional spatial distribution of a generic parameter ξ having similar second-moment statistics and distributions (i.e. histograms), but different degrees of spatial correlation: weak correlation (top right) and strong correlation (bottom right).

Knowledge of the spatial behavior of soil properties is thus very important in uncertainty-based geotechnical analysis and design, at least because: (a) geotechnical design is based on site characterization, which

objective is the description of the spatial variation of compositional and mechanical parameters of soils; (b) the values of many geotechnical parameters depend on in-situ state factors (e.g. stress level, overconsolidation ratio, etc.) which are related to spatial location; (c) quantification of the magnitude of spatial variability can contribute significantly to reducing the degree of conservatism in design by allowing the inclusion of the spatial averaging effect in geotechnical uncertainty models.

The goal of spatial variability analyses is to assess quantitatively how much and in which way a given property changes along pre-determined spatial orientations. The vast majority of spatial variability analyses rely on the calculation of spatial statistical properties of data sets. Higher-level analyses require the results of lower-level analyses, and allow consideration of additional effects and parameters which improve the quality of the results. The trade-off for the simplicity of lower-level analyses lies in the limited generality of results as well as in the imperfect modeling of the behavior of geotechnical systems. A detailed insight into such techniques, along with a comparative example of different levels of analysis for slope stability evaluation, is provided by Uzielli et al. (2006a).

4.1 Spatial averaging of variability

An implicit manifestation of spatial correlation which is commonly encountered in geotechnical engineering practice is that the representative value of any soil property depends on the volume concerned in the problem to be solved. With reference to a given soil unit and to a specific problem, the geotechnical engineer is trained to define the design values of relevant parameters on the basis of the magnitude of the volume of soil governing the design. Any laboratory or in-situ geotechnical measurement includes some degree of spatial averaging in practice, as tests

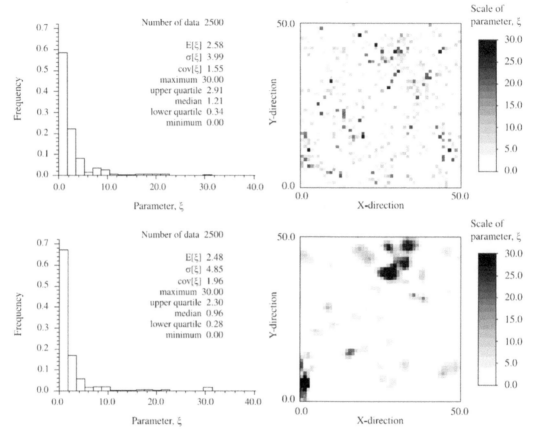

Figure 5. Comparative representation of spatial data with similar statistical distributions (top and bottom right) but different magnitudes of spatial correlation: weak correlation (top right) and strong correlation (bottom right) (from El-Ramly et al. 2002).

are never indicative of point properties, but, rather, are used to represent volumes of soil. The spatial averaging effect results in a reduction of the effect of spatial variability (and, hence, of excessive conservatism) on the computed performance because the variability (in statistical terms, variance) is averaged over a volume, and only the averaged contribution to the uncertainty is of interest as it is representative of the 'real' physical behavior.

4.2 Random field modeling of spatial variability

Among the available techniques to investigate spatial variability, *random field theory* has been most frequently referred to in the geotechnical literature. A *random field* is essentially a set of values which are associated to a one- or multi-dimensional space. Values in a random field are usually spatially correlated, meaning that spatially adjacent values can be expected (on the basis of statistical and probability theory) not

to differ as much as values that are further apart. It can be described (in the second-moment sense) by its mean, standard deviation (or coefficient of variation) and scale of fluctuation, as well as by an autocorrelation function, which describes the way in which a given property is correlated in space.

The scale of fluctuation is a concise indicator of the spatial extension of correlation. Within separation distances smaller than the scale of fluctuation, the deviations from the trend function are expected to show significant correlation. When the separation distance between two sample points exceeds the scale of fluctuation, it can be assumed that little correlation exists between the fluctuations in the measurements. Though a dependence on soil type has been noted (Uzielli et al. 2005a; 2005b), the scale of fluctuation is not an inherent property of a soil parameter and, when it exists, can be estimated using a variety of methods such as autocorrelation model fitting, calculation of the variance reduction function and semivariogram fitting. These methods, which basically rely on the statistical investigation of spatial correlation properties of the random field, are described in detail in Uzielli et al. (2006a). The scale of fluctuation is also useful to approximate the variance reduction function as shown in Eq (1).

In random field analysis of dependent random variables, it is not conceptually correct to estimate the coefficient of variation on the raw data set as the presence of a spatial trend, (for instance due to effective overburden stress or overconsolidation) would introduce bias in such estimate. Phoon & Kulhawy (1999a) defined the *coefficient of variation of inherent variability* more rigorously as the ratio of the standard deviation of the residuals of data detrending to the mean value of the spatial trend. Very few estimates of the 'real' coefficient of variation of inherent variability are available in the literature; these are reported in Uzielli et al. (2006a). Calculation of the COV as described above provides a more realistic assessment of the dispersion of values around a spatial trend as discussed in Section 3.2.

Uzielli et al. (2006a) provided an extensive list of references to geotechnical spatial variability analyses. Recent contributions include Jaksa (2006), Cherubini et al. (2006), Chiasson & Wang (2006) and Uzielli et al. (2006b).

4.3 *Advanced analysis of geotechnical systems*

The description of a random field in the second-moment sense through a mean, standard deviation, a scale of fluctuation and a spatial correlation function is useful to characterize a spatially variable soil property. However, some possible limitations in this approach should be recognized. For instance, if spatial variability of soil properties is included in an engineering model, stresses and/or displacements which would not appear in the homogeneous case (i.e. in which variability is not addressed) could be present. Random field theory alone is unable to model the influence of spatial variability in the *behavior* of geotechnical systems.

A number of studies have focused, in recent years, on the combined utilization of random fields, non-linear finite element analysis and Monte Carlo simulation for investigating the behavior and reliability of geotechnical systems when the variability of soil properties which are relevant to the main presumable failure mechanisms is considered. Some important general observations can be made on the basis of their results. First, when soils are modelled as spatially variable, the modelled failure mechanisms are quite different – and significantly more complex – than in the case of deterministic soil properties. Second, there generally exists a *critical correlation distance* which corresponds to a minimum-reliability state. Third, phenomena governed by highly non-linear constitutive laws are affected the most by spatial variations in soil properties.

Popescu et al. (2005), for instance, investigated the differential settlements and bearing capacity of a rigid strip foundation on an overconsolidated clay layer. The undrained strength of the clay was modeled as a non-Gaussian random field. The deformation modulus was assumed to be perfectly correlated to undrained shear strength. Ranges for the probabilistic descriptors of the random field were assumed from the literature. Uniform and differential settlements were computed using non-linear finite elements in a Monte Carlo simulation framework. Figure 6a shows the contours of maximum shear strain for a uniform soil deposit with undrained strength of $100 \, \text{kPa}$ and for a prescribed normalized vertical displacement at center of foundation $\delta / B = 0.1$. In this case the failure mechanism is symmetric and well-defined. The results of the analyses indicated that different sample realizations of soil properties corresponded to fundamentally different failure surfaces. Figure 6b shows an example of a sample realization in which the spatial distribution of undrained strength is not symmetric with respect to the foundation. Hence, as could be expected, the configuration at failure, shown in Figure 6c, involves a rotation as well as a vertical settlement. The repeated finite-element analysis allows appreciation of compound kinematisms (settlements and rotations) of the footings, which could not be inferred from bearing capacity calculations involving non-variable parameters. It was observed that failure surfaces are not mere variations around the deterministic failure surface; thus, no 'average' failure mechanisms could be identified. Another notable result was the observed significant reduction in the values of bearing capacity spatially in the heterogeneous case in comparison with the deterministic model. Figure 6d shows that the normalized pressure which causes a given level of normalized settlement is always higher in the deterministic case. A number of other studies implementing integrated, statistics-based methods for the advanced analysis of geotechnical systems are presented and discussed in Uzielli et al. (2006a).

5 IDENTIFICATION OF PHYSICALLY HOMOGENEOUS SOIL VOLUMES

Second-moment statistics of geotechnical properties available in the literature are generally associated to soil types (e.g. 'clayey soils'). The identification of physically homogeneous units from tested soil volumes is an essential prerequisite for meaningful soil characterization and, consequently, for reliable design. From a qualitative point of view, if analyses are performed on data sets from, say, interbedded stratigraphic units, their results can be expected to be less confidently related to a specific soil type.

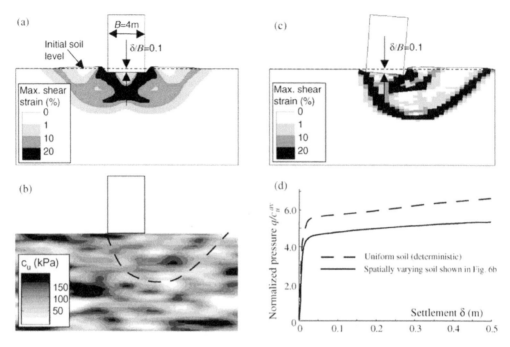

Figure 6. Selected results of investigation on homogeneous and spatially random foundation soil (Popescu et al. 2005).

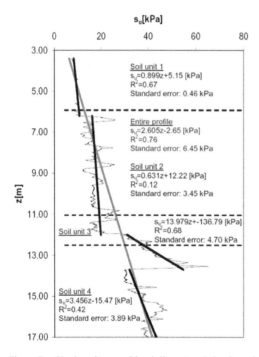

Figure 7. Single-unit vs. multi-unit linear trends in view of the assignment of design values of undrained shear strength (data from Lacasse et al. 2007).

Correspondingly, from a quantitative perspective, resulting statistics (and derived design parameters) can be expected to be less precise and accurate, and inferences made for engineering purposes less refined. Figure 7, for instance, illustrates the effect of subdividing the data from a CPTU sounding into homogeneous units on linear detrending of undrained shear strength values. The gray line represents the linear trend which is obtained by generalized least squares regression on the entire profile; the black lines correspond to the trends in each individual homogeneous soil layer. It may be seen that the unit-level approach provides a set of linear trends which are significantly more respondent to the undrained strength profile. The fact that the determination coefficient is highest for the entire-profile approach indicates that the linear regression model explain a higher quota of the variation of s_u with depth than that the unit-level trends, for which inherent variability plays a more relevant role. Hence, design values based on such approach (for instance, taken as the linear trend minus one standard deviation) would be, depending on the depth, beneficially less over-conservative or less under-conservative.

Examples of both subjective and objective assessment of physical homogeneity are available in the geotechnical literature. A purely subjective assessment relying uniquely on expert judgment may not provide optimum and repeatable results. A purely objective assessment based exclusively on numerical criteria

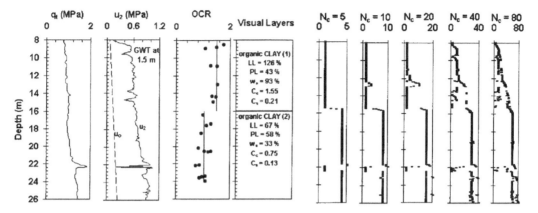

Figure 8.   Identification of stratigraphic boundaries from cluster analysis – comparative configurations for variable number of clusters on same profile of corrected cone tip resistance (Hegazy & Mayne 2002).

is not feasible in practice, as at least the preliminary selection and assessment of data quality and the definition of relevant parameters require geotechnical expertise. Hence, in practice, homogeneity assessment is always both subjective and objective to some degree.

In geotechnical analyses, physical homogeneity can be assessed in terms of either soil *composition* or soil *behavior*. These do not display a one-to-one correspondence, as soils which may be homogeneous in terms of composition may not be so in terms of mechanical behavior due to inherent properties of the geomaterials, in-situ state and imperfections in classification systems. Uzielli & Mayne (2008) presented an example of CPTU-based classification in which the Robertson (1990) charts provide heterogeneous categorization for a highly plastic calcareous clay due to strong cementation. Hence, it is essential to define the criteria adopted for physical homogeneity assessment and to report them when presenting results. These aspects are discussed in greater detail in Uzielli et al. (2006a).

Statistical-based procedures have been proposed in the geotechnical literature with the aim of providing increasingly repeatable and objective criteria for the assessment of soil homogeneity. Two examples are illustrated in the following. The aim is not to replace or downplay the importance of judgment; rather, to provide a more solid framework for the rational application of geotechnical expertise.

Hegazy & Mayne (2002) proposed a method for soil stratigraphy delineation by cluster analysis of piezocone data. The objectives of cluster analysis applied to CPT data were essentially: (a) objectively define similar groups in a soil profile; (b) delineate layer boundaries; and (c) allocate the lenses and outliers within sub-layers. Cluster analysis, allows identification of stratigraphic boundaries at various levels of resolution: if a configuration with a low number of clusters is referred to (e.g. 5 cluster-configuration in

Figure 8), the main boundaries are identified; with increasing number of clusters, greater resolution is achieved, by which less pronounced discontinuities are captured. Other applications of statistical clustering to CPT data include Facciorusso & Uzielli (2004).

Uzielli (2004) proposed a statistical moving window procedure to identify physically homogeneous soil units. Each moving window is made up of two semi-windows of equal height above and below a centre point. A suitable range of moving window height could be established as 0.75–1.50 m on the basis of soil type, test measurement interval and the type of soil failure (bearing capacity, shear) induced by the testing method. At each center point, user-established statistics (e.g. range, COV) are calculated for data lying in the interval corresponding to the upper and lower limits of the moving window. Homogeneous soil units are essentially identified by delineating soundings into sections where the values of the aforementioned statistics do not exceed preset thresholds. The magnitudes of such thresholds represent the degree of required homogeneity. A minimum width for homogeneous units can be preset by the user to ensure statistical numerosity of samples. Figure 9 shows the application of the method to cone penetration testing data from the heavily interbedded stratigraphy of Venice lagoon soils (Uzielli et al. 2008). Applications of the moving window procedure also include Uzielli et al. (2004; 2005a; 2005b).

The performance of the homogeneous soil unit identification methods should always be assessed critically using geotechnical knowledge and statistical procedures. For instance, a variety of soil classification charts based on in-situ tests are available in the geotechnical literature; these provide an effective means for subjective assessment of the homogeneity of a data set. If data points from a soil unit plot as a well-defined cluster (with possible limited outliers, if

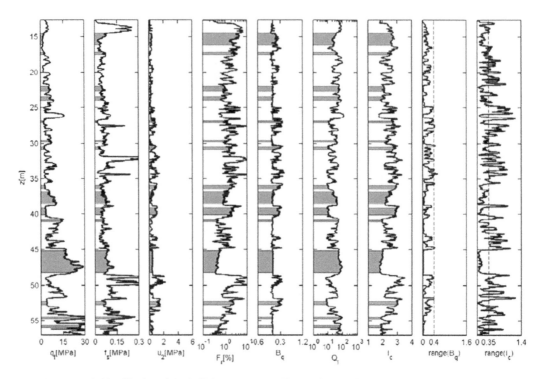

Figure 9. Example identification of physically homogeneous soil layers (in gray) from CPTU test data in Venice lagoon soils (Uzielli et al. 2008).

tolerated by the user) in such charts, homogeneity can be expected. Figure 10 illustrates an example application to dilatometer data using the chart by Marchetti & Crapps (1981), which is useful for DMT-based classification. Examples for CPT data are available in Uzielli (2004) and Uzielli et al. (2004; 2005a; 2005b). The performance of the moving window procedure can also be assessed objectively. Uzielli et al. (2005a) employed a statistical check to assess the performance of the moving window statistical methodology illustrated previously by calculating the coefficient of variation of the soil behavior classification index I_c (which is an efficient descriptor of soil behavior) in each identified homogeneous soil unit.

6   GEOSTATISTICAL ESTIMATION OF SOIL PROPERTIES

In geotechnical site characterization or for design purposes, it is often desirable or necessary to acquire a multi-dimensional spatial map of parameters of interest. As measurements are almost invariably 'too few' or 'not where they would be most useful', it becomes necessary to estimate soil properties at specific locations where observations are not available. The term *kriging* encloses a set of geostatistical techniques which allow

such inferential estimation. Kriging techniques are essentially weighted, moving average statistical interpolation procedures which minimize the estimated variance of interpolated values with the weighted averages of their neighbors. The input information required for kriging includes: available data values and their spatial measurement locations; information regarding the spatial correlation structure of the soil property of interest; and the spatial locations of target points, where estimates of the soil property are desired. The weighting factors and the variance are computed using the information regarding the spatial correlation structure of the available data. Since spatial correlation is related to distance, the weights depend on the spatial location of the points of interest for estimation. Formal aspects of kriging can be found, for instance, in Journel & Huijbregts (1978), Davis (1986) and Carr (1995). Two examples from the literature are summarized in the following.

Figure 11a (Lacasse & Nadim 1996) shows the locations of cone penetration test soundings in the neighborhood of a circular-shaped shallow foundation which was to be designed; Figure 11b reports the superimposed profiles of cone resistance in the soundings. It was of interest to estimate the values of cone resistance in other spatial   locations under the design location of the foundation itself. Figure 12

184

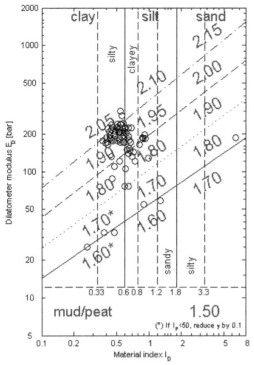

Figure 10. Assessment of the statistical moving window method: plotting data from a physically homogeneous soil unit on Marchetti and Crapps' (1981) I_D–E_D chart.

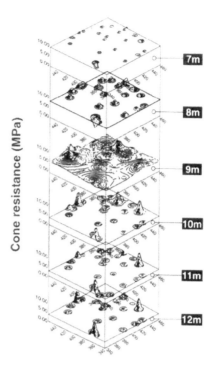

Figure 12. Contours of cone penetration resistance at each meter as obtained by geostatistical kriging (Lacasse & Nadim 1996).

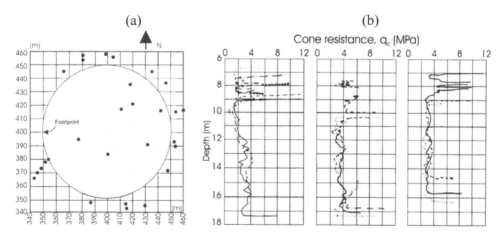

Figure 11. Input data for geostatistical analyses: (a) locations of cone penetration test soundings; (b) superimposed profiles of cone resistance (Lacasse & Nadim 1996).

presents the contours of cone penetration resistance at each meter as obtained by kriging. The 3D graphic representation provides improved insight into the possible spatial variation of cone resistance and the most likely values beneath the foundation. The results of the analysis enabled designers to determine more reliably the position of a clay layer and to use higher shear strength in design, thus reducing conservatism.

Baise et al. (2006) presented an integrated statistical method for characterizing geologic deposits for

185

liquefaction potential using sample-based liquefaction probability values. The method consists of three steps, namely: (1) statistical characterization of samples; (2) evaluation of the spatial correlation; and (3) estimation of the distribution of high liquefaction probability values. In presence of spatial correlation, kriging was used to evaluate the spatial clustering of high liquefaction probability in view of a regional liquefaction potential characterization. An example output is shown in Figure 13.

7 CORRELATION, STATISTICAL DEPENDENCE AND STATIONARITY ASSESSMENT

When dealing with more than one random variable, uncertainties in one may be associated with uncertainties in another, i.e. the uncertainties in the two variables may not be independent. Such dependency (which may be very hard to identify and estimate) can be critical to obtaining proper numerical results in engineering applications making use of well-known probabilistic techniques such as First-Order Second-Moment approximation (FOSM) and First-Order Reliability Method (FORM). Neglecting correlation may result in overdesign or underdesign, depending on the nature of the correlation.

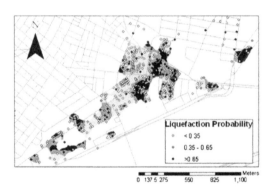

Figure 13.   Interpolated map of liquefaction probability in an artificial fill obtained by kriging (Baise et al. 2006).

There are two general approaches to testing for statistical dependence: *parametric* and *nonparametric*. Parametric approaches require the formulation of hypotheses regarding the nature and distribution of the data set under investigation. Non-parametric approaches, on the contrary, do not make such basic assumptions. Consequently, the latter are more widely applicable than parametric tests which often require normality in the data. While more widely applicable, the trade-off is that non-parametric tests are less powerful than parametric tests.

The most common measure of statistical dependence among random variables is *Pearson's linear correlation coefficient*. This measures the degree to which one uncertain quantity varies *linearly* with another. The correlation coefficient is non-dimensional, and varies in the range $[-1, +1]$. A higher bound implies a strict linear relation of positive slope (e.g. Figure 14), while the lower bound attests for a strict linear relation of negative slope. The higher the magnitude, the more closely the data fall on a straight line. Uzielli et al. (2006a) presented several literature examples of Pearson's correlation coefficient between different soil properties at selected sites. The clause of linearity should not be overlooked when interpreting the meaning of correlation: two uncertain quantities may be strongly related to one another, but the resulting correlation coefficient may have negligible value if the relationship is non-linear (e.g. Figure 14).

To overcome the sensitivity of correlation parameters on linearity, it is necessary to refer to procedures which are capable of measuring the strength of the dependence between two variables regardless of the type of relation. A number of statistical tests are available to perform such investigation. *Kendall's tau test* (e.g. Daniel 1990) involves the calculation of the test statistic, τ_{ken}, which measures the degree of concordance (or discordance) between two random variables, i.e. the strength of the trend by which one variable increases (or decreases) if the other variable increases (or decreases). Hence, linearity of the relationship is no longer a constraint. The values of τ_{ken} range from -1 to $+1$, indicating, respectively, perfect negative and positive correlation; values close to zero indicate weak dependence. Critical values of τ_{ken} for rejecting the

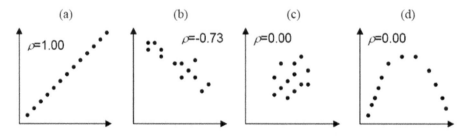

Figure 14.   Examples of correlation coefficient between two variables.

null hypothesis of independence are available in tabulated form (e.g. Daniel 1990) or, for larger data sets, assuming that the standardized τ_{ken} statistic is normally distributed, using standard normal distribution tables for a desired confidence level. Kendall's tau test can also be used to establish, when in doubt, whether a random variable should be addressed as being dependent or independent, i.e. to check whether there is a significant spatial trend which would result in dependency. Application of Kendall's test to geotechnical data can be found in Jaksa (1995) and Uzielli et al. (2004).

7.1 Weak stationarity

Stationarity is often an important prerequisite for spatial variability analyses of soil properties because many statistical procedures employed therein are based on the assumption that data samples consist of at least *weakly stationary* observations. A process is said to be stationary in the second-order sense (or weakly stationary) if: a) its mean is constant (i.e. there are no trends in the data); b) its variance is constant; and c) the correlation between the values at any two points at distinct spatial locations, depends only on the interval in the spatial distance between the points, and not on the specific spatial location of the points themselves. In the case of laboratory or in-situ geotechnical testing, stationarity can generally only be identified in a weak or second-order sense because of the limitations in sample size. Weak stationarity is discussed in greater detail in Uzielli et al. (2006a).

As statistical independence implies stationarity (the converse is not true), procedures designed for the assessment of the former could be (and have often been) used for assessing the latter. However, the matter should be handled with caution. Classical parametric and non-parametric statistical independence tests such as Kendall's tau test are based on the assumption of spatially uncorrelated input data. This assumption is antithetical to the fundamental premise in geotechnical spatial variability analysis, namely the existence of a correlation structure in spatially varying soil properties. Due to the resulting bias in autocorrelation coefficients in the correlated case, the application of such tests may result in unconservative assessments, i.e. non-stationary sets may be erroneously classified as weakly stationary.

The MBSR test, proposed by Uzielli (2004) as an evolution of a procedure proposed by Phoon et al. (2003), explicitly includes the hypothesis of correlation in data, and allows overcoming of the aforementioned bias. The test consists essentially in the comparison between the maximum value B_{max} of a 'Bartlett statistic', calculated from a moving sampling window, and a critical value B_{crit}. The latter is calculated from *profile factors* which depend on: the degree of spatial correlation of a given soil property of interest (parameterized by the *scale of fluctuation*); the

measurement interval of available data; the spatial length of the soil record of length; and the spatial extension of the statistical sampling window assigned by the investigator. Critical values at 5% level of significance for several autocorrelation models, among which single exponential, cosine exponential, second-order Markov and squared exponential, were provided by Phoon et al. (2003). The null hypothesis of stationarity in the variance is rejected at 5% level of significance if $B_{max} > B_{crit}$. The MBSR method is described in detail in Uzielli et al. (2006a). Example applications of the method are provided in Uzielli et al. (2004; 2005a; 2005b).

8 SOIL CLASSIFICATION

Soil classification based on laboratory and in-situ testing is invariably affected by considerable uncertainty. In most general terms, the main sources of uncertainty are testing uncertainty (including the effects of sample disturbance) and (especially in the case of in-situ testing, where samples are seldom retrieved) the uncertainty due to the lack of direct information regarding compositional properties of penetrated soils. The two sources of uncertainty can be assumed to be independent, and can be investigated separately. The effect of testing uncertainty is illustrated using an example in Section 9.2. The lack of perfect correlation between mechanical characteristics of soils (which depend on the mode of failure induced by a specific testing technique) and their composition is the main contributor to uncertainty in classification. Soil classification from CPT testing, for instance, relies on the mechanical response of soils to cone intrusion at least in terms of bearing capacity (for cone resistance) and shear failure (for sleeve friction). Due to the inherent complexity of geomaterials and to the variability of in-situ conditions from one testing campaign to another, mechanical properties cannot be related univocally to compositional characteristics. Hence, sets of measurements yielding the same numerical values may pertain to soil volumes which are compositionally different.

Quantitative uncertainty estimation in soil classification is deemed more useful to the geotechnical engineer than qualitative uncertainty assessment, as the most widely used classification methods based on in-situ tests (e.g. Robertson's 1990 chart-based system) operate in a quantitative sense. Moreover, the results of soil classification are used in several engineering applications, for instance liquefaction susceptibility evaluation (e.g. Robertson & Wride 1998).

Zhang & Tumay (2003 and previous) proposed two uncertainty-based classification approaches (one *probabilistic*, based on probability density functions, and one *possibilistic*, based on fuzzy sets) making use of CPT data. The approaches aim to quantify

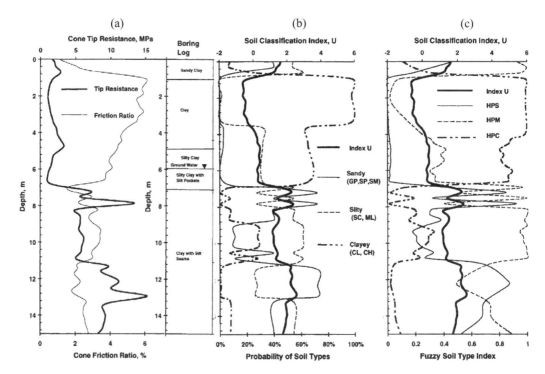

(a)   (b)   (c)

Figure 15.   National Geotechnical Experimentation Site Texas A&M University: (a) CPT measurements; (b) probabilistic classification; (c) possibilistic classification (Zhang & Tumay 2003).

the likelihood (in probabilistic and possibilistic terms, respectively) that a soil whose mechanical properties have been measured belongs to a given soil type (in the compositional sense). The imperfect relation between soil composition and mechanical response to penetration was modelled statistically, through numerous comparisons of: (a) the value of a specifically defined soil classification index from couples of cone resistance and sleeve friction measurements; and (b) a compositional classification based on the USCS system. Subsequently, the value of such coefficient for each couple of measurements is associated with a measure of the degree of probability (or possibility) of being classified as one of the following three categories: (a) highly probable clayey soils [HPC], comprising USCS soil types GP, SP and SM; (b) highly probable mixed soils [HPM], comprising USCS soil types SC and ML; and (c) highly probable mixed soils [HPC], comprising USCS soil types CH and CL. The *membership functions* used in the possibilistic approach quantify the degree of membership (from 0: no membership to 1: full membership) of a given value of the soil classification index in the HPS, HPM and HPC fuzzy sets. For any set of CPT measurements, it is possible to describe continuous profiles in terms of soil behavior, quantifying, respectively, the *probability* and the *degree of possibility* that, at

each depth, the penetrated soil behaves like a cohesionless, mixed or cohesive soil. Figure 15a reports the cone tip resistance and sleeve friction measurements as well as the descriptions pertaining to an adjacent boring. Figure 15b and Figure 15c illustrate the results of probabilistic and possibilistic classification of data. It may be observed that the probabilistic and possibilistic approaches provide information which is qualitatively consistent but quantitatively different. For instance, at each measurement depth the three probability values invariably add up to 100%; this is not a necessary condition for possibilistic classification.

9   ESTIMATION OF EPISTEMIC
     UNCERTAINTY

The procedures for the quantification of aleatory uncertainty, the assessment of statistical independence and stationarity, geostatistical kriging and the identification of homogeneous soil units reported in the previous sections implicitly neglect epistemic uncertainty, i.e. they assume that: (a) measured values reflect the true values of the parameter of interest; (b) the statistics calculated from samples are perfectly representative of the real populations; and (c) the geotechnical calculation models employed are perfect

188

parameterizations of the relations between measurements and design values. None of these hypotheses is strictly acceptable.

The objective estimation of epistemic uncertainty is more difficult than that of aleatory uncertainty. While a variety of well established mathematical procedures (e.g. random field modeling, geostatistical kriging, wavelet theory, maximum likelihood methods) are available to pursue the estimation of inherent variability, a frequentist investigation on epistemic uncertainty requires data in sufficient number and of sufficient quality. Such conditions are very seldom met in routine geotechnical practice.

9.1 Statistical estimation uncertainty

Any quantitative geotechnical variability investigation must rely on sets (in statistical terms, *samples*) of measured data which are limited in size. *Sample statistics* are unable to represent the 'true' *population statistics* perfectly, in the sense that: (a) they may be biased; and (b) that there is some degree of uncertainty in their estimation. In other words, they are affected by *statistical estimation uncertainty*. If a data set consisting of n elements is assumed to be *statistically independent*, the expected value of the sample mean is equal to the (unknown) population mean; hence, the sample mean is an *unbiased estimator* of the population mean. However, the sample mean has a variance and, consequently, a standard deviation. The latter is given by $s_\xi/\sqrt{n}$, in which s_ξ is the sample standard deviation and n is the numerosity of the data sample. The coefficient of variation of statistical estimation error is thus defined as

$$COV_{se} = \frac{s_\xi}{m_\xi \sqrt{n}} \qquad (6)$$

Confidence intervals for sample statistics can also be calculated (see e.g. Ayyub & McCuen, 2003).

Statistical estimation uncertainty is epistemic in nature though its magnitude also depends on the COV of aleatory uncertainty of the random variable of interest. It can be effectively quantified using statistical theory, and generally reduced by increasing the number of available data though the standard deviation could increase with additional observations.

9.2 Measurement uncertainty

Measurement uncertainty results from a combination of several sources including the uncertainty associated with systematic testing error (i.e. equipment and operator/procedural effects), and random testing error which is not assignable to specific testing parameters. Measurement uncertainty can in principle be reduced by improving the quality of testing equipment,

| precise but inaccurate measurements | accurate but imprecise measurements | precise and accurate measurements |

Figure 16. Target analogy for the conceptual distinction between accuracy and precision (adapted from Orchant et al. 1988).

controlling testing conditions and ensuring strict adherence to test procedures.

Past research has made use of a variety of terms to qualify measurement uncertainty. Here, *accuracy* is defined as 'the closeness of agreement between any one measured value and an accepted reference value.' Orchant et al. (1988) defined *precision* as 'the closeness of agreement between randomly selected individual measurements or test results'. Hence, precision describes the repeatability of a test. The conceptual distinction between accuracy and precision may be appreciated in the 'target analogy' in Figure 16.

Measurement uncertainty is generally characterized statistically by a *measurement bias* (i.e. the possible consistent overestimation or underestimation of the real value of the object parameter, conceptually affine to accuracy), and a *measurement dispersion* (i.e. the scatter of measurements on presumably homogeneous soil volumes, conceptually related to precision). The above definitions, as well as results of past research focusing on the estimation of the magnitudes of the sources of laboratory and in-situ testing variability suggest that reliable direct quantification of measurement epistemic uncertainty requires repeated, comparative tests in replicate conditions (e.g. Orchant et al. 1988; Phoon & Kulhawy 1999a; 2005). Moreover, documentation on equipment and procedural controls during in situ testing is usually not detailed sufficiently to allow for a quantitative evaluation of measurement errors.

Literature values for design coefficients of variation of measurement uncertainty for a number of design parameters are shown in Table 4; coefficients of variation of measurement uncertainty for in-situ testing methods are given in Table 3. In some cases, values are assigned to a large extent subjectively on the basis of experience and judgment. In other cases, they are derived objectively by statistical estimation.

Neglecting measurement uncertainty is acceptable for tests which have been shown to be largely operator-independent and to have very low random measurement errors such as the CPTU or DMT (see Table 3). Results of tests with high measurement and random uncertainty (such as the SPT, which also has a

Table 3. Coefficients of variation of measurement uncertainty for in-situ testing methods (Kulhawy & Trautmann, 1996).

Test	Equipment	Oper./proc.	Random	Total[a]	Range[b]
Standard penetration test (SPT)	0.05[c]–0.75[d]	0.05[c]–0.75[d]	0.12–0.15	0.14[c]–1.00[d]	0.15–0.45
Mechanical cone penetration test (CPT)	0.05	0.10[e]–0.15[f]	0.10[e]–0.15[f]	0.15[e]–0.22[f]	0.15–0.25
Electric cone penetration test (ECPT)	0.03	0.05	0.05[e]–0.10[f]	0.07[e]–0.12[f]	0.05–0.15
Vane shear test (VST)	0.05	0.08	0.10	0.14	0.10–0.20
Dilatometer test (DMT)	0.05	0.05	0.08	0.11	0.05–0.15
Pressuremeter test, pre-bored (PMT)	0.05	0.12	0.10	0.16	0.10–0.20[g]
Self-boring pressuremeter test (SBPMT)	0.08	0.15	0.08	0.19	0.15–0.25[g]

[a] $COV(total)^2 = COV(equipment)^2 + COV(operator/procedure)^2 + COV(random)^2$.
[b]   Because of statistical estimation uncertainty and subjective judgment involved in estimating COVs, ranges represent plausible magnitudes of measurement uncertainty for field tests.
[c,d] Best- to worst-case scenarios, respectively, for SPT.
[e,f] Tip and side resistances, respectively.
[g]   It is likely that results may differ for p_0, p_f and p_L, but data are insufficient to clarify this issue.

Table 4. Literature values for design coefficients of variation of measurement uncertainty for design parameters (Phoon 2004; Lacasse, personal communication).

Property[a]	Test[b]	Soil type	Point COV	Spatial avg. COV[c]
s_u(UC)	Direct (lab)	Clay	0.20–0.55	0.10–0.40
s_u(UU)	Direct (lab)	Clay	0.10–0.35	0.07–0.25
s_u(CIUC)	Direct (lab)	Clay	0.20–0.45	0.10–0.30
s_u(field)	VST	Clay	0.15–0.50	0.15–0.50
s_u(UU)	q_t	Clay	0.30–0.40[e]	0.30–0.35[e]
s_u(CIUC)	q_t	Clay	0.35–0.50[e]	0.35–0.40[e]
s_u(UU)	N_{SPT}	Clay	0.40–0.60	0.40–0.55
s_u^d	K_D	Clay	0.30–0.55	0.30–0.55
s_u(field)	PI	Clay	0.30–0.55[e]	–
s_u(CAUC)	Direct (lab)	Clay	0.15–0.20	–
s_u(CAUE)	Direct (lab)	Clay	0.20–0.30	–
s_u(DSS)	Direct (lab)	Clay	0.10–0.15	–
ϕ'	Direct (lab)	Clay, sand	0.07–0.20	0.06–0.20
ϕ'(TC)	q_t	Sand	0.10–0.15[e]	0.10[e]
ϕ_{cv}	PI	Clay	0.15–0.20[e]	0.15–0.20[e]
K_0	Direct (SBPMT)	Clay	0.20–0.45	0.15–0.45
K_0	Direct (SBPMT)	Sand	0.25–0.55	0.20–0.55
K_0	K_D	Clay	0.35–0.50[e]	0.35–0.50[e]
K_0	N_{SPT}	Clay	0.40–0.75[e]	–
E_{PMT}	Direct (PMT)	Sand	0.20–0.70	0.15–0.70
E_D	Direct (DMT)	Sand	0.15–0.70	0.10–0.70
E_{PMT}	N_{SPT}	Clay	0.85–0.95	0.85–0.95
E_D	N_{SPT}	Silt	0.40–0.60	0.35–0.55

[a] s_u: undrained shear strength; UU: unconsolidated undrained triaxial compression test; UC: unconfined compression test; CIUC: consolidated isotropic undrained triaxial compression test; CAUC: consolidated anisotropic undrained triaxial compression test; CAUE: consolidated anisotropic undrained triaxial extension test; DSS: direct simple shear test; s_u(field): corrected su from vane shear test; ϕ': effective stress friction angle; TC: triaxial compression; ϕ_{cv}: constant volume effective friction angle; K_0: in-situ horizontal stress coefficient; E_{PMT}: pressuremeter modulus; E_D: dilatometer modulus.
[b]   VST: vane shear test; q_t: corrected cone tip resistance from piezocone testing; N_{SPT}: standard penetration test blow count; K_D: dilatometer horizontal stress index; PI: plasticity index.
[c]   Spatial averaging COV for an averaging distance of 5 m.
[d]   Mixture of s_u from UU, UC and VST.
[e]   COV is a function of the mean (see Phoon & Kulhawy 1999b).

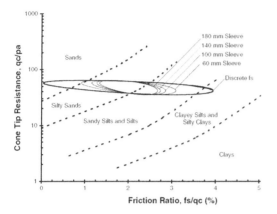

Figure 17. Effect of varying sleeve length on soil classification of simulated CPT data (Saussus et al. 2004).

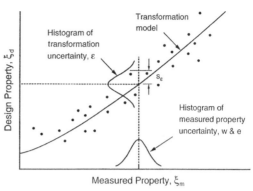

Figure 18. Probabilistic characterization of a geotechnical transformation model (Phoon & Kulhawy 1999b).

very large measurement interval, resulting in samples which are smaller in size and are thus affected by higher statistical estimation uncertainty) are most often not reliable inputs for such methods. The effects of measurement uncertainty can be assessed statistically. For instance, regarding the influence of measurement uncertainty on soil classification, Saussus et al. (2004) showed, using simulated profiles and analyzing the spatial correlation of CPT-related parameters, that CPT sleeve friction measurement introduces unnecessary redundancy due to the length of the standard friction sleeve compared to the measurement interval. Consequently, the smoothing and filtering of friction data affects, in the context of soil classification, the uncertainty in friction ratio values. Figure 17 shows an example result of the study, highlighting the significant effect of friction sleeve length on the classification of an artificially simulated sample realization of CPT measurements. It can be seen that the uncertainty in friction ratio increases with decreasing sleeve length. While studies of this type are not likely to be conducted on a routine basis, they are of extreme importance as they provide an idea of the relevant effect of testing equipment on important phases of geotechnical characterization and design.

9.3 Evaluation of transformation uncertainty

The direct measurement from a geotechnical test typically is not directly applicable for characterization or design purposes. Hence, a transformation (or calculation) model is required to relate test measurements to design parameters. In the process of model characterization, some degree of uncertainty is introduced. This is true whether the model is obtained by empirical fitting (because of inherent variability and epistemic uncertainty) or by theoretical analysis (because of the simplification of the physical reality).

An example of transformation model is the cone factor model, by which design undrained shear strength is estimated from CPT cone tip resistance (appropriately corrected in the case of piezocone data). The cone factor is calibrated for a specific site by regression analysis on CPT data and the undrained shear strength measured from laboratory tests. It should be highlighted, for this specific example, that due to shear strength anisotropy (different laboratory tests induce different types of failure in test samples) the cone factor is related to a specific test type, e.g. triaxial compression, triaxial extension, direct simple shear, etc.

Transformation uncertainty can be characterized statistically as shown in Figure 18. The transformation model is evaluated using regression analysis. In such an approach, the set of residuals of detrending are modeled as a zero-mean random variable. Transformation uncertainty is taken as the representative dispersion statistic (e.g. standard deviation or COV) of such random variable.

In practice, model characterization is not easily achieved, as reliable model statistics can only be evaluated by: (a) realistically large-scale prototype tests; (b) a sufficiently large and representative database; and (c) reasonably high-quality testing in which extraneous uncertainties are well controlled. Generally, available testing data are insufficient in quantity and quality to perform robust statistical assessment of model error in most geotechnical calculation models. Nonetheless, model uncertainty is of great importance in uncertainty-based design approaches such as reliability-based design, as its magnitude often exceeds that of other uncertainty components. Hence, efforts should be direct towards the compilation of a database of model statistics. Examples are given in Phoon & Kulhawy (1999b; 2005), Juang et al. (2005) and Uzielli et al. (2006b).

10 CLOSING REMARKS

Recognizing the existence of variability in geotechnical data for characterization and design purposes is the first step towards the consistent modeling, processing and assessment of the associated quantitative uncertainties. Statistical theory lies at the base of numerous methods and procedures which allow practical implementation of such uncertainty-based approaches by geotechnical engineers. The main benefits of such perspective for characterization and design are perhaps a more rational assessment of the degree of conservatism and risk and an improvement in the costs of investigation and design. This paper, far from aiming to be a self-standing contribution, aims to stimulate and accompany the growing diffusion of uncertainty-based approaches in geotechnical research and practice.

ACKNOWLEDGEMENTS

The author is grateful to the Norwegian Geotechnical Institute (NGI) for permitting use of the data for Figure 1, Figure 2 and Figure 7.

REFERENCES

Ang, A.H.S. & Tang, W.H. 1975. *Probability concepts in engineering planning and design*. New York: John Wiley & Sons.

Baecher, G.B. & Christian, J.T. 2003. *Reliability and statistics in geotechnical engineering*. New York: John Wiley & Sons.

Baise, L.G., Higgins, R.B. & Brankman, C.M. 2006. Liquefaction hazard mapping – statistical and spatial characterization of susceptible units. *Journal of Geotechnical and Geoenvironmental Engineering*, ASCE 132(6): 705–715.

Carr, J.R. 1995. *Numerical analysis for the geological sciences*. Englewood Cliffs: Prentice Hall.

Cherubini, C., Vessia, G. & Pula, W. 2006. Statistical soil characterization of Italian sites for reliability analyses. In T.S. Tan, K.K. Phoon, D.W. Hight & S. Leroueil (eds.), *Proceedings of the 2nd International Workshop on Characterisation and Engineering Properties of Natural Soils. Singapore, November 29–December 1, 2006*. The Netherlands: Taylor & Francis.

Chiasson, P. & Wang, Y.J. 2006. Spatial variability of sensitive Champlain sea clay and an application of stochastic slope stability analysis of a cut. In T.S. Tan, K.K. Phoon, D.W. Hight & S. Leroueil (eds.), *Proceedings of the 2nd International Workshop on Characterisation and Engineering Properties of Natural Soils. Singapore, November 29–December 1, 2006*. The Netherlands: Taylor & Francis.

Christian, J.T. 2004. Geotechnical engineering reliability: how well do we know what we are doing? *Journal of Geotechnical and Geoenvironmental Engineering*, ASCE 130(10): 985–1003.

Dai, S.H. & Wang, M.O. 1992. *Reliability analysis in engineering applications*. New York: Van Nostrand Reinhold.

Daniel, W.W. 1990. *Applied nonparametric statistics – 2nd edition*. Boston: PWS-Kent.

Davis, J.C. 1986. *Statistics and data analysis in geology – 2nd edition*. New York: John Wiley & Sons.

Duncan, J.M. 2000. Factors of safety and reliability in geotechnical engineering. *Journal of Geotechnical and Geoenvironmental Engineering*, ASCE 126(4): 307–316.

El-Ramly, H., Morgenstern, N.R. & Cruden, D.M. 2002. Probabilistic slope stability analysis for practice. *Canadian Geotechnical Journal* 39: 665–683.

Facciorusso, J. & Uzielli, M. 2004. Stratigraphic profiling by cluster analysis and fuzzy soil classification. In A. Viana da Fonseca & P.W. Mayne (eds.), *Proceedings of the 2nd International Conference on Geotechnical Site Characterization ISC-2, Porto, 19–22 September 2004*: 905–912. Rotterdam: Millpress.

Harr, M.E. 1987. *Reliability-based design in civil engineering*. New York: McGraw-Hill.

Hegazy, Y.A. & Mayne, P.W. 2002. Objective site characterization using clustering of piezocone data. *Journal of Geotechnical and Geoenvironmental Engineering*, ASCE, 128(12): 986–996.

Hight, D.W. & Leroueil, S. 2003. Characterisation of soils for engineering practice. In T.S. Tan, K.K. Phoon, D.W. Hight & S. Leroueil (eds.), *Characterisation and engineering properties of natural soils; Proceedings of the first International Workshop on Characterisation and Engineering Properties of Natural Soils. Singapore, December 2–4, 2002*: 255–360. Lisse: Swets & Zeitlinger.

Jaksa, M.B. 1995. The influence of spatial variability on the geotechnical design properties of a stiff, overconsolidated clay. *Ph.D. thesis*. University of Adelaide.

Jaksa, M.B. 2006. Modeling the natural variability of an overconsolidated clay in Adelaide, South Australia. In T.S. Tan, K.K. Phoon, D.W. Hight & S. Leroueil (eds.), *Proceedings of the 2nd International Workshop on Characterisation and Engineering Properties of Natural Soils. Singapore, November 29–December 1, 2006*. The Netherlands: Taylor & Francis.

Journel, A.G. & Huijbregts, C.J. 1978. *Mining geostatistics*. London: Academic Press.

Juang, C.H., Yang, S.H. & Yuan, H. 2005. Model uncertainty of shear wave velocity-based method for liquefaction potential evaluation. *Journal of Geotechnical and Geoenvironmental Engineering*, ASCE 131(10): 1274–1282.

Kulhawy, F.H. & Trautmann, C.H. 1996. Estimation of in-situ test uncertainty. In C.D. Shackleford, P.P. Nelson and M.J.S. Roth (eds.), *Uncertainty in the Geologic Environment: From Theory to Practice*, Geotechnical Special Publication No. 58: 269–286. New York: ASCE.

Lacasse, S. & Nadim, F. 1996. Uncertainties in characterising soil properties. In C.D. Shackleford, P.P. Nelson and M.J.S. Roth (eds.), *Uncertainty in the Geologic Environment: From Theory to Practice*, Geotechnical Special Publication No. 58: 49–75. New York: ASCE.

Lacasse, S., Guttormsen, T., Nadim, F., Rahim, A. & Lunne, T. 2007. Use of statistical methods for selecting design soil parameters. *Proceedings of the 6th International Conference on Offshore Site Investigation and Geotechnics, London, September 11–13, 2007*. London: Society for Underwater Technology.

Marchetti, S. & Crapps, D.K. 1981. Flat dilatometer manual. *Internal Report*. G.P.E. Inc.

Nadim, F. & Lacasse, S. 1999. Probabilistic slope stability evaluation. In *Proceedings of the 18th Annual Seminar on Geotechnical Risk Management, Hong Kong, May 14, 1989*: 177–186.

Orchant, C.J., Kulhawy, F.H. & Trautmann, C.H. 1988. Reliability-based foundation design for transmission line structures: critical evaluation of in-situ test methods. *Report EL-5507(2)*. Palo Alto: Electric Power Research Institute.

Phoon, K.K. & Kulhawy, F.W. 1999a. Characterisation of geotechnical variability. *Canadian Geotechnical Journal* 36: 612–624.

Phoon, K.K. & Kulhawy, F.W. 1999b. Evaluation of geotechnical property variability. *Canadian Geotechnical Journal* 36: 625–639.

Phoon, K.K. & Kulhawy, F.H. 2005. Characterisation of model uncertainties for laterally loaded rigid drilled shafts. *Géotechnique* 55(1): 45–54.

Phoon, K.K., Quek, S.T. & An, P. 2003c. Identification of statistically homogeneous soil layers using modified Bartlett statistics. *Journal of Geotechnical and Geoenvironmental Engineering*, ASCE 129(7): 649–659.

Popescu, R., Deodatis, G. & Nobahar, A. 2005. Effects of random heterogeneity of soil properties on bearing capacity. *Probabilistic Engineering Mechanics* 20: 324–341.

Réthati, L. 1988. *Probabilistic solutions in geotechnics*. New York: Elsevier.

Robertson, P.K. 1990. Soil classification using the cone penetration test. *Canadian Geotechnical Journal* 27: 151–158.

Robertson, P.K. & Wride, C.E. 1998. Evaluating cyclic liquefaction potential using the cone penetration test. *Canadian Geotechnical Journal* 35: 442–459.

Snedecor, G.W. & Cochran, W.C. 1989. *Statistical methods*. Ames: University of Iowa Press.

Uzielli, M. 2004. Variability of stress-normalized CPT parameters and application to seismic liquefaction initiation analysis. *Ph.D. thesis*. University of Florence, Italy.

Uzielli, M., Vannucchi, G. & Phoon, K.K. 2004. Assessment of weak stationarity using normalised cone tip resistance. *Proceedings of the ASCE Joint Specialty Conference on Probabilistic Mechanics and Structural Reliability, Albuquerque, New Mexico, July 26–28, 2004* (CD ROM).

Uzielli, M., Vannucchi, G. & Phoon, K.K. 2005a. Random field characterisation of stress-normalised cone penetration testing parameters. *Géotechnique* 55(1): 3–20.

Uzielli, M., Vannucchi, G. & Phoon, K.K. 2005b. Investigation of correlation structures and weak stationarity using the CPT soil behaviour classification index. In G. Augusti, G.I. Schuëller & M. Ciampoli (eds.), *Safety and Reliability of Engineering Systems and Structures – Proceedings of ICOSSAR 2005, Rome, 19–23 June, 2005*. Rotterdam: Millpress (CD ROM).

Uzielli, M., Lacasse, S., Nadim, F. & Phoon, K.K. 2006a. Soil variability analysis for geotechnical practice. In T.S. Tan, K.K. Phoon, D.W. Hight & S. Leroueil (eds.), *Proceedings of the Second International Workshop on Characterisation and Engineering Properties of Natural Soils*. Singapore, November 29–December 1, 2006. The Netherlands: Taylor & Francis.

Uzielli, M., Lacasse, S., Nadim, F. & Lunne, T. 2006b. Uncertainty-based analysis of Troll marine clay. In T.S. Tan, K.K. Phoon, D.W. Hight & S. Leroueil (eds.), *Proceedings of the 2nd International Workshop on Characterisation and Engineering Properties of Natural Soils. Singapore, November 29–December 1, 2006*. The Netherlands: Taylor & Francis.

Uzielli, M. & Mayne, P.W. 2008. Comparative CPT-based classification of Cooper Marl. *Proceedings of the 3rd International Conference on Site Characterization, Taiwan, April 1–4, 2008*. The Netherlands: Taylor & Francis.

Uzielli, M., Simonini, P. & Cola, S. 2008. Statistical identification of homogeneous soil units for Venice lagoon soils. *Proceedings of the 3rd International Conference on Site Characterization, Taiwan, April 1–4, 2008*. The Netherlands: Taylor & Francis.

Vanmarcke, E.H. 1983. *Random Fields: analysis and synthesis*. Cambridge: MIT Press.

Zhang, Z. & Tumay, M.T. 2003. Non-traditional approaches in soil classification derived from the cone penetration test. In G.A. Fenton & E.H. Vanmarcke (eds.), *Probabilistic Site Characterisation at the National Geotechnical Experimentation Sites*, Geotechnical Special Publication No. 121: 101–149. Reston: ASCE.

Geotechnical and Geophysical Site Characterization – Huang & Mayne (eds)
© 2008 Taylor & Francis Group, London, ISBN 978-0-415-46936-4

Characterization of residual soils

A. Viana da Fonseca
Faculdade de Engenharia, University of Porto, Portugal

R.Q. Coutinho
Federal University of Pernambuco, Recife, Brazil

ABSTRACT: This paper presents an overview, in the perspective of the authors, of the factors that condition the assessment of mechanical characteristics of residual soils, especially when these are derived from the interpretation of in situ testing techniques. Specific, although none exclusive, very important factors, such as microstructure, cohesive-frictional characteristics, stiffness non-linearity, small and large strain anisotropy, weathering and destructuration, condition of saturation, consolidation/permeability characteristics and rate dependency are herein discussed in the light of known parametrical correlation proposals, developed for transported soils. Results of some directional studies, from international and local (from the authors' view) experience, are presented, both in saprolitic, as in lateritic soils. The conceptual characterization of these unusual soils properties, are discussed in the perspective of the use of in situ tests for the design of geotechnical structures.

1 INTRODUCTION TO THE SINGULARIRY OF RESIDUAL SOILS

The assessment of mechanical characteristics of natural soils implies new techniques or, better, new interpretation methods (Lumb, 1962, Mello, 1972, Collins, 1985, Lacerda et al. 1985, Novais Ferreira, 1985, Vargas, 1985, Vaughan, 1985, Wesley, 1990) capable of measuring soil properties conditioned by effects of some very important factors, such as microstructure, stiffness non-linearity, small and large strain anisotropy, weathering and destructuration, consolidation characteristics and rate dependency (Schnaid, 2005). No saturated conditions prevail in most of soils and will have to be dealt carefully. Interpretation methods for the knowledge of bonded geomaterials, such as residual soils, are as much challenge as the complexity of their behaviour. Cross-correlation of different and multiple measurements from different tests is one preferential solution: more measurements in one test the better!...

Residual soils, particularly from igneous rocks, are geologically formed by upper layer of heterogeneous soil masses of variable thickness, overlaying more or less weathered rocks (Blight, 1997). In fact, variable weathering processes, temperature, drainage, and topography, have reduced the rocks in-place to form overburden residual soils that range from clay topsoil to sandy silts and silty sands that grade with depth back into saprolite and partially-weathered rocks

(Martin, 1977). Igneous rocks, like granite, are composed mainly of quartz, feldspar and mica. Quartz is resistant to chemical decomposition, while feldspar and mica are transformed mainly to clay minerals during the weathering process. As weathering proceeds, the stress release as a result of the removal of the over-lying material accelerates the rate of exfoliation (stress release jointing) and the wetting and drying processes in the underlying fresh rock. These processes increase the surface area of the rock on which chemical weathering proceeds, which leads to deep weathering profiles (Irfan 1996, Ng & Leung, 2006). Micropetrographic studies evidence that the process of weathering gradually reduces the contents of feldspar in both volcanic and granitic rocks. The amount of clay minerals, microfractures and voids increases with the degree of weathering. As expected, the quartz remains fairly constant throughout the weathering processes (Viana da Fonseca, 2003, Viana da Fonseca et al. 1994, 2006, Ng & Leung, 2006). As reported, attempts have been made (Massey et al. 1989, Irfan, 1996) to correlate relationships between the degrees of weathering and engineering properties of weathered rocks with reasonable success. However, for practical design and construction purposes these correlations are not sufficiently stable and accurate.

Depending on the degree of alteration, some soils do not keep features of the parent rock, while others are strongly influenced by relict structures inherited from the parent rock (Rocha Filho et al. 1985, Costa

195

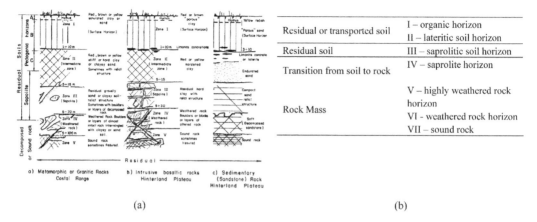

Residual or transported soil	I – organic horizon
	II – lateritic soil horizon
Residual soil	III – saprolitic soil horizon
Transition from soil to rock	IV – saprolite horizon
Rock Mass	V – highly weathered rock horizon
	VI - weathered rock horizon
	VII – sound rock

(a)   (b)

Figure 1.   (a) Typical profiles of Brazilian residual soils (some of five zones may be absent in particular cases) (Vargas, 1985); (b) Weathering profile according to Pastore (1992), cited Lacerda & Almeida (1995).

Filho et al. 1989). The relict structure of the parent rock can be left within the derived soil materials with evidence of bonding or dissolved bond features, as well as cracks and fissures from the original fractured rock mass (Mayne and Brown, 2003).

In the weathering profile (the sequence of horizons that is formed) one can possibly found materials varying from sound rock, weathered rock, soil which retain rock characteristics (called young residual soil or saprolitic soil) to the horizon where no remaining rock characteristics can be seen (structure and mineralogy – knows as mature residual soil or lateritic soil). In this top layer the presence of transported soil may also occur (for instance colluvium), which may be difficult to distinguish from the true residual soils. Figure 1a and b present a proposal for the classification profiles.

The horizon II (lateritic soil) is usually formed under hot and humid conditions involving a high permeability profile which often results in a bond structure with high contents of oxides and hydroxides of iron and aluminum (laterization). It is noteworthy that not all soils belonging in this horizon have undergone enough pedogenetic evolution for laterization. The horizon III (saprolitic soil) can show a high level of heterogeneity both vertical and laterally as well as a complex structural arrangement which retain rock characteristics. The texture and mineralogy of these soils can vary considerably, function of the degree of weathering and leaching that are submitted. In a tropical region the weathering profile very often shows a very narrow or inexistent horizon V, while in temperate climates this zone can be reasonably thick.

Mitchell and Coutinho (1991) (see also Lacerda & Almeida, 1995, and Clayton & Serratrice, 1998) presented a general view of many soils which are called unusual soils (see Schnaid et al. 2004; Coutinho et al. 2004a) where is included bonded soils – granitic

saprolitic soil and laterite soils, unsaturated – collapsible soils. Bonding and structure are important components of shear strength, in general residual soils present a cohesive-frictional nature (characterized by c' and ϕ'). Anisotropy derived from relic structures of the parent rock can be also a characteristic of a residual soil. In those conditions, the structure formed during the weathering evolution / process can become very sensitive to external loads, requiring adequate sample technique in order to preserve that relic structure.

Usually, the resulting void ratio and density of the soil are not usually directly related to stress history, unlike the case of sedimentary clayey soils (Vaughan, 1985, Vaughan et al. 1988, Geological Society, 1990, Viana da Fonseca, 2003). The presence of some kind of bonding, even weak, usually implies the existence of a peak shear strength envelope, showing a cohesion intercept and a yield stress which marks a discontinuity in stress strain behaviour. Examples of observed yield stresses in residual soils are found in Sandroni (1981) – from gneiss, in Vaughan et al. (1988) – from basalt, in Coutinho et al. (1997, 1998) – from gneiss, in Viana da Fonseca (1998, 2003), Viana da Fonseca & Almeida e Sousa (2002), Viana da Fonseca et al. (1997a, 1998b, 2006) from granite, or, in Machado & Vilar (2003) – sandstone and magmatic rocks.

Structure in natural soil has two "faces": the "fabric" that represents the spatial arrangement of soil particles and inter-particle contacts, and "bonding" between particles, which can be progressively destroyed during plastic straining (given place for the term "destructuration"). Most – if not all! – geomaterials are structured, but naturally bonded soils are dominated by this effect on their mechanical response (Leroueil & Vaughan, 1990). Here cohesive component due to cementation can dominate soil shear strength, at engineering applications involving low stress levels (Schnaid, 2005) or in specific stress-paths where this component is relevant (cuts or slopes, for instance).

Residual masses present strong heterogeneity, changing gradually their characteristics laterally and vertically (with depth), especially regarding their mechanical properties. As a sign of this, it is common to have to adopt shallow foundations – footings and mats – and driven piles or drilled and bored piles, in very limited areas, depending upon the consistency of the overburden soils and the depth to parent rock. An accurate mapping of the spatial variability of the mechanical properties, necessary for geotechnical design, is very challenging being this improved recently by the use of geophysical methods (Viana da Fonseca et al. 2006). This will be explored later in this text. Several in situ testing methodologies, such as SPT, CPT, DMT, PMT and SBPT, and geophysical survey, surface and borehole seismic tests, electrical resistivity and GPR, have been used to give different insight to these particular soils. This will be the main goal of these lines.

The background investigation on the effects of bonded structure is well reference by Schnaid (2005) and is has been the aim of the authors preocupation during the last years (Viana da Fonseca, 2003; Viana da Fonseca et al. 2006; Coutinho et al. 1997, 1998, 2000; Coutinho et al. 2004a). Most of this research has been focusing in laboratory tests carried out on natural specimens retrieved from the field, but, as stated by Schnaid (2005) difficulties in testing natural soils are twofold: disturbance to the structure that can occur during the sampling process and spatial variability inherent to natural deposits emerging from both the degree of cementing and the nature of the particles (e.g. Stokoe & Santamarina, 2000). A significant work to understand this important factor has been made with artificially cemented, but they definitely cannot reproduce the deposition process and the distinctive structure of natural soils (Schnaid, 2005). To distinguish features of behaviour emerging from bonded structure from those related to changes in state, constitutive laws conceived for the unbonded material are modified accordingly to introduce the bond component (from the very early days – Leroueil & Vaughan, 1990, to the very recent – Pinyol et al. 2007).

Water table is in many cases deep in the profile; hence the soils are generally in unsaturated conditions. In this case, the role of matrix suction and its effect on the soil behavior has to be recognized and considered in the interpretation of in situ tests. The main difference between saturated soils and unsaturated soils is the existence of a negative pressure in the water of the pores; largely know as suction, which tends to increase resistance and rigidity of the soil. The mechanical behavior of an unsaturated soil is currently evaluated according to four variables – net mean stress $(p - u_a)$, deviator stress q, suction s $(u_a - u_w)$ and specifc volume v, where u_a is the air pressure and u_w the pore water pressure according to several models early developed (e.g. Fredlund, 1979; Alonso et al. 1990; Wheeler and Sivakumar, 1995; and Futai et al. 1999).

Residual soils derived from a wide variety of parent rocks can be collapsable. Residual collapsing soils have a metastable state characterized by honeycomb structure and partially saturated moisture condition that can develop after a parent rock has been thoroughly decomposed or while the decomposition is happening (Vargas, 1973). Commonly, metastable residual soils form under condition of heavy concentrated rainfalls in short periods of time, long dry periods, high temperature, high evaporation rates, and flat slopes so that, leaching of material can occur. There are two mechanisms of bonding in the metastable soil structure: soil water suction and cementation by clay or other types of fines particles. The clay fraction of these soils is usually composed of kaolinite, though gibbsite and halloysite are also common. Residual collapsible soils usually have low activity and low plasticity. Colluvial deposits (or mature residual soils) can become collapsible in environments where the climate is characterized by alternating wet and dry seasons that cause a continuous process of leaching of the soluble salts and colloidal particles much like residuals soils (from Mitchell and Coutinho, 1991).

1.1 Cementation in natural soils – reactive force (Contact-Level)

There are many mechanisms leading to cementation. Some agents lithify the soil around particles and at contacts, while other processes change the initial physical-chemical structure (Mitchell 1993). Cementation is a natural consequence of aging and the ensuing diagenetic effects in soils. Most natural soils have some degree of inter-particle bonding. The stress-strain behavior, stiffness, strength and the volume change tendency of soils can be drastically affected by the degree of cementation. Santamarina (2001) defines two regions that can be identified: the low-confinement cementation controlled region and the high-confinement stress-controlled region. To facilitate comparing this cementation-reactive force to other forces, the tensile force required to break the cement at a contact is computed in Santamarina (2001). The author emphasises that the stress-strain behaviour, the strength and the volume change tendency of soils can be drastically affected by the degree of cementation (Clough et al. 1981; Lade and Overton 1989; Airey and Fahey 1991; Reddy and Saxena 1993; Cuccovillo and Coop 1997). Two regions – quoting Santamarina (2001) – can be identified: the "cementation-controlled" region at low confinement, and the "stress-controlled" region at high

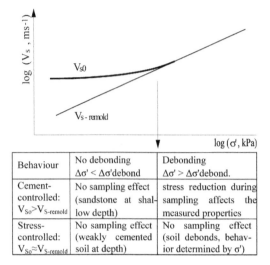

Behaviour	No debonding $\Delta\sigma' < \Delta\sigma'$debond	Debonding $\Delta\sigma' > \Delta\sigma'$debond.
Cement-controlled: $V_{So} > V_{S\text{-remold}}$	No sampling effect (sandstone at shallow depth)	stress reduction during sampling affects the measured properties
Stress-controlled: $V_{So} \approx V_{S\text{-remold}}$	No sampling effect (weakly cemented soil at depth)	No sampling effect (soil debonds, behavior determined by σ')

Figure 2. Skeletal forces vs. cementation strength – sampling and debonding (Santamarina 2001).

confinement. In the cementation-controlled region, the small-strain shear stiffness can increase by an order of magnitude, the strength is cementation controlled, the buckling of chains is hindered (lower initial volume contraction), and the soil tends to brake in blocks (immediately after breaking, the inter block porosity is null, hence shear tends to cause high dilation, even if the cemented soil within the blocks has high void ratio). Fundamentally, bonded soils are evolutive soils, with the mechanical properties changing irreversibly – even the most elemental, such as elastic modulus – with the stress-strain level (Viana da Fonseca, 1998, 2003). The cementation thickness can be related to the shear modulus of the soil Gs by considering a modified Hertzian formulation (Santamarina, 2001). The probability of debonding increases if the stress reduction $\Delta\sigma'$ approaches or exceeds the soil capacity $\Delta\sigma'$debond. The upper bound for stress reduction is the in situ state of stress $\Delta\sigma' \approx \sigma'_0$ (details and data in Ladd and Lambe 1963). While $\hat{G}_g$ and σ_{ten} cannot be readily evaluated (where σten is the tensile strength of the bonding agent and Gg is the shear modulus of the mineral that makes the particles), this model illustrate the importance of the in situ shear wave velocity V_{so} in determining whether a soil will experience sampling effects due to stress reduction. Figure 2 summarizes these observations into four regions, and provides a framework for organizing available data and further studies on sampling effects (sampling disturbance is reviewed in Jamiolkowski et al. 1985). Although premature the potential impact of sampling on the bases of the in-situ shear wave velocity V_{S0} may be pursue.

The relative relevance of cementation and confinement can be identified by comparing the shear wave velocity in situ V_{S0} with the velocity in a remoulded specimen $V_{s\text{-remold}}$ that is subjected to the same state of stress (Santamarina, 2001). In general, one should suspect cementation if $V_{S0} > V_{S\text{-remold}}$, as the opposite. De-bonding during sampling affects the behaviour of both sands and clays, or, even most, in residual soils, and it can cause important differences between the soil response measured in the laboratory and in the field (Tatsuoka and Shibuya 1992; Leroueil, 2001; Stokoe and Santamarina 2000). This is the base of one of the processes, explored later in the text, for measuring damage in sampling.

The values of shear modulus (G_{max}, or G_0, a function of V_S) obtained from in situ seismic tests, such as Cross-Hole, may be close to the values obtained from laboratory resonant column tests or triaxial tests with bender elements (Giachetti, 2001, Viana da Fonseca et al. 2006, Ferreira et al. 2007). Measured G_{max} values are larger than the values predicted by some correlations methods developed for transported soils, such as those proposed by Hardin (1978), both in saprolitic soils (Viana da Fonseca et al. 2004) or, more, in lateritic soils, by as much as 300% (Barros & Hachich, 1996). Suction, which will be dealt later in the text, and cementation (the iron and aluminium are usually present in lateritic soils) are claimed to be responsible for this difference. Empirical relationships such as the one proposed by Ohsaki & Iwasaki (1973), relating G_{max} and SPT usually underestimate G_{max} in the lateritic soil (Viana da Fonseca, 1996, 2001, Barros & Hachich, 1996, Viana da Fonseca et al. 1997a, 1998a, 2003, 2004, 2006).

1.2 Cementation in natural soils – fundaments of a cohesive pattern of behaviour

Cohesion is not as a simple concept as most may feel. On this purpose Santamarina (1997) points out the dangerous oxymoron of the concept "Cohesive Soil", concluding: the "common sense" behind "cohesive soils" biases the "perceived reality" so that the "known uncertainty" "almost always" becomes "pretty ugly".

As referred by Santamarina (1997) and restated by Locat et al. (2003), cohesion in a soil can come from different possible sources. Six are pointed out by the latest authors. The first one is due to electrostatic forces providing contact strength, i.e., van der Waals attraction and double layer forces, related to ionic concentration of the pore fluid (only in cohesive soils). The second is cementation, that is a chemical bounding, i.e., the cementation due to lithification of soil around particles and at contacts, and physical-chemical processes that are created by diagenesis or weathering and can be found in both cohesive or granular soils and can be generated during or after the formation of the soil. The third one is adhesion of clay particles around some

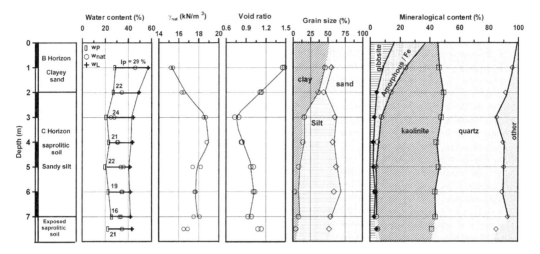

Figure 3. Index tests data and soil mineralogy of a residual tropical soil from Ouro Preto – Brazil (Futai et al. 2004).

larger silt or sand size particles, also call clay bounding. The fourth one is contact cementation develop with time and pressure. The fifth one results from the inter-action of organic matter with particles, mostly fibers, which can attract particles to from larger strings or aggregates. Another process, called arching will be considered as part of the friction component. The sixth source of cohesion is caused by suction (or negative pore pressure, in partial saturation conditions) which can result into an apparent cohesion. Some of the above sources are well known, and we would therefore like to expand on: adhesion and contact cementation. The first cited author notes however that some of cohesive signs are non real sources, but consequent reflexes of other phenomena: undrained shear – when the rate of loading exceeds the rate of pore-water pressure dissipation; dilatancy – the tendency of a soil to dilate that is direct related to density and decreases with increasing confinement; and, particle eccentricity – that show post peak behaviour and shear banding even in loose specimens.

Cohesion is very much "unstable" as the compatibility of deformations and progressive failure dictate that, for large strain, only critical state soil parameters should be used in ultimate analysis (capacity). Due to the brittleness of cementation, only friction remains in the critical state (Lambe and Whitman 1969, Atkinson 1993, and Wood, 1990, as cited by Santamarina, 1997). This sensitivity will, as much as the progressive behaviour discussed for small strain shear modulus – or the shear wave velocity, indict a very significant reflexive attitude when dealing with characterization based on in situ tests (each and all with certain levels of inducing actions for their execution) or lab test, very much dependent of sampling action.

2 BACKGROUND CHARACTERISTICS

2.1 Index Properties

2.1.1 Some typical profiles
Results of index tests and mineralogy data of a residual tropical soil from Ouro Preto – Brazil are summarized in Figure 3. The amount of clay is greater in horizon B and comparison grain size analyses with and without defloculant has revealed that all clay is flocculated. Figure 4 presents results of a complete borehole performed in the unsaturated gneissic residual soil together with some properties of the materials. In general, the mature residual soil (horizon B) is composed of porous sandy clay, with the clay particles in flocculated condition (aggregated). The young residual soil (horizon C/Cr) is a silty sandy with mica, but still preserving some characteristics of parent rock (foliation of around 30° with horizontal). The rock (biotite gneiss) presents a high degree of fracturing with low RQD values in some depths.

2.1.2 Classification based on mechanical indices from in situ tests results
Classification charts based on the piezocone (CPTU) results, such as those proposed by Roberston (1990), may reflect considerable dispersion in the material type, as this evaluation may be conditioned by unreliable pore pressure measurements. An example of a typical trendy classification of a saprolitic residual soil from Porto is presented in Figure 5 (Viana da Fonseca et al. 2006). This material is indexed to cemented and aged silty clays to clayey sands. Lab tests over recoiled samples have confirmed it mainly as clayey silty sand, when the analysis is made by classical wetting sieving and sedimentation with no use of chemical defloculation (Viana da Fonseca et al. 2004).

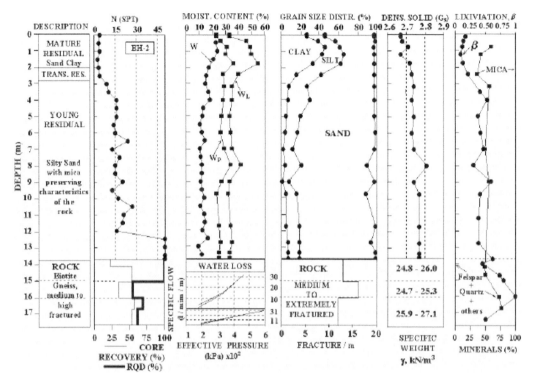

Figure 4. Result of a borehole (SPT/RQD) and characteristic properties of an unsaturated gneissic residual soil from Pernambuco – Brazil (Coutinho et al. 1998, 2000).

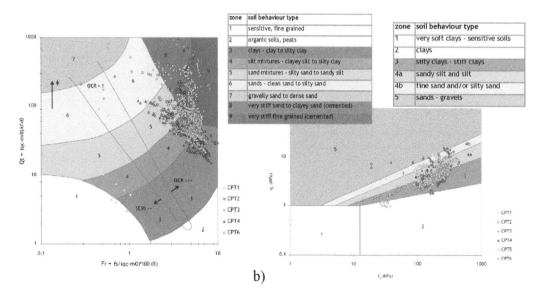

Figure 5. Soil behaviour classification charts: Robertson's (1990), Eslami & Fellenius's (1997) (Viana da Fonseca et al. 2006).

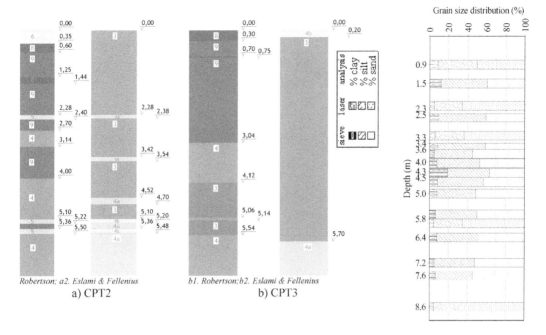

Figure 6. Soil behaviour classification profiles (Costa, 2005), based on the proposals of Robertson (1990) and Eslami & Fellenius (1997), and compared with lab analysis.

However, grain distribution deferred from laser beam analyser, reveal that silty sand class prevails, with very low clay fraction. Another classification based on the CPTU results introduced by Eslami and Fellenius (1997), directed to pile design but also useful for geotechnical purposes, Figure 5. The same results are presented in Figure 6 in two specif CPT profiles (Costa 2005, Viana da Fonseca et al. 2006), generating classification profiles throughout depth, by applying both proposals and confirming that this residual soil profile is very much heterogeneous.

The most surprising trend is the prevailing incidence of clayey matrices, which is not in accordance to the grain size distribution defined in laboratory tests. This diatomic behaviour was also explored by other classifications, such as those proposed by Marchetti (1980) and Zhang & Tumay (1999), and turned out to be also divergent in the percentage of fines. Figure 7 illustrates the discrepancies observed for one of the profiles – that of CPTU8, DMT6 and DMT8 (very close to each other).

There is clearly an unsuitable classification, which evidences the specificity of these soils and the compulsory need to adapt most of the classical proposals, developed for transported soils, to residual profiles.

This fact has been identified in other sites, a good example being the one described by Mayne and Brown (2003) referring to Piedmont residual soils. In this case, soils are comprised by fine sandy silts to silty

fine sands, resulting from weathering of gneiss and schist. Classical classifications, such as the Unified Soils System, categorize these soils unsatisfactorily, since the definition of fine and coarse-grained groups are very much sensitive for such an extensive grain size distribution. The material index (ID) from flat dilatometer tests consistently gives values that are characteristic of silty materials ($0.6 \leq ID \leq 1.8$) when performed in Piedmont residuum (e.g., Harris & Mayne, 1994; Wang & Borden, 1996). The strata seem to change in random fashion, thus suggesting high variability over short distances, but, as referred by the cited authors, this is illusionary due to the fact that the mean grain size of the Piedmont residuum is close to the 75-micron criterion that separates fine-grained from coarse-grained fractions (0.075 mm, or the No. 200 sieve size). In fact, the Piedmont residuum acts more as a dual soil type (SM-ML), exhibiting characteristics of both fine-grained soils (undrained) and coarse-grained soils (drained) when subject to loading. These contradictions in ID indices, such as stress history ratios, when comparing the results coming from lab tests over undisturbed samples and those inferred from in situ tests (CPTU, DMT…) results, may be due to the specificity of the high degree of non-linearity, typical of these cemented soils, in very distinct patterns from the corresponding transported soils (with similar density and moisture content). The authors conclude that great care should be taken in using empirical

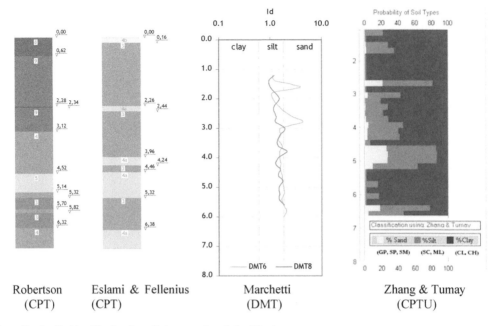

Figure 7. Profile identification from distinct tests based classifications.

correlations in non-textbook geomaterials, being necessary to verify design parametrical findings, by performing regional inter-crossing of experimental data (with emphasis to the use of comparison of calibration chambers and geotechnical structures prototypes tests results).

Mayne and Brown (2003) have used the results of seismic piezocone tests from Piedmont residual soils at Opelika, to plot in graph with the normalized cone tip resistance, $Q = (qt - \sigma_{vo})/\sigma'_{vo}$, versus the small-strain shear modulus normalized to cone tip stress, G_{max}/q_t. Based on this chart, the Piedmont residuum categorizes as a sand to silt mixture. In contrast, the same SCPTU data plotting the small strain stiffness directly with net cone resistance index the same soil to clayey group. Taking account that fissures in the residual soil are believed to reflect the relict discontinuities of the original fractured rock formation (Finke & Mayne, 1999), the measurement of positive u1 and negative u2 pressures at the shoulder position are characteristic of stiff fissured clays (Mayne et al. 1990; Lunne et al. 1997). The very fast dissipation records observed in these residual materials, however, are not indicative of clay materials (e.g., Burns & Mayne, 1998). All these signs are, therefore, a clear proof that the application of such classifications in these soils should dealt with great care. In fact, Mayne and Brown (2003) point out some relevant points that arise from a significant number of cone penetration tests (CPTs). The phenomenon of positive u1 readings with negative u2 response has been previously observed in fissured overconsolidated

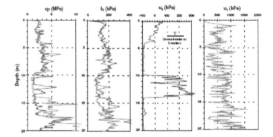

Figure 8. CPTu with midface u1 and shoulder u2 porewater pressures (Mayne & Brown, 2003).

clays; the Piedmont residuum has relict features of the parent rock, including remnant bonds from the intact rock, as well as the discontinuities and cracks of the rock mass (Sowers, 1994): the midface u1 response is positive (see illustration in Figure 8), because it is dominated by the destructuration of the residual intact bonding of the original rock continuum, while the shoulder u2 readings are negative because they reflect shear-induced porewater pressures and remnant rock discontinuities within the matrix (e.g., Finke et al. 1999; Mayne & Schneider, 2001).

Mayne & Brwon (2003) emphasize that in the "vadose" zone above the groundwater table, the penetration porewater pressures at midface have been observed to be either positive or near zero, while at the shoulder, the values can be positive, zero, or negative, attributing these variances to the transient

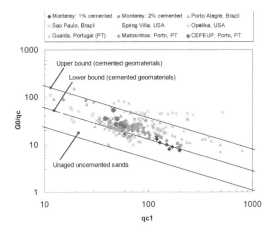

Figure 9. Relationship between G_0 and q_c for residual soils (completed from Schnaid et al. 2004).

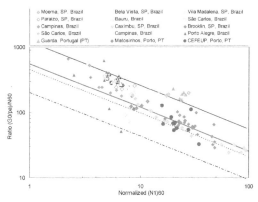

Figure 10. Relationship between G_0 and N_{60} for residual soils (completed, Schnaid et al. 2004).

capillary conditions due to degree of saturation, partial or full, in the residual fine-grained soils, depending on the humidity, infiltration, and prior rainfall activities around the actual time of testing; the importance and relevance of unsaturated soil mechanics in these residual silty soils can be appreciated, yet not implemented and well-studied in this geology.

Bearing in mind that soil classification using CPTU data is indirect and relies on empirical charts developed for strata interpretation, Schnaid et al. (2004) make the point that u_2 measurements cannot always be considered useful to ensure a proper soil classification in unusual materials. Since classification charts should rely on at least two independent measurements, in the absence of pore pressure measurements, the authors suggest that q_c should be compared with the small strain stiffness G_0. The G_0/q_c ratio provides a measure of the ratio of elastic stiffness to ultimate strength and may therefore be expected to increase with ageing and cementation, primarily because the effect of these on G_0 is stronger than on q_c. Other authors, such as Bellotti et al. (1994), Rix & Stokoe (1992), Lunne et al. (1997) and Fahey et al. (2003) have reported, in sands, some new insights by correlating G_0/q_c versus q_{c1}, where q_{c1} is defined as:

$$q_{c1} = \left(\frac{q_c}{p_a}\right)\sqrt{\frac{p_a}{\sigma'_{v0}}} \qquad (1)$$

where p_a is the atmospheric pressure.

Once profiles of q_c and G_0 are determined, these values can be directly used to evaluate the possible effects of stress history, degree of cementation and ageing for a given profile, as already recognised by Eslaamizaad & Robertson (1997). Data points are shown in Figure 9 for CPT tests carried out in residual soils (artificially cemented Monterey soils are

also included). The plotted data completes the synthesis done by Schnaid et al. (2004), including the values obtained for the FEUP experimental site – CEFEUP, Porto, PT (Viana da Fonseca et al. 2006). Since residual soils always exhibit some bond structure, the data fall outside and above the band proposed by Eslaamizaad & Robertson for uncemented soils.

The variation of G_0 with q_c observed in the range of sand deposits was expressed by upper and lower bounds. The upper bound for uncemented material can be assumed as a lower bound for cemented soils and a tentative new upper bound for cemented materials can be expressed as (Schnaid et al. 2004):

$$\left.\begin{aligned} G_0 &= 800\sqrt[3]{q_c\sigma'_v p_a} \quad \text{upper bound : cemented} \\ G_0 &= 280\sqrt[3]{q_c\sigma'_v p_a} \quad \text{lower bound : cemented} \\ & \qquad\qquad\qquad\quad \text{upper bound : uncemented} \\ G_0 &= 110\sqrt[3]{q_c\sigma'_v p_a} \quad \text{lower bound uncemented} \end{aligned}\right\} \quad (2)$$

As for the CPT, Schnaid et al. (2004) point out that SPT N values can also be combined with seismic measurements of G_0 to assist in the assessment of the presence of bonding structure and its variation with depth. Such a combination is provided on Figure 10, which plots G_0/N_{60} against $(N_1)_{60}$ in residual soils, where $(N_1)_{60} = N_{60}(p_a/\sigma'_{vo})^{0.5}$ and is analogous to q_{c1} on Figure 11. Values of FEUP experimental site are also included. The bond structure is seen to have a marked effect on the behaviour of residual soils, with normalised stiffness (G_0/N_{60}) values considerably higher than those observed in cohesionless materials. A guideline formulation to compute G_0 from SPT tests may be given by the equations:

$$\frac{(G_0/p_a)}{N_{60}} = \alpha N_{60}\sqrt{\frac{p_a}{\sigma'_{vo}}} \quad \text{or} \quad \frac{(G_0/p_a)}{N_{60}} = \alpha (N_1)_{60} \quad (3)$$

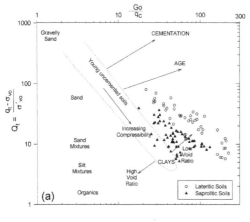

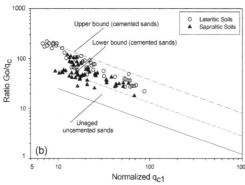

Figure 11. Relationship between Go and qc for three sites in Brazil: (a) Robertson et al (1995) chart (b) Schnaid et al (2004) chart (modified from Giacheti & De Mio, 2008).

where α is a dimensionless number that depends on the level of cementation and age as well as on the soil compressibility and suction. The variation of G_0 with N can also be expressed by upper and lower boundaries (Schnaid et al. 2004):

$$
\left.
\begin{aligned}
G_0 &= 1200\sqrt[3]{N_{60}\sigma'_v p_a^2} && \text{upper bound : cemented} \\
G_0 &= 450\sqrt[3]{N_{60}\sigma'_v p_a^2} && \text{lower bound : cemented} \\
&&& \text{upper bound : uncemented} \\
G_0 &= 200\sqrt[3]{N_{60}\sigma'_v p_a^2} && \text{lower bound uncemented}
\end{aligned}
\right\} \quad (4)
$$

Giacheti & De Mio (2008), in a paper for this conference, present SCPT test results from three relatively well-studied tropical research sites (Bauru, São Carlos and Campinas) in the State of São Paulo, Brazil. All the SCPT data for the three sites were plotted on the Robertson et al (1995) and Schnaid et al (2004) chart in Figure 11. Robertson et al (1995) chart considers normalized cone penetration $[Q_t = (q_t - \sigma_{vo})/\sigma'_{vo}]$ versus

the ratio Go/q_t (Figure 11a) and Schnaid et al (2004) chart plotting Go/q_c ratio *versus* q_{c1} (Figure 11b).

SCPT test allows calculation of the Go/q_c ratio simplifying interpretation and reducing soil variability. The interpretation of SCPT data indicated that the bonded structure of tropical soils produces Go/q_c ratios that are systematically higher than those measured in cohesionless soils. It was also observed that lateritic soils achieve a higher Go/q_c ratio than saprolitic soils.

The high elastic stiffness and the low point resistance at the surficial soil layers is caused by laterization process which enriches the soil with iron and aluminum and their associated oxides, the high concentration of oxides and hydroxides of iron and aluminum bonds support a highly porous structure. Relating an elastic stiffness to an ultimate strength is an interesting approach to help identify tropical soils since the low strain modulus from seismic tests reflects the weakly cemented structure of lateritic soils while the penetration brakes down all cementation. The results are in agreement with the propositions of Schnaid et al. (2004).

Within the Federal District (Brasília – Brazil) extensive areas (more than 80% of the total area) are covered by a weathered latosoil of the tertiary-quaternary age. This latosoil has been extensively subjected to a laterization process and it presents a variable thickness throughout the District, varying from few centimeters to around 40 meters. There is a predominance of the clay mineral caulinite, and oxides and hydroxides of iron and aluminum. The variability of the characteristics of this latosoil depends on several factors, such as the topography, the vegetal cover, and the parent rock. In localized points of the Federal District the top latosoil overlays a saprolitic/residual soil with a strong anisotropic mechanical behavior (Cunha and Camapum de Carvalho 1997) and high (SPT) penetration resistance, which originated from a weathered, folded and foliate slate, the typical parent rock of the region. Figure 12 shows typical soil profile with results from natural water content in different periods of the year. Typical N (SPT) results are also presented.

To evaluate the influence of the seasonal variation of in situ tests results, a linear correlation with suction was attempted; the best option is to correlate N with the suction normalized by the void ratio (pF/e), proving that there is no direct relation to suction (pF).

2.2 Mechanical properties: deformability and strength parameters (saturated soil)

These soils exhibit clearly yield loccus. This is defined as a stress or stress state at which there is a discontinuity in stress-strain behavior, and a decrease in stiffness. This yield stress is similar to that of an overconsolidated sedimentary soil, except that it is caused by chemical bonding between particles instead of

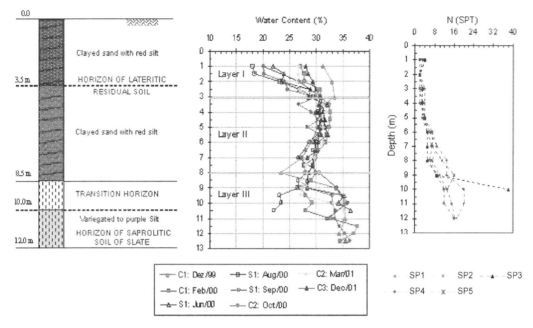

Figure 12. Typical soil profile with results from natural water content in different periods of the years 1999 to 2001, with SPT, University of Brasília, Brazil (Mota et al. 2007).

precompression. The deformation of residual soils *in situ* will be small unless the yield stress is exceeded. Once the yield stress has been exceeded bonding is progressively destroyed as strains increase. The slope of a void ratio vs log stress plot in one-dimensional or isotropic stress compression test after yield is a function of the yield stress and initial porosity of the soil – Figure 13 – rather than being an intrinsic function of the grading and mineralogy of the soil (Vaughan et al. 1988; GSEGWP, 1990; Mitchell & Coutinho, 1991). Usually the compressibility of the soils is measured in laboratory in an oedometer test (a one-dimensional compression test), as shown in Figure 13. The parameter C_{CS} for bonded soil has been found to be a function of initial void ratio, rather than of grading and plasticity, as in unbounded soils (Vaughan, 1988; Lacerda & Almeida, 1995).

The results of oedometer tests presented in Figure 13b reveal three distinct zones typical of cemented soils (Vaughan, 1988) and corresponding to different states and ranges of the compressibility index: (i) "stable", $C_r = 0.007–0.018$; (ii) "metastable", $C_{cs} = 0.129–0.289$; and (iii) "granular or de-structured", $C_c = 0.101–0.172$. In the first state, the soil generally preserves its natural cemented structure, while in the third state this structure is completely destroyed; the intermediate state is associated with progressive de-structuring, resulting from the gradual breakage of interparticle bridges by compression. While the first zone can be likened to an

overconsolidated state in transported soils, the second, although apparently similar to a normally consolidated state, is really a transition state characterised by higher values of compressibility than the same material would have under remoulded conditions (granular state). Results for the remoulded soil are included in the figure. A virtual preconsolidation stress, σ'_{vP}, can be defined as the value which separates the first two zones, where most of the stress states generated by current loading conditions are situated.

Figure 14 shows data from various sources, including results of young and mature residual soils from Pernambuco and Rio de Janeiro, Brazil (Coutinho et al. 1998 and Futai et al. 2007); the young residuals soils showing a correlation: $C_c = 0.41 (e_0 - 0.2) + 0.01$.

The residual soils present a cohesive-frictional nature with its shear strength significantly influenced by: (1) the presence of bonding between particles, that gives both a component of strength and of stiffness; (2) widely variable void ratio which is a function of the weathering process and is not related to stress history; and (3) partial saturation, possibly to considerable depth (GSEGWP, 1990). The measurement of shear strength of residual soils requires samples of high quality, and test specimens should be large enough to include large particles fabric elements. Sandroni, 1977, in Lacerda & Almeida (1995), studied gneissic residual soils, and found that the mineralogy of the sand fraction, including the proportions of mica and feldspar, correlates well with the strength of the

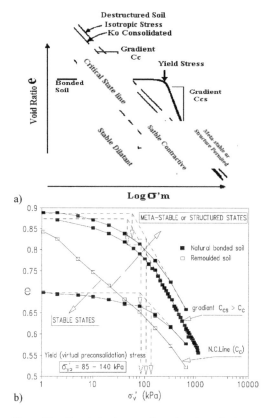

a)

b)

Figure 13. Correlation between Log σ'_m and e_0: a) adapted GSEGWP (1990); and, b) data from Viana da Fonseca (1996, 1998).

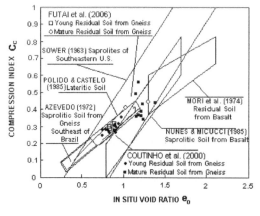

Figure 14. Correlation between C_c and e_0 (modified from Coutinho et al. 1998).

soil, which tends to be higher as the feldspar content increases and the mica content decreases. Figure 15 shows how the mineralogy of the coarse fraction affects the shear strength of saprolitic gneissic soil, including other results of young residual soils from Pernambuco (Coutinho et al. 1998, 2000) and Rio de Janeiro (Futai et al. 2006), Brazil, which are in agreement with the early results. Strength parameters from other types of residual soils are also presented.

The residual strength is reached along a surface of discontinuity after large shear displacements. It is characterized by a residual friction angle, ϕ'_r without any cohesion. The residual friction angle depends on the percentage of particles that can be reoriented during shearing. As plate-shape particles are found among clay minerals most of the time, the residual friction angle has often been related to the clay fraction, CF, or Plasticity Index of soils. It is worth noting that these relationships do not directly take into account the shape of the particles (for instance presence of mica in silty/sandy soils) and / or the aggregate structure present in tropical clayey soils and may consequently be misleading. Also, for soils for which the particles cannot be reoriented the residual friction angle is not significantly lower than the critical state friction angle (Leroueil & Hight, 2003, Lacerda, 2004 and Coutinho & Silva, 2005; Lacerda & Fonseca, 2003; Fonseca & Lacerda, 2004; Fonseca, 2006). Figure 16 shows results of ϕ'_r versus clay fraction (%) for many types of soils, where C represents colluvium and R / S represents residual soil and FB Barreiras Formation. All other points are tests on sands, kaolin and bentonite.

The soils from "Barreiras" Formation and some residual soils are in agreement with the correlation; however, despite the high clay content, colluvial material or the mature residual granite (R5) due to its aggregate structure present a high residual strength and, despite the low clay content, the saprolitic gneiss (S1 to S4) due to high presence of mineral mica present a low residual strength.

Undistorbed samples – using blocks and high quality piston samplers – have been carefully tested in the University of Porto, comprising CK_0D triaxial tests – in compression with bender element (BE) readings and in extension – with local strain measurements, resonant column (RC) and oedometer tests (Viana da Fonseca, 2003, Greening et al. 2003, Viana da Fonseca et al. 2006). To illustrate some of these results, stress-path triaxial tests and corresponding stress-strain curves are shown in Figure 17. At rest coefficient K_0 was taken as 0.50. Regional experience indicates even lower values (Viana da Fonseca and Almeida e Sousa, 2001).

From tests results, the cohesive-frictional nature of strength is clear ($\phi' = 45.8°$; $c' = 4.5\,kPa$), being also notorious the anisotropic stress-strain answer, very much evident when comparing the curves in

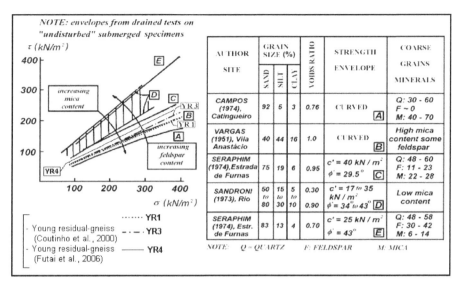

	GRAIN SIZE (%)			VOIDS RATIO	STRENGTH ENVELOPE	COARSE GRAINS MINERALS
AUTHOR SITE	SAND	SILT	CLAY			
CAMPOS (1974), Catingueiro	92	5	3	0.76	CURVED **A**	Q: 30 - 60 F ~ 0 M: 40 - 70
VARGAS (1951), Vila Anastácio	40	44	16	1.0	CURVED **B**	High mica content some feldspar
SERAPHIM (1974),Estrada de Furnas	75	19	6	0.95	c' = 40 kN / m² ϕ' = 29.5° **C**	Q: 48 - 60 F: 11 - 23 M: 22 - 28
SANDRONI (1973), Rio	50 to 80	15 to 30	5 to 10	0.30 0.90	c' = 17 to 35 kN / m² ϕ'= 34° to 43° **D**	Low mica content
SERAPHIM (1974), Estr. de Furnas	83	13	4	0.70	c' = 25 kN / m² ϕ' = 43° **E**	Q: 48 - 58 F: 30 - 42 M: 6 - 14

NOTE: Q = QUARTZ   F: FELDSPAR   M: MICA

NOTE: envelopes from drained tests on "undisturbed" submerged specimens

- Young residual-gneiss (Coutinho et al., 2000) ······ YR1 – · – YR3
- Young residual-gneiss (Futai et al., 2006) —— YR4

Material	Grain size (%)			Strength envelope (saturated condition)	Mineralogy
	Sand	Silt	Clay		
Young residual – gneiss - PE (Coutinho et al. 2000)	83 60-68	11 19-23	4 7-12	ϕ'= 29.3°, c' = 2.9 kPa ϕ'= 30.1°, c' = 2.2 kPa	M = 40-50%, Q and others ~ 50-60%
Young residual – gneiss - RJ (Futai et al. 2006)	38	50	12	ϕ'= 30°, c' = 10 kPa (average)	kaolinite= 40%, Q= 45%, gibbsite and others
Residual – calcareous - PE (Silva et al. 2004)	20	39	40	ϕ'= 31.6°, c' = 11.3 kPa	calcite = 70%, ilite, kaolinite and others = 30%
Mature residual – granite - PE (Lafayette, 2006)	24	16	60	ϕ'= 31.3°, c' = 7.6 kPa	kaolinite, ilite and Q
Mature residual – granite - PE (Silva, 2007)	28	23	49	ϕ'= 26.3°, c' = 9.7 kPa	kaolinite, Q (predominant) and M
Young residual – Porto granite (Viana da Fonseca, 1996,2003)	52-69	17-32	4-8	ϕ'= 37-38°, c'= 9-12kPa ϕ'$_{cv}$=31.6°	M ~ 23%, Q ~ 38%, F ~ 14%, K ~ 24%
Young residual – Porto granite (Viana da Fonseca, 2006)	50-95	10-41	1-8	ϕ'= 46 °, c'= 5kPa ϕ'$_{cv}$=33°	M~12-45%,Q~1-25%, F ~ 2%, K ~ 54-90%

Figure 15.   Influence of mineralogy in strength of residuals soils from gneiss (Sandroni, 1977, modified by Coutinho et al. 2004a).

compression and extension stress-paths. This has been confirmed during several research programs in residual soil from Porto granite – an inherently homogeneous rock, which reveals the induced anisotropy in a soil characterized by low at rest stress state (Viana da Fonseca, 1998, 2003, Viana da Fonseca et al. 2003, 2006).

Specimens tested under drained shear show a transition from brittle/dilatant behaviour to a ductile/compressive response, as confining stress increases (Figure 18, from Viana da Fonseca et al. 1997a, 1998b, Viana da Fonseca & Almeida e Sousa, 2002). As resumed by Santamarina (2001), dilatant

behaviour in some very porous relic structured soils, for low confining stresses, are definitely due the formation of grooms of very small cement particles crushing and pushing out the bigger grains, while disaggregating, and evoluting to critical state… Under high confinement stresses this will not occur and the energy necessary to disrupt the structure and disaggregate the bonds rapidly prevails by the volumetric component. We have, therefore, different patterns in the two stress states.

Bonding condition gives rise to tensile strength, explaining the cohesive-frictional nature of these residual soils. Mohr-Coulomb shear strength is

207

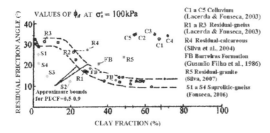

Figure 16. Residual friction angles of colluvium and residual soil (modified from Lacerda, 2004; Skempton 1985, after Lacerda & Fonseca, 2003).

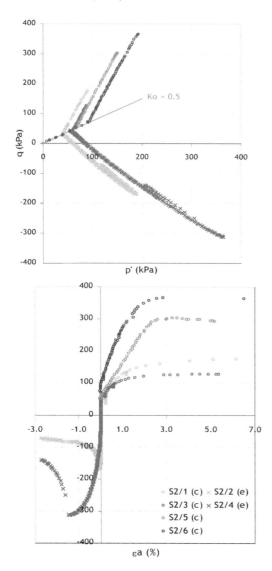

Figure 17. Triaxial tests: stress-path and stress-strain curves (compression and extension).

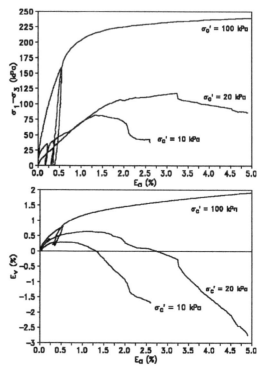

Figure 18. CID triaxial tests under three distinct consolidation stresses (Viana Fonseca et al. 1997).

markedly defined by the binomial ϕ' and c' and this cohesive intercept will be as much relevant as the loading stress path is as less dominated by volumetric compression as possible: Figure 19, taken from the work of Rios Silva (2007) shows markedly the highest evidence of the cohesive tensile component when a compression path with decreasing of mean effective stress prevails, in opposition to others with increasing of mean effective stress; this has notorious significance in the differences of derived geotechnical parameters obtained for in situ expansion tests (such as pressuremeters) versus compressive tests, such as penetrating tools!

One of the most important and unresolved question is whether the yield locus is isotropic or anisotropic, and if this is, or not, centred in the K_0 stress axis. Viana da Fonseca et al. (1997a) reported some results of isotropic consolidation tests with local measurement of axial and radial strain, which provided values of the virtual isotropic preconsolidation stress slightly lower than the one deduced from the oedometer tests, taking $K_0 = 0.38$ (Figure 20a).

The shape of the yield curves of natural clays has also been studied by many of authors (e.g, Tavenas and Leroueil, 1977; Graham et al. 1983; Smith et al. 1992; Diaz-Rodriguez et al. 1992) and it has been observed

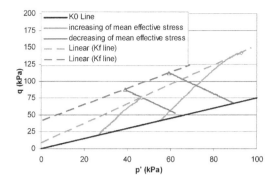

Figure 19. CK$_0$D triaxial tests under distinct stress-paths (Rios Silva, 2007).

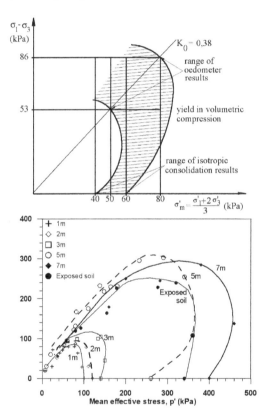

Figure 20. Yield surfaces for volumetric compression ($\sigma1$ = vertical stress, $\sigma3$ = horizontal stress) – Viana da Fonseca et al. (1997a); b) Limit state curve for saturated condition (Futai et al. 2006).

that the yield locus is anisotropic centred in the K$_0$ stress axis due to the conditions prevailing during their deposition. In bonded soils this not yet well known, as limited data are available. Leroueil and Vaughan (1990) (see also, Leroueil & Hight, 2003) suggested

that, on the basis of the data available at that time, residual soils and soft rocks may have yield curves centred on the hydrostatic axis. Machado & Vilar (2003) working with young residual soil from a sandstone obtained yield curves also centred along the p axis. Schnaid (2005) pointed out that factors controlling the shape of the limit state curve in bonded soils has not yet fully identified, due to lack of experimental data in natural samples. Futai et al. (2006) however showed that yield curves of tropical soils under saturated conditions may be isotropic or anisotropic with respect to the hydrostatic axis, depending on the degree of weathering, the nature of the mother rock, and diagenesis. Futai et al. (2006) shows limit state curves from gneissic mature and young residual soils determined by a number of tests chosen in order to explore different regions of the q:p′ space. The results obtained at all depths are presented in Figure 20b. The expansion of the limit state curves with the increase of depth is quite clear in the figure. It is shown that limit state curves for soils from depths of 1.0 and 2.0 m in horizon B are centrated on the hydrostatic axis. Limit state curves of soils from horizon C (depths 3–5 m) are not centred on the hydrostatic axis, which may be due to the remaining 'mother' rock anisotropy, showing similar shapes observed for natural clays that are anisotropic due to the K$_{onc}$ stress conditions prevailing during their deposition (Diaz-Rodriguez et al. 1992).

2.3 Mechanical properties in unsaturated conditions

Unsaturated soils in their various categories (residual soils, collapsible soils, expansive soils and so forth) can be found almost everywhere in the world. Their importance in the geotechnical field came to be recognized with the development of the arid, semi-arid and tropical regions causing engineers to work on soils showing geotechnical behavior different than those predicted according to the principle of effective stress (PES) which has been developed for saturated soils and temperate climate. Fredlund (2006) illustrates the progression from the development of theories and formulations to practical engineering protocols for a variety of unsaturated soil mechanics problems (e.g., seepage, shear strength, and volume change). The ground surface climate is a prime factor controlling the depth to the groundwater table and therefore, the thickness of the unsaturated soil zone. The zone between the ground surface and the water table is generally referred to as the unsaturated zone. The water degree of saturation of the soil can range from 100% to zero. The changes in soil suction result in distinct zones of saturation. The zones of saturation have been defined in situ as well in the laboratory (i.e., through the soil-water characteristic curve): capillary finge – boundary effect; two phase fluid flow – transition; dry (vapour transport to water) – residual.

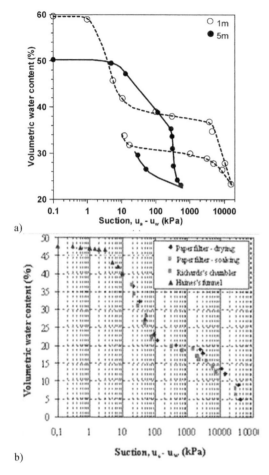

a)

b)

Figure 21.   Soil water retention curve: (a) Futai et al. (2007); and (b) Lafayette (2006).

Several approaches can be used to obtain a soil-water characteristic curve (Fredlund, 2006).

Figure 21a (Futai et al. 2006) compares the soil-water retention curves of two gneissic residual soil (mature and young) horizons B and C, measured using the suction plate, for suction lower than 30 kPa, pressure plate (suction between 30 kPa to 500 kPa) and the filter paper technique (Chandler & Gutierrez, 1986) for higher suction. The differences between the two soils regarding grain size distribution; mineralogical composition and microstructure directly influence the water retention capacity. Porosimetry measurements appear to confirm the results of the grain size analysis, i.e., horizon B possess smaller pores and higher clay content than horizon C. The overall analysis of grain size, soil microscopy and porosimetry suggests a meta-stable structure for the horizon B soil comprising micro and macro pores. Figure 21b (Lafayette, 2006), shows the soil-water retention curve of a mature

granite residual soil, measured using Haines'funnel, Richards's chamber and the filter paper technique.

The bimodal retention curve presented by the two mature residual soils is typical of tropical highly weathered soils (Camapum de Carvalho & Leroueil, 2004) containing aggregated particles uncemented or cemented by iron oxides linked by clay bridges. According to the same authors, these soils have two air entry points, corresponding to macro and micro pores (see also Feuerharmel et al. 2007). The suction developed in an unsaturated soil can affect the stress strain response, the shear strength and the yield stress-curve, and the permeability of the soil.

In a general form, the unsaturated shear strength equation can be written as follows:

$$\tau = c' + (\sigma_n - u_a)\tan\varnothing' + (u_a - u_w)f_1 \qquad (5)$$

where τ = shear strength; c' = effective cohesion intercept; σ_n = total normal stress on the failure plane at failure; $\varnothing'$ = effective angle of internal friction; and f_1 = soil property function defining the relationship between shear strength and soil suction.

The soil properties, c', and, are presented as saturated soil constant but the soil property, f_1, varies in response to the amount of water filling the voids of the soil (i.e., it is a function of matric suction). There is curvature to the shear strength envelope with respect to matric suction and the curvature can be related to the SWCC. In many soils the shear strength reaches an ultimate value; some particular soils have shown that the shear strength drops off to a lower value at high value of suction, i.e., greater than residual soil suction (Fredlund, 2006; Vilar, 2007). Some empirical functions have been proposed to deal with the non linearity of the shear envelope of unsaturated soils. For instance, Vilar (2007) has proposed a hyperbolic equation.

Thinking in practical solutions, Vilar (2007) proposed an expedite method to predict the shear strength of unsaturated soils, using data from saturated samples and from air dried samples or, alternatively, from samples tested at a known suction that is larger than the maximum expected suction in the problem.

Futai et al. (2006) present values of cohesion intercept c and friction angle ϕ as a function of suction are shown in Figure 22 and it is noticed that both c and ϕ increase with the suction, although the latter increases just slightly in the range s = 100–300 kPa for the 1.0 m deep specimen. The increase of the cohesion intercept c with suction is well known (Fredlund, 2006; Vilar, 2007) but the increase in ϕ with the suction is less common. In the same figure also is showed results from a mature granite residual soil (Lafayete, 2006) presenting some similarity in the results. Based on these works, it is to be expected that, depending on mineralogical composition, the friction angle can vary with suction.

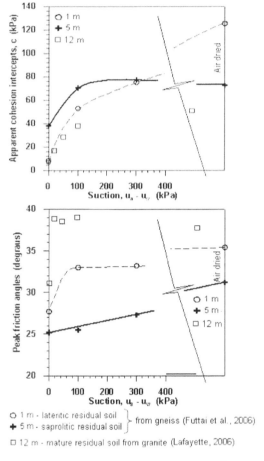

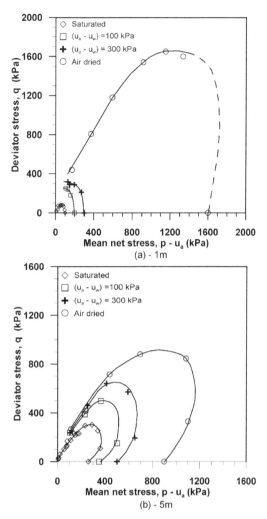

Figure 22. Cohesion intercept c and friction angle φ versus suction (Futai et al. 2007, with results from Lafayette, 2006).

Figure 23. Yield curves under constant suction values (Futai et al. 2004).

The effect of increasing suction is to enlarge the yield curves, with maintaining the shape. Yield curves not centred in the hydrostatic axis have been obtained for compacted and artificially cemented unsaturated soils (Cui & Delage, 1996; Maâtouk et al. 1995; Leroueil & Barbosa, 2000). Machado & Vilar (2003) obtained yield curves centred in the hydrostatic axis for an intact sandstone young residual soil sandy, probably due to the isotropic characteristic of the mother rock.

Futai et al. (2007) determined yield curves for saturated conditions, for suction values equal to 100 kPa and 300 kPa and in air dried conditions. Yield curves under different suctions of specimens of 1.0 m and 5.0 m depths are shown in Figure 22. It is observed that suction has a strong effect on the increase in size of the yield curves maintaining the shape. The yield a strong effect on the increase in size of the yield curves maintaining the shape.

The yield curves of the 1.0 m specimens (Figure 23a), appear to be isotropic, as previously observed in saturated specimens of the 2.0 m horizon B (Futai et al. 2004). The 5.0 m deep specimens presented anisotropic yield curves, as observed (Futai et al. 2004) in saturated specimens from other depths at horizon C. Cui & Delage (1996) also obtained for a compacted silt soil anisotropic yield curves of the same shape, with suctions in the range 200–1500 kPa. Machado et al. (2002) obtained isotropic yield curves for an intact sandstone residual soil collected at a depth of 8 m and subjected to suctions nature of the s = 0, 100 and 200 kPa. This isotropic yield curves could be a consequence of isotropy of parent rock.

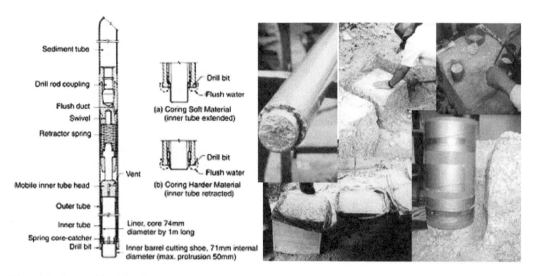

Figure 24. Retractable triple-tube core-barrel (Mazier) (GCO 1987) and block sampling (Viana da Fonseca & Ferreira, 2002, Ferreira, 2003, 2008).

2.4 *Sampling a fount of uncertainty*

Sampling quality is a key parameter in characterizing feasible soil properties through laboratory testing. In order to minimize the disturbance introduced by sampling process, innovative techniques have been proposed for retrieving high quality samples. Hight (2000) examines the effects of sampling on the behaviour of soft clays, stiff clays, and sands, describing improvements that have been made to more common methods of sampling, which have enabled higher quality samples to be obtained, and describing how the quality of these samples can be assessed. The less confiable and most elaborated and reliable samplers for clays and sands were thoroughly analysed in the very complete state-of-the-art of Hight, 2000. Residual soils are too much variable to index them to clays or sands, or intermediate materials, but certainly their behaviour is very much dependent on the macro and micro interparticle bonded structure, which has to be preserved both when they are weak (very sensitive to induced strains) or stronger (for less weathered profiles). Conventional rotary sampling and block sampling are considered as suitable sampling techniques to obtain sufficiently intact samples for determining shear strength and stiffness of soils derived from in situ rock decomposition (GCO 1987). In Hong Kong or in Portugal, the Mazier core-barrel is becoming common for soil sampling (Figure 25; GCO 1987). When a soil sample with the least possible disturbance is required, the block sampling technique can be applied to retrieve the soil sample. Block samples are obtained by cutting exposed soil in trial pits or excavations (Figure 24).

As described by Ng & Leung (2007a), Mazier sampler is a triple-tube core-barrel fitted with a retractable

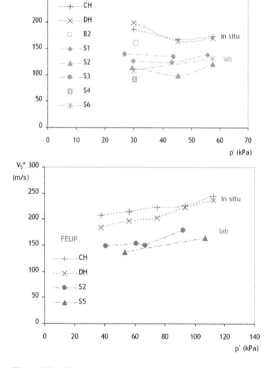

Figure 25. Normalized shear wave velocities, for different types of samplers used in residual soil from granite in to experimental sites in Porto (Ferreira et al. 2002).

shoe. The cutting shoe and connected inner barrel projects ahead of the bit when drilling in soil and retracts when the drilling pressure increases in harder materials. This greatly reduces the possibility of any drilling fluid coming into contact with the core at or just above the point of cutting and hence minimizes sample disturbance. Moreover, the Mazier core-barrel has an inner plastic liner that protects a cored sample during transportation to a laboratory. The core diameter is 74 mm, which is compatible with some laboratory triaxial testing apparatus. Local experience in Hong Kong regards the Mazier sampling technique as the most suitable sampling method available for weathered granular materials at depths. However, comparable studies between samples recoiled by this method and blocks, developed in the University of Porto (Ferreira et al. 2002), have proved that this technique is very sensitive to the executions process, mostly for the injection procedure (pressure, flow volume and type of fluid), putting in risk the natural structure for the more weathered and granular profiles. Some results are presented below. In the above referred paper and in a more recent one (Ng & Leung, 2007b), the authors recognize the limitations associated to this technique, emphasizing the importance the control of shear-wave velocities in controling the relative higher quality of the block specimens, with average 27% higher values than those of the Mazier specimens.

The definition of a reliable methodology for assessing sampling quality is a very important topic. Among the most referenced techniques, such as fabric inspection, measurement of initial effective stress, measurement of strains during reconsolidation (details in Hight, 2000), the comparison of seismic wave velocities in situ and in laboratory specimens has seen increasingly accepted as most promising, specially in natural structured soils, such as residual masses. The characterization of experimental sites on residual soil from granite in Porto has enabled an objective approach to the proposed methodology of assessment of sampling quality in the soils in question. In this context, extensive geological and geotechnical surveying was undertaken, which included a series of Cross-Hole and Down-Hole tests in boreholes where samples were previously obtained, by means of different sampling techniques. In situ and laboratory tests results have enabled comparisons between seismic wave velocities measured in all different conditions involved: in the field; on undisturbed high quality samples (block sampling); on disturbed samples – tube sampling – and remoulded samples – reconstituted in laboratory (Ferreira et al. 2002). The main sampling effects were identified, in terms of the tools, techniques and procedures currently employed (Figure 25).

A posteriori, Viana da Fonseca et al. (2006) presented new comparative analysis of in situ seismic V_S obtained in CH tests and measured in the laboratory,

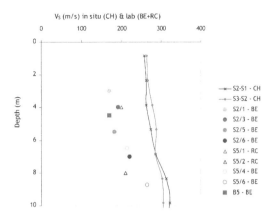

Figure 26. V_S profiles from in situ (crosshole, CH) and triaxial tests with Bender Elements over specimens obtained by different sampling techniques – Porto saprolitic soil (Ferreira, 2008).

with bender elements (BE) or in the resonant column (RC) – these turned to be very much convergent with values form BE, as reported by Ferreira et al. 2007. These are compared in Figure 26, where the similarity of V_S trends in depth from both in situ and laboratory tests is evident and the differences encountered may well be mainly due to the disturbances associated with the sampling processes. These residual soils are particularly sensitive to sampling, since their behaviour is strongly controlled by the structure inherited from the parent rock.

These issues relating to sensitivity to sampling processes were discussed in detail in Viana da Fonseca and Ferreira (2002) and Viana da Fonseca et al. (2006). In terms of the samples recoiled using tube samples, it has been observed that the geometric characteristics of the sampler were determinant, both in terms of the cutting edge taper angle as well as of the area ratio. The good performance of the Shelby sampler is worth mentioning, which is most likely associated to the difference between the internal diameter of the sampling tube and the final diameter of the tested specimens, requiring the trimming of the sample to the appropriate size, after its extrusion from the sampler. This process, despite being quite delicate, enables to eliminate the peripheral areas of the sample which have experienced greater distortions during both sampling and extrusion, pointing out to the relevance of the laboratory sample preparation techniques. Finally, samples recoiled by Mazier process are surprisingly highly affected. Its disturbance is probably related with the "forced" application of high water injection pressures at the cutting shoe, during the sampling procedure with the equipment, which is, in these geomaterials, heavily aggravating. Ageing may also explain why velocity of body waves of natural deposits of some age differ from that of same soil in laboratory with the same state of effective stress and void ratio.

Other trends are presented by Ng et al. (2000) using suspension P-S velocity logging and crosshole seismic measurements (Ng & Wang 2001), in Hong Kong, showing that the shear wave velocity profiles appear to be generally consistent with the SPT 'N' profiles, which is in opposition with the implication of cementation in these residual materials. This may be a consequence of a granular matrix stronger than the matric fabric.

3 IN SITU TESTING FOR MAPPING AND CLASSIFICATION OF RESIDUAL SOILS

3.1 Specific trends and taking advantage of the direct capacity of field techniques

Howie (2004) give emphasis to seismic wave velocities in identifying the presence of fabric and structure that causes residual soils to behave differently from transported soils. Some points are remarked by the author: (i) How can we use or adapt existing correlations developed on the basis of chamber testing on textbook soils? (ii) Under what circumstances is it valid to transfer existing correlations to nontextbook soils? (iii) What is an appropriate test on which to base correlations? Is it useful to carry out SPTs or CPTs? Can we interpret the results? What is the effect of grain size, compressibility, etc. on the test being carried out? How do we select valid parameters for correlation? Is c' a fundamental soil parameter? (iv) What place should the SPT have in site characterization of such cohesive-frictional soils?

Apart from these points, other have been considered very important and reported by Coutinho et al. (2004a): the fact that there is a highly variation of the hydraulic conductivity of residual soil with suction, when these are in unsaturated condition, there is the possibility of occurrence of partial drainage during CPTs and SPTs tests (Schnaid et al. 2004); or, the widely variable void ratio which is a function of the weathering process and not being related to stress history.

For Schnaid et al (2004), the challenge in the field of *in situ* tests includes: (i) the evaluation of the applicability of existing theoretical and empirical approaches in order to extend the experience of 'standard' clays and sands to other geomaterials; (ii) the development interpretative methods that incorporate new constitutive models whenever required; and (iii) the gathering experimental data that justifies the applicability of proposed interpretation methods to engineering applications.

The applicability and potential of existing techniques, implies a critical appraisal on how results can

be compiled to obtain a ground model and appropriate geotechnical parameters.

From the existing field techniques can be broadly divided into (Schnaid et al, 2004): (i) non-destructive or semi-destructive tests that are carried out with minimal overall disturbance of soil structure and little modification of the initial mean effective stress during the installation process. The non-destructive group comprises seismic techniques, selfboring pressuremeter and plate loading tests, a set of tools that is generally suitable for rigorous interpretation of test data under a number of simplified assumptions; (ii) invasive, destructive tests were inherent disturbance is imparted by the penetration or installation of the probe into the ground. Invasive-destructive techniques comprise SPT, CPT and dilatometer. These penetration tools are robust, easy to use and relatively inexpensive, but the mechanism associated to the installation process is often fairly complex and therefore a rigorous interpretation is only possible in few cases.

3.2 Geophysical survey – persecuting heterogeneity

Geophysical methods (namely: seismic, electrical and electromagnetic) have been playing an increasingly more determinant role in near-surface site investigation and characterization for geotechnical purposes, specifically in mapping the geological and/or geotechnical spatial variability of rock/soil mass formations.

Some comparative advantages: flexibility (air, underwater, surface and borehole, in-situ and laboratory; 1D, 2D or 3D mapping; adjustable resolution and data processing procedures; fast coverage of large areas), well established theoretical basis; competitive cost and effectiveness. The synergetic integration of results from different in-situ geophysical and geotechnical as well as laboratory methods/tests in the analytic/interpretative process may reduce significantly the inherent uncertainties associated to modelling the studied geo-system.

The common geophysical in-situ methods may be classified in: (i) Seismic: P- and/or S-waves, surface (conventional and tomographic refraction; reflection; surface waves – SASW, MASW, Passive seismic...), borehole(s) (down-hole; up-hole; VSP; PS-logging; conventional and tomographic cross-hole) and combined surface/borehole(s) variants; (ii) Electrical: surface (1D profiling, 1D sounding, 2D and 3D imaging), borehole(s) (1D profiling; 2D and 3D imaging); (iii) Electromagnetic: surface (GPR, VLF), borehole(s) (GPR); (iv) Other mixed techniques (geotechnical/geophysical) are SCPT, RSCPT and SDMT.

There are some experimental sites in residual soils, where geophysical borehole and surface methods have been used to compare their performance for mapping characterization. In the University of Porto (CEFEUP), this has been made exhaustively in Porto residual soil from granite (Carvalho et al. 2004, 2007, Almeida et al. 2004, Lopes et al. 2004, Viana da Fonseca et al. 2006). These included, namely: P- and S-wave conventional (RC) and tomographic (RT) refraction, high resolution shallow reflection, cross-hole (CH), down-hole (DH), 2D electrical resistivity imaging and ground probing radar (GPR). Direct and derived results from the applied methods and techniques were compared between them, as well as with some of the available geological and geotechnical information. The results are synthesized below. Recent developments in the interpretation of P- and S-wave seismic cross-hole data are presented in this conference (Carvalho et al. 2008).

Geophysical methods have been increasingly carried out throughout the world: namely in Hong Kong using PS-logging (Kwong 1998, Ng et al. 2000), or in Portugal and Brazil using cross-hole (Ng & Wang 2001) and down-hole (Wong et al. 1998) seismic measurements.

3.2.1 Intrusive (borehole) techniques

Boreholes have been used thoroughly for S and P-wave seismic CH and DH surveys. V_S and V_P variability with depth are very well defined by them, and tomographic (RT) refraction sections have been used for a more complete imaging. The fact that the P waves CH detected clearly the presence of water around 1.5 m to 2 m below the piezometer measurement (11.5–12.0 m deep) was associated to the high sensitivity of P waves to total saturation and in less degree to the presence of moist. Conversely, the water level presence distinctive sign may be the transitory decrease in both P and S-wave velocity starting at 10.5 m (Figure 27).

3.2.2 Surface methods

Conventional and tomographic P- and S-wave seismic refraction, shallow reflection, (Figure 28a & b) pole-pole electric resistivity imaging and GPR (Figure 28c) were successfully used for the evaluation of the generic site geological/geotechnical 2D variability (Carvalho et al. 2004, Almeida et al. 2004, Viana da Fonseca

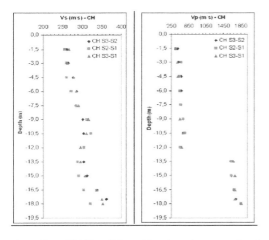

Figure 27. CH VS and VP profiles, 3 sections, water level 11.5–12.0 m deep (Viana da Fonseca et al. 2006).

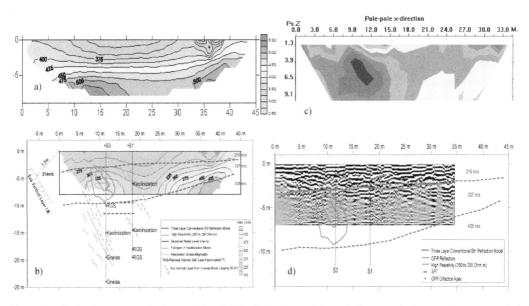

Figure 28. Refraction tomography P-wave (a) and SH velocity (b) models overlaid by geophysical results and geological model used for seismic reflection; c) Electrical resistivity image; d) Processed radargram with interpreted events overlaying the resistivity and CH model and N-SPT values in boreholes S3 and S1 (Carvalho et al. 2004, Almeida et al. 2004).

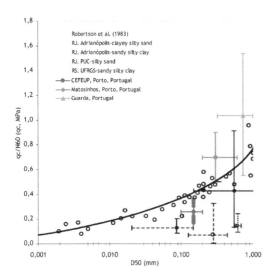

Figure 29. Ranges of q_c/N versus D_{50} on Brazilian residual soils, with experimental site results (Danziger et al. 1998, Viana da Fonseca et al. 2004, 2006; Coutinho et al. 2004).

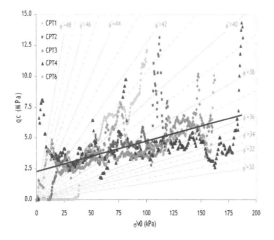

Figure 30. q_c vs σ'_{v0}, and the angle of shearing resistance, ϕ' – adaptation of Robertson and Campanella (1983) to CEFEUP results (Viana da Fonseca et al. 2006).

et al. 2006). Those methods yielded non-contradictory results globally consistent with the existing geological map and compatible with a convergent integrating "final" model (Figs. 28 to 30).

From the experience accumulated with these thorough campaigns in residual soils from granite, some aspects should be mentioned, namely:

(i) tomographic surface refraction is adequate for an average 2D distribution of P- and S-waves velocities and derived elastic parameters (shear modulus, G_0, Poisson ratio, σ) as well as for geological mapping based in a previous model. Possible hindrance: to attain the desired depths may be necessary rather long seismic lines;

(ii) conventional cross-hole allows obtaining reliable 1D profiles of P- and S-waves velocities and deriving the related elastic parameters. Tomographic cross-hole allows obtaining reliable and higher resolution 2D variation profiles of P- and S-waves velocities and derived elastic parameters, when compared to surface refraction;

(iii) Varying soil degree of saturation, above the full saturation level, seems to play an important role in the S-wave CH profiles due to the influence of capillary forces / suction effects on the soil effective stress condition and associated shear strength. Changes in density due to varying water content seem to be comparatively much less influent in S-wave velocity;

(iv) When supposedly soil full saturation depth is attained, high frequency events appear in the horizontal components of the CH;

(v) One of the relevant conclusions is the high spatial variability correlation presented by seismic and electrical imaging section models as well as the very consistent similar horizontal interface pattern common to interpretative models from seismic stacked section, conventional refraction and GPR radargram. In addition, the SH wave velocity fields (cross-hole, reflection and refraction) and resistivity model supports the local geological evidence.

(vi) The resulting refraction, mostly tomographic, and electrical models point out to the adequacy of both methods in mapping the local underground heterogeneities, both horizontally and vertically, inside the more or less weathered granitic mass as well as the boundary with the gneissic migmatite.

SH-wave seismic reflection is a method that usually allows a very good and detailed spatial image of the near surface. The difficulty of acquiring good quality S-wave data in noisy environment, due to the difficulty in producing a sufficient high-energy signal, leads to the concentration of the energy in the direct and refracted wave, avoiding hyperbolic reflection. The SH-wave seismic reflection method was applied with success in ISC'2 test site, in the campus of the University of Porto – FEUP (Lopes et al. 2004), allowing to a good definition of weathering profile, recognizing layers of thickness <1 m (for short spacing between geophones), and showing correctly the lateral variability to depths that depend on the length of the acquisition line.

Another very interesting technique, the Surface Wave Method (SWM) – a multichannel surface wave test based in a one-dimensional wave propagation theory – allows to obtain a V_S soil profile until depths that depend of site specific characteristics, the equipment and acquisition scheme. This test has also proved to be very robust in noisy conditions. The presence of lateral variability can cause misinterpretation of surface wave data as the dispersion properties of a signal contain information about all the area crossed by the acquisition line. Due to the relevance of identifying the errors that depend of the theoretical model in surface wave data, Strobbia & Foti (2006) developed the MOPA procedure that allows to check their presence and importance. Performing MOPA prior to final data processing allows to verify the quality of data and to decide the better approach for processing. In multistation data the presence of lateral variability can be assessed dividing the acquired seismogram in several parts and processing them separately. Even thought the data acquired in the University of Porto test site gave results similar to the obtained crosshole V_S (Carvalho et al. 2004), the careful processing of the data permitted to identify the presence of a lateral variability (details in Lopes et al. 2004). The processing of the surface wave acquisition splitting the multistation data in smaller seismograms permitted to observe an increase of the velocity in E direction (Figure 31) also identified with other seismic techniques by Carvalho et al. (2004).

3.3 Geomechanical characterization

3.3.1 In-situ tests in residual soils and correlations between their results

The appropriate field tests should consider firstly the nature of the construction and the proposed methods of analysis and secondly the nature of the ground. The best in-situ tests for each given project and ground conditions will then be the ones which give the required information regarding the understanding of geological and geotechnical ground conditions and the relevant geotechnical properties and behaviour of the ground such as their constitutive laws to be used by the proposed methods of analysis used in the structural design (Gomes Correia et al. 2004).

The difficulties of sampling residual soils, as referred above, which cause a number of problems for the characterization of stress-strain behaviour of soils through laboratory tests, make in situ tests very important tools in geotechnical practice. The most common tests are by far, the dynamic penetration tests (SPT, dynamic probing – ex. DPSH), and other more limited in penetration capacities – CPT and DMT. Other, more time consuming, such as PMT or PLT (plate load tests), and the utmost technique of self boring pressuremeter (SBPT) are becoming more frequent as they give a more fundamental parametrical information (Gomes Correia et al. 2004; Fahey, 1998, 2001, Fahey et al. 2003, 2007). More recently a special attention is being put to seismic tests for the evaluation of initial shear modulus (G_0), regarded as a highly

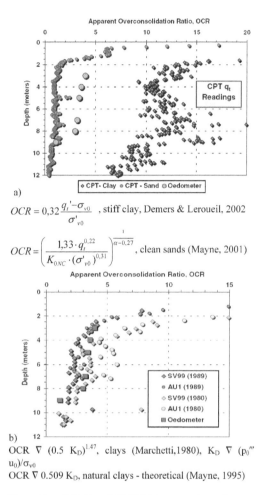

$$OCR = 0{,}32\frac{q_t{}'-\sigma_{v0}}{\sigma'_{v0}}$$, stiff clay, Demers & Leroueil, 2002

$$OCR = \left(\frac{1{,}33 \cdot q_t^{0{,}22}}{K_{0NC} \cdot (\sigma'_{v0})^{0{,}31}}\right)^{\frac{1}{\alpha-0{,}27}}$$, clean sands (Mayne, 2001)

OCR ∇ (0.5 K_D)$^{1.47}$, clays (Marchetti,1980), K_D ∇ (p_0''' u_0)/σ_{v0}

OCR ∇ 0.509 K_D, natural clays - theoretical (Mayne, 1995)

Figure 31. CPT (a) and DMT (b) interpretations of apparent (virtual) OCR in Piedmont residual silts compared with oedometer (Mayne and Brown, 2003).

important benchmark parameter. Although in situ tests suffer serious limitations in terms of interpretation of their results, they nevertheless make a valuable contribution to geomechanical characterization.

Profile specificities and correlations between in situ tests parameters

Apart from evaluation of natural macro-spatial variability, fabric of residual soils due to preserved relic heritage may be fairly homogeneous in geotechnical terms, being demonstrated by the results obtained with continuous sampling from drilling, observed from SPT sampler and, rarely, from high quality samplers (Viana da Fonseca et al. 2004).

The cross-comparison of SPT, CPTU, DMT, PMT and several CH, DH, SASW and CSWS, has been allowing the derivation of several correlations between geomechanical parameters, base for classification, evaluation of at rest stress state, deformability and strength properties, for geotechnical engineering purposes. As it has been made since the early days in geotechnical research, these studies have a purpose to allow for the use of any or some of these tests to get what others do more fundamentally and with more confidence, avoiding some expensive and time consuming tests whenever requested for the success of a project. When data are combined there is more scope for rational interpretation and, for this reason, emphasis has been placed on correlations with mechanical properties that are based on the combination of independent measurements (Schnaid, 2005).

CPT – SPT correlations
The usefulness of correlating results from SPT and CPT tests, also as a sign indexation to granular and physical characteristics, has leaded researchers to the evaluation of the dependence of q_c/N_{60} ratio on the mean grain size, D_{50}, as suggested by Robertson and Campanella (1983). For instance in the case of Porto residuals soils, such as the reported by Viana da Fonseca et al. (2006), this ratio varied from 0.17 to 0.36 ($D_{50} = 0.15$ mm, Figure 32). Different parent rocks generally produce different correlations for the same

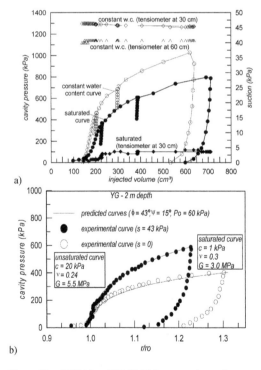

a)

b)

Figure 32. CPT (a) and DMT (b) interpretations of apparent (virtual) OCR in Piedmont residual silts compared with oedometer (Mayne and Brown, 2003).

particle size distribution, due to intrinsic heterogeneity (Danzinger et al. 1998). The Brazilian data show a general trend of lower values for the relation q_c/N_{60} with D_{50} than that expressed by Robertson & Campanella's (1983) average line (Figure 29). Data from FEUP experimental site (Viana da Fonseca et al. 2004, 2006) are in close agreement with those results, but somewhat contradictory with Porto silty sand data (Viana da Fonseca, 1996, 2003) or saprolitic granite soils from Guarda, much coarser matrix, plotted on the same graph (Rodrigues & Lemos, 2004).

The results obtained and presented on Figure 29 clearly show that, in the case of saprolitic soils from Guarda and the former Porto sites, q_c/N relations are conspicuously higher than those proposed for the granular sedimentary soils. This fact should be related to the greater sensitivity of the q_c parameter of the CPT test, than the value of N of the SPT test, concerning the cohesive part of the resistance, due to the existence of weak inter-particulate bonding and significant quartz coarse grains. It is indeed reasonable to accept that grain size distribution plays an important part in controlling stress-strain behaviour, since the coarser grain size of Guarda's saprolitic granite soils exhibits a higher q_c/N ratio than the saprolitic granite soils from Porto, whose grain size is finer. The Brazilian residual soils and those from FEUP site have a q_c/N ratio that is lower than that predicted for sedimentary soils. Both theses soils have a similar mineralogical nature due to the original rock: the Brazilian rock being made of gneiss and sandstone, and the FEUP rock made of granite at the interface of gneiss and schist. These findings lend further support to the idea that grain size properties do not in themselves explain the behaviour of the residual soils, meaning that other parameters must be incorporated into the analysis of the behaviour of these soils, namely, weathering indices, chemical and mineralogical ones.

An important aspect is the link between the drainage conditions during cone penetration and that expected in the design problem (Lunne et al. 1997). Takesue et al. (1995) showed this aspect for a volcanic soil, pointing out that the change in drainage, function of the penetration rate, has a larger effect on the sleeve friction than on the cone resistance. This is consistent with the fact that cone resistance is a total stress measurement contrarily to sleeve friction that is controlled by the effective stresses. The drainage also affects the relationship between CPT and SPT. In fact SPT is the summation of cutting shoe resistance and friction along the outside wall (and to a less extend along the inside wall) of the SPT.

It should be emphasized that there is no absolute certainty that some of these correlations may be affected by a non systematic correction of depth (Skempton, 1986), which is discussed recently as not well fundament (Cavalcante et al. 2004, or Odebrecht et al. 2004), and may explain some incongruence if – and that may be quite common in most of the cases – some of these correlations were developed from tests near surface.

PMT versus CPT/SPT correlations

Viana da Fonseca et al. (2001, 2003, 2006) made a synthesis for two experimental sites on Porto granite saprolitic soils; there are some derived ratios between PMT and SPT or CPT parameters. Some correlations are included in the Table 1.

Ratios between distinct values of Young's moduli inferred from the investigations conducted have the obvious interest of fulfilling the needs of geotechnical designers to obtain data from different origins for each specific purpose.

Viana da Fonseca et al. (2001, 2003) reported some interesting correlations from the data available at the experimental sites: (i) values of Young's moduli determined directly, with no empirical treatment, or even, no deriving assumptions; (ii) common constant ratios that are assumed to correlate SPT (DP) or CPT parameters with Young's modulus, comparing them with transported soils; (iii) relative values of moduli can be summarized in the way that is expressed in Table 2a, while some relations could be pointed out between in situ tests, as expressed in Table 2b.

Table 1. Ratios between in situ tests parameters (Viana da Fonseca et al. 2001, 2003, 2006).

N_{60}/p_l (MPa)	14.6	E_m/p_l	10–12
N_{60}/E_m (MPa^{-1})	1.4	E_{mur}/E_m	1.4–1.9
q_c/p_l^*	4–6	E_D/E_{PMT}	$\cong 1.5$

N_{60} – number of blow for a energy ratio of 60%; q_c – CPT resistance; E_m, E_{mur} – pressuremeter modulus ("elastic" and un-reload); p_f and p_l – Creep and net limit pressure of PMT

Table 2a. Ratios between defomability (Young and Shear) modulus (Viana da Fonseca et al. 1998a, 2001, 2003; Topa Gomes et al. 2008).

$\frac{E_0(CH)}{E_{s1\%}(PLT)}$	$\frac{E_0(CH)}{E_{ur}(PLT)}$	$\frac{E_0(CH)}{E_m}$	$\frac{G_0(CH)}{G_{ur}(SBP)}$
$\cong 8$–15	$\cong 2$–3	$\cong 20$–30	$\cong 1,7$–3

CH – values obtained in cross-hole tests; PLT – values obtained from plate load tests

Table 2b. Average ratios between Young's modulus and in situ "gross" tests.

$\frac{E_0(CH)}{N_{60}(SPT)}$	$\frac{E_0(CH)}{q_c(CPT)}$	$\frac{E_0(CH)}{q_d(DPL)}$	$\frac{E_0(CH)}{p_l(PMT)}$
$\cong 10$ (MPa)	$\cong 30$	$\cong 50$	$\cong 8$

CH , PLT – ibidem; N60, qc, qd, pl – resistances values

Interesting to refer that for most designs, Sabatini et al. (2002) state that the elastic modulus corresponding to 25 percent of failure stress, E_{25}, may be used. In Piedmont residual soils, the use of the dilatometer modulus, E_D, as equal to E_{25} has shown to provide reasonably accurate predictions of settlement (Mayne & Frost, 1988). However, the specific evaluation of E/E_0 reduction value associated with this FS equal to 4 (E_{25}) was derived equal to 0.34.

3.3.2 *Strength from in situ tests – particularities due to the cohesive-frictional nature of residual soils*

Empirical correlations are very useful for the purpose of common design of foundations and other simple geotechnical structures and have been derived from an accumulated database, described in Carvalho et al. 2004, Gomes Correia et al. 2004, Viana da Fonseca et al. (2004, 2006). SPT and CPT can be used to predict the peak angle of shear resistance in granular soils when normalised stress-level of $p_a = 100$ kPa, by:

$$(N_1)_{60} = \frac{N_{60}}{(\sigma'_v / p_a)^{0,5}} \qquad q_{c1} = \frac{q_c}{(\sigma'_v / p_a)^{0,5}} \qquad (6)$$

Hatanaka and Uchida (1996) obtained the following equation for the evaluation of the angle of shearing for peak, also corroborated by Mayne (2001) for residual silty sand in Atlanta and Georgia:

$$\phi'_p = \left[15,4(N_1)_{60}\right]^{0,5} + 20° \qquad (7)$$

Robertson and Campanela (1983) and Kulhawy and Mayne (1990) recommended, respectively, for unaged, uncemented quartz sands, the following correlation with CPT:

$$\phi'_p = \arctan\left[0,1 + 0,38 \cdot \log(q_c / \sigma'_{v0})\right] \qquad (8)$$

$$\phi'_p = 17,6° + 11,0 \cdot \log(q_{c1}) \qquad (9)$$

In residual soils from Porto, values of $(N_1)_{60}$, taken from the SPT tests, allowed to derive the angle of shearing resistance from Décourt's (1989) or Hatanaka and Uchida (1996)

Proposal, adapting for an average state-of-practice with 60% efficiency in Portugal; this last being expressed by:

$$\phi'_p = \sqrt{15.4 \cdot (N_1)_{60}} + 20° \qquad (10)$$

ranging from 35° to 41°, with an average of 38°. The relation between q_c from CPT and σ'_{v0} (expressed in equation 7) is presented in Figure 30. CPT results reveal a moderate increase of q_c in depth. Robertson and Campanella's proposal tends to lead to higher values of ϕ', especially at lower depths, than those obtained from triaxial tests, since the effective cohesive component is not considered.

This reflects the simultaneous sensitivity of q_c towards frictional and cohesive components. In the present case, the CPT results are rather constant in depth, crossing a wide range of friction angles (35–42°) with higher incidence at 37°, which is much lower than the one obtained in the laboratory tests. Being attributed to the behaviour of a cohesive-frictional soil, where the lower confinement levels are dominated by the cohesive component, while the higher are mostly governed by friction.

Theoretically, peak and post peak resistance can be obtained by pressuremeter tests (PMT/SBPT). Due to the influence of disturbance during installation, peak strength is unreliable from PMT. For drained conditions the angle of shearing resistance can be estimated by the approach proposed by Hughes *et al.* (1977), which assumes the material has a linear Mohr-Coulomb plastic behaviour and zero cohesion intercept, with a constant dilation angle (ψ), for deformations up to around 10%. The pressure-expansion data, when plotted as log p (cavity pressure) versus log ε (cavity strain), tends to be linear, with a slope, S. This value of S may be used to determine the values of ϕ' and ψ by the following expressions as follow:

$$\sin\phi' = \frac{S}{1+(S-1)\sin\phi'_{cv}} \qquad (11a)$$

$$\sin\psi = S+(S-1)\sin\phi'_{cv} \qquad (11b)$$

Residual soils are, however, characterized by a cohesive-frictional strength, for what any of these procedures cannot be used by themselves for strength evaluation purposes.

The usual way of interpreting pressuremeter tests by plotting data in ways that a single parameter can be extracted from a specific part of the experimental pressure-expansion curve is not feasible under these conditions (such as the undrained strength in clay and friction angle in sand). An obvious alternative is to reproduce the whole pressuremeter curve, varying the constitutive parameters in order to obtain a best fit of the experimental data as advocated by many authors (Jefferies, 1988; Ferreira & Robertson, 1992; Fahey & Carter, 1993; Schnaid et al. 2000). Since different sets of parameters can produce an equally good fit to measured data, engineering judgement is always required and independent data from laboratory tests is generally desirable. The fitting process is part of the definition / check of initially estimated parameters.

A new cavity expansion model that incorporates the effects of structure and structure degradation into cylindrical cavity expansion theory was introduced by Mántaras & Schnaid (2002) and Schnaid & Mántaras (2003). The Euler Method is applied to solve simultaneously two differential equations that lead to the continuous variations of strains, stresses and volume changes produced by cavity expansion. Despite the mathematical complexity, an explicit expression for the pressure-expansion relation is derived without any restriction imposed on the magnitude of deformations. The propose approach consists of a unified solution for interpretation of pressuremeter tests. A case study presented by Schnaid & Mántaras (2004) illustrated the applicability of the proposed approach.

The extension of the cavity expansion theory to accommodate the framework of unsaturated soil behaviour in the interpretation of pressuremeter tests is discussed by Schnaid & Coutinho (2005). Fontaine et al. (2005) present experimental results and the analytical interpretation of several Menard pressuremeter tests on an unsaturated soil matrix. The authors show specifically deals with the curve fitting methodology to derive the parameters of the PMT test.

Other approaches have been proposed, starting from a set of results derived from laboratory tests and cross-hole seismic tests, and applying a "distorted" hyperbolic model, incorporated into numerical programs, suchs as CAMFE (Fahey & Carter, 1993), aiming to find the set of parameters that may give the best fit to the observed results. A large number of variables may be involved, making the finding of a unique set of parameters to match each curve quite complex. This wil be referred below.

Cohesive-frictional shear strength derived from multi-size loading tests

A simple approach, although limited to shallow foundations, is the recourse to multi-size loading tests (PLT, for instances). Punching type failures in these residual soils, give very little definition of an inflexion zone of the pressure-settlement curve, but log-log plotting can be a better way to detect it (Viana da Fonseca et al. 1998a).

This has been used preliminarily to interpret the results of loading tests (Viana da Fonseca, 1996) and the values for the failure load were obtained: $q_f = 700\,kPa$ (B = 0.30 m); $q_f = 821\,kPa$ (B = 0.60 m); $q_f = 950\,kPa$ (B = 1.20 m). Failure load is associated to a "punching" mechanism, which can be distinct from the ultimate equilibrium defined by limit state analysis (Terzaghi-Meyerhof-Vésic). If we use the values of those failure loads (Viana da Fonseca et al. 1997b), in the bearing capacity formulation, taking account of the water level position, three equations are obtained. These can be optimised to get the two strength parameters range. The derived values were: $c' = 6.3–6.9\,kPa$ and $\phi' = 36.5°–37.0°$, revealing a fair agreement with the results obtained in extensive laboratory testing of undisturbed samples (Viana da

Fonseca, 1998) and *in situ* testing (Viana da Fonseca et al. 1998a).

DMT to evaluate the cohesive-frictional nature of residual soils

The use of Marchetti's dilatometer test (DMT) on residual soils has not been very much exploited, with the exception of a few singular cases. Some consistent research has been made in Portugal (Cruz et al. 2004; Cruz & Viana da Fonseca, 2006) to study its efficiency in such soils, trying to define specific correlations that may explain their mechanical behaviour. One of the goals of these studies has been to establish correlations for deriving the strength parameters due to cemented structure, revealed by the presence of an effective cohesive intercept, c', taking the Mohr-Coulomb criterion.

Since the test allows the determination of two basic independent parameters (P_0 and P_1), it may lead to the evaluation a pair of strength parameters, both the angle of shear resistance and the effective cohesive intercept. The main goal of the methodology is the establishment of correlations that allows for the evaluation of the increment of resistance resulting from the cemented structure, associated to the effective cohesive intercept, c'. Cruz et al. (2004) resume the application of such a specific methodology on residual soils from granite in five experimental sites located nearby the city of Porto.

The mechanical characterization the soils has been based on "in situ" (DMT, CPT/CPTU and PLT) tests and laboratory (triaxial, CK_0D and CID) tests over good quality undisturbed samples. The determination of reference values of c' is mostly based on triaxial tests. In the study presented by Cruz et al. 2004, effective cohesion values ranged from 5 to 25 kPa, for values of the angle of shearing resistance in the range of 32° to 42°. The evaluation of the effects of cemented structure was based on the results of DMT and DMT+CPT tests, namely from lateral stress index, K_D, "virtual (or apparent) overconsolidation ratio", v(A)OCR (DMT) and the ratio M/q_c, from DMT and CPT. These results were then compared with the cohesive intercept obtained from triaxial and PLT tests, fitting very well. The concept of overconsolidation ratios does not have the same meaning for sedimentary and residual soils, although the presence of a naturally cemented structure gives rise to similar behaviour. Pre-consolidation stress, which may be represented by the second yield (y_2) in Burland's (1989) concept, is called "virtual" (or "apparent") pre-consolidation stress and the relation between this and the vertical rest stress, 'v(A)OCR', has the same meaning as other well established general concept: "yield stress ratio = YSR. Thus, vOCR derived from the DMT test on residual soils reflects the strength resulting from the cementation structure, normalised in relation to the effective

vertical stress. It is important to note that OCR evaluation is I_D and K_D dependent (that is P_0 and P_1 dependent), allowing to be confident on the determination of both angle of shear resistance and effective cohesive intercept. By the same principle of the singularity of the dual parametric evaluation in DMT, Cruz et al. (2004) present another alternative way to derive the pair of c' and ϕ', which is based on the ratio M/q_c. This value has been emphasised by Marchetti (1997) as a useful tool for the definition of OCR on granular soils, given the greater sensitivity of the M parameter to variations in compaction or cementation, when compared to the tip resistance, q_c. Behind this approach is the same principle of the superlative conditioning of traction resistance in stiffness parameters as in cohesive intercept, when compared to frictional resistance. As a corollary of this methodology, the authors compare c' with preconsolidation pressure, σ'_p, obtained via DMT, point out the ability of the test to feel the cementation structure.

Indeed, the increase in strength due to the cemented structure is associated to an effective cohesion intercept c', which is not related to the presence of clayey/fine material. Cruz & Viana da Fonseca (2006) proposed the following equation:

$$c' = 0.367 \, v(A)OCR + 3.08 \qquad (12)$$

Once c' is obtained, it is reasonable to expect that it can be used to correct the overestimation of ϕ', when correlations for transported soils are taken. Thus, considering the difference between this derived value (ϕ'_{DMT}) and the real frictional component ($\phi'_{triaxial}$) and comparing it with c', evaluated in triaxial conditions over undisturbed samples, Cruz & Viana da Fonseca (2006) demonstrated a very good correlation. This methodology has been tested in growing number of sites with quite fair results (ex.: Viana da Fonseca et al. 2008, in this conference).

As it is detailed in these works, the angle of shear resistance determined from DMT and CPTU tests by mean of Marchetti's (1997) and Robertson and Campanella's (1983) correlations in regional residual soils, range from 35° to 45°, which are globally higher than those determined under controlled conditions in triaxial tests on undisturbed samples, if the cohesive fraction is not considered. Taking into account the low influence that sampling has on the evaluation of the angle of shear resistance, the registered difference on the value of ϕ' is attributed to the effects of the cementation structure on qc and K_D parameters. From available data, the cohesion obtained from DMT, c'DMT, ranged from 5 to 15 kPa, which were confirmed by triaxial testing.

The subject of evaluation of the utmost index, the apparent or virtual overconsolidation ratio (v(A)OCR), has been explored with detail in Piedmont residuum (Figure 31). Its determination in lab consolidation tests has been erroneous, due to inadequate sampling techniques, as standard thin-walled Shelby tubes (Mayne and Brown, 2003). Values deduced with lab oedometer and triaxial tests indicate that these soils are apparently lightly overconsolidated ($1 \leq v(A)OCR < 4$) with supporting trends given by midface CPT porewater pressures (u_1), flat dilatometer tests, and shear wave correlations. In contrast, which is dichotomous, measured cone tip stresses (qt) and shoulder porewater pressures (u_2) implicate soil behaviour that is characteristic of stiff fissured clay with high $v(A)OCR$. Figure 31 resumes some of these surprising results.

Unsaturated versus saturated conditions
The main difference between saturated soils and unsaturated soils is the existence of a negative pressure in the water of the pores; largely know as suction, which tends to increase resistance and rigidity of the soil. This finding can be observed in the results from pressuremeters tests on a residual granite soil showed by Schnaid et al. (2004) (Figure 32).

The first test was performed at an *in situ* pressuremeter data and theoretical cavity expansion curves (Figure 32b), in which the strength parameters measured in the laboratory are combined to the *in situ* horizontal stress assessed from pressuremeter tests to produce theoretical pressure-expansion curves for tests carried out under unsaturated and soaked conditions.

3.3.3 Deformability properties of residual soils
Maximum stiffness and serviceability
An important subject that conditions the interpretation of geomechanical characterization methods is the well recognised elastic linear behaviour at very small and small strains and a non-linear answer at medium to high strains. This non-linear pre-failure behaviour complicates in situ test interpretation and may conflict with simplifying assumptions made in the past. It is then crucial to define and identify the type of modulus that will be adopted.

The small strain shear modulus G_0 is related with the Young's modulus E_0 by:

$$G_0 = E_0/[2(1+v)] \tag{13}$$

This modulus, if properly normalised with respect to void ratio and effective stress, is in practical terms independent of the type of loading, number of loading cycles, strain rate and stress/strain history. It is a fundamental parameter of the ground, considered as a benchmark value, which reveals its true elastic behaviour.

There are several relations derived from experimental studies between small strain shear modulus versus effective stress state. In fact, as stated by Cho

et al. (2006): "the small-strain shear stiffness (G_0 or G_{max}) of a soil reflects the nature of interparticle contacts, such as the Hertzian deformation of contacting smooth spherical particles; the resulting nonlinear load–deformation response determines the stressdependent shear wave velocity". The most direct and congruent relation takes the direct dependence between the two variables (taking into account the direct relation: $G_0 = \rho \cdot V_s^2$):

$$V_s = \alpha \cdot \left(\frac{p'}{p_a} \right)^\beta \tag{14}$$

However, taking into account the advantage of normalizing for the density influence, on order to better identify the strict influence of cementation factors there are several relations derived from experimental studies between small strain shear modulus versus void ratio and effective stress, since the very early proposals by Hardin and Richard (1963) – equations (14) and (15) – to more recent approaches, such as those reported by Jamiolkowski et al. (1995) – equation (16):

$$G_0 = S \cdot p_a^{1-n} \cdot F(e) \cdot p'^n \tag{15}$$

where p_a is a reference stress, generally assumed equal to $100\,kPa$; p' is the effective mean stress; S and n are experimental constants; $F(e)$ the void ratio function generally adopted as:

$$F(e) = \frac{(C-e)^2}{1+e} \tag{16}$$

C is a function of the shape and nature of grains.

$$F(e) = e^{-x} \tag{17}$$

It should be stated that when the reference pressure (p_a) is not used in formulae, then the value of S will expressed in pressure unities. As expressed by Gomes Correia et al. (2004), on top of the evaluation of the small strain shear modulus, it is now well established that some more relevant information can be obtained by the following category of tests: evaluation of anisotropy by using polarised shear waves; estimation of K_0; evaluation of material damping; evaluation of modulus degradation curve with strain; and, evaluation of undrained behaviour and susceptibility of in situ materials to static or cyclic liquefaction.

Comparison between in situ and laboratory measured values of Vs velocity offers insight into the quality of the undisturbed samples, as will be referred later in this text. Time effects have to considered (Gomes Correia et al. 2004).

In granular materials, there are several proposals for natural alluvial sands, aged and cemented (ex: Ishihara, 1986), while for residual soils some specific proposals have been presented, such as those reported

Table 3. Stiffness vs stress state parameters for residual soil (adapted frrom Schnaid, 2005).

Soil (locals), references	S	n
Alluvial sands (…), Ishihara (1982)	7.9 to 14.3	0.40
Saprolite from granite (Matosinhos, Porto, Portugal), Viana da Fonseca (1996, 2003)	110	0.02 $p' < 100\,\text{kPa}$
Saprolite from granite (CEFEUP, Porto, Portugal), Viana da Fonseca et al. (2004)	65	0.07
Saprolite from gneiss (Caximbu, Sao Paulo, Brazil), Barros (1997)	60 to 100	0.30 $p' < 100\,\text{kPa}$
Saprolite from granite (Guarda, Portugal), Rodrigues & Lemos (2004)	35 to 60	0.35
Competely decomp. tuff (Hong Kong), Ng & Leung (2007b)	37 to 51	0.20–0.26
Cachoeirinha lateritic soil (Porto Alegre, Brazil). Consoli et al. (1998) and Viana da Fonseca et al. (2008)	79	0.18
Passo Fundo lateritic soil (Porto Alegre, Brazil), Viana da Fonseca et al. (2008)	181	0

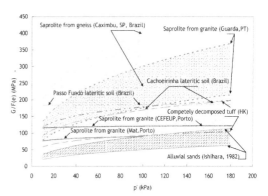

Figure 33. Comparison between observed and reference proposals of G_0 variation with effective stress (reference to equation 18 and parameters in Table 3).

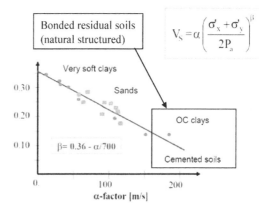

Figure 34. Vs: skeletal stiffness – state of stress (adapted from Santamarina, 2005).

for saprolitic soil of granite in Portugal (Porto) (Viana da Fonseca, 1996; Barros, 1997; Rodrigues & Lemos, 2004; Viana da Fonseca et al. 2006):

$$\frac{G_0(\text{MPa})}{F(e)} = S \cdot \left[p'_0\,(\text{kPa})\right]^n \qquad (18)$$

$F(e)$ is a void ratio function calculated by:

$$F(e) = \frac{(2.17 - e)^2}{1 + e} \qquad (19)$$

These results show that the constant value of the small strain shear modulus expression is much higher for residual soils. Schnaid (2005) present a synthesis that is now enlarged in Table 3, where parameter S is much higher than the value adopted for cohesionless soils, as result of local weathering conditions, while the exponent n, is in general, very low, reflecting that the influence of the mean effective stress is substantially lower. These different values of power n are conse-quence of different types of bonding between grains affecting the Hertz type of behaviour existing in partic-ulate materials (Biarez et al. 1999, Viana da Fonseca, 2003, Viana da Fonseca et al. 2006, Schnaid, 2005) – Figure 33.

Interesting to realize that this behaviour is very con-gruent with the general pattern of relations between shear wave (Vs) and effective stress state (p'/p_a), as suggested by Santamarina (2005) – Figure 34. In this figure, and taking reference to representative data included in Table 3, the region of cemented soils is associated to the behaviour of residual soils, with very low values of the stress depnednce exponent "β" (with similar meaning of "n", in equation 17).

G_0 derived from in situ tests in routine analysis
Penetration Tests (SPT, CPT…) are common in-situ test for geotechnical investigation. Correlations between penetration parameters and stiffness (these well expressed by the very small strain stiffness param-eters, such as G_0 from CH seismic survey) are very sensitive to stress-strain levels, very much dependent on the degree of cementation. Diverse dependence of the initial or dynamic stiffness values and strength parameters from penetration tests, with confinement

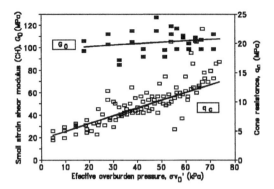

Figure 35. Relationship between G_0 and q_c for residual soils (Viana da Fonseca, 1996).

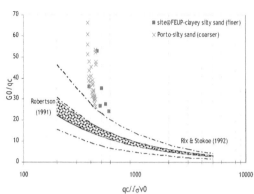

Figure 36. G_0/q_c versus $q_c/\sqrt{\sigma'_{v0}}$ from in-situ tests at FEUP experimental site (Viana da Fonseca et al, 2004, 2006) versus reference curves for uncemented and unaged soils.

stress is a consequence of such progressive destructuring (Figure 35), which is another consequence of the effect of the duality between cement-control and stress control zones (considerations in the introduction to this paper).

From the empirical correlations between SPT values versus shear wave velocity, the classical proposal of Seed et al. (1986) for sands is:

$$Vs = 69 \cdot N_{60}^{0,17} \cdot Z^{0,2} \cdot F_A \cdot F_G \qquad (20)$$

where: Vs is the shear wave velocity (m/s), N_{60} is the SPT value (Er = 60%), Z is the depth (m), F_G is a geological factor (clays = 1; sands = 1,086), F_A is the age factor (Holocene = 1; Pleistocene = 1,303), ρ is the total mass density.

Following Stroud's (1988) suggestion, a simple and very useful power law between G_0 and N_{60} is:

$$G_0(\mathrm{MPa}) = C \cdot N_{60}{}^n \qquad (21)$$

For saprolitic soils of Porto granite the constant values were the following: $C \cong 60$; $n = 0.2$–0.3 (Viana da Fonseca, 2003, Viana da Fonseca et al. 2004, 2006), depending on the relative content of clay, silt and sand. The variation of G_0 with effective mean stress (p'_0) is very low when compared with other parameters, such as N_{60}. Linear – or close – correlations between G_0 and N_{60} for relevant values of p'_0 on shallow foundations strongly underestimate elastic stiffness on cemented soils when using laws developed for transported masses (Stroud, 1988).

Pinto & Abramento (1997) present also some very ilustrative correlations between SPT values and G_0, in gneissic residula soils in SãoPaulo, Brazil, with constant values vry similar to those obtained in Porto: $C \cong 62.8$; $n = 0.3$.

The same trends are expressed in the correlations between q_c and G_0, being quite different in uncemented and unaged cohesionless soils (Robertson, 1991, or Rix and Stokoe, 1992) and in residual soils,

such as those deduced in cemented residual soils in Figure 36. These late follow a completely different trend.

Another way of doing this representation is by means of the proposals on Schnaid (1999, 2005). This has obvious correspondence to the results expressed above, for G_0/N_{60} vs $(N_1)_{60}$ and G_0/q_c vs q_{c1}, in Figures 9 and 10. Cementation and ageing have different influences over q_c and G_0, which allows for the identification of "unusual" soils such as highly compressible sands, cemented and aged soils and clays with either high and low void ratio. As stated above in section 2.1.2, this particular relation between stiffness and penetration resistances (both in q_c and N_{60}) constitutes a step towards classification, as well as another way of correlating these values for parametrical deduction.

Study of stiffness anisotropy

It is well documented that shear wave velocity of soil depends on stresses in the direction of wave propagation and particle motion and is independent of stress normal to the shear plane (Roesler 1979).

The inherent stiffness anisotropy (degree of anisotropy under isotropic stress state) of completely decomposed granite (CDG) was not proved obvious from the limited set of test results of Ng and Leung (2006). However, under the influence of a higher vertical effective stress, Ghv were found to be on average higher than Ghh by about 22% at each of the stress states (p' = q = 80, 160, 400 kPa), showing stress-induced anisotropy.

Ng & Leung (2007a & b) show multidirectional shear wave velocity measurements taken on a Mazier and Block specimen, and the measured degree of inherent stiffness anisotropy of completely decomposed tuff (CDT) is lower in Mazier specimens than in block specimens due to a higher degree of sample disturbance. The average values of Ghh/Ghv of the block and Mazier specimens were 1.48 and 1.36, respectively. The shear moduli, G_{hh}, G_{hv} and G_{vh}, of the block specimen are consistently higher than those of Mazier specimen.

The degree of inherent stiffness anisotropy of the block specimen of natural CDT ($G_{hh}/G_{hv} = 1.48$) is of a similar order to some natural clays such as London clay, possibly attributed to some geologic processes such as metamorphism of the parent rock mass of CDT. When this happens it is very difficult to identify induced anisotropy, apart of the ineherent factor.

Effects of suction on anisotropic stiffness
The results of the analysis of this factor appeared to prove that an increase in the matric suction can be considered as equivalent to an increase in the mean effective stress, and hence the shear wave velocities increase (Mancuso et al. 2000, Ng and Leung, 2006). The rate of increase of the shear wave velocities reduces as the matric suction exceeds the air-entry value of the material ($u_a-u_w = 50$ kPa). Air-water menisci (or contractile skins) form at the inter-particle contacts as the soil specimens start to desaturate at matric suctions exceeding the air-entry value. This results in an increase in the normal force acting on the particles and thus the shear wave velocities increase. Upon further increase of the matric suction, the meniscus radius reduces, which limits the increase of the shear wave velocities (Mancuso et al. 2000).

The essential isotropy of the matric suction will fundamentally affect the sensitivity of the measured ratio of stiffness anisotropy, expressed as G_{hh}/G_{hv}, as a sign to reveal the stress induced anisotropy solely due to effective components along v and h directions.

Degree of non-linearity – Degradation of maximum sher modulus (G_0)
For practical purposes, it is necessary to extrapolate the results of small strains to the range of strain of engineering significance, generally 0.001% to 0.5%. This need arises from the recognition that the displacements of well designed civil engineering structures are generally quite small and overpredicted when using soil parameters that are inferred from conventional soil tests in theoretical settlement solutions (Burland, 1989; Tatsuoka et al. 1997; Simpsom 2001; Jardine et al. 2001). There are empirical settlement solutions based on the measured parameter that should not be demerited or specific purposes and conditions (Menard, 1962, Schmertmann, 1970), but should be restricted to them (Schnaid, 2005).

A very important skilfulness of recent research progresses is the possibility of deriving more general stress-strain responses from the stiffness degradation laws ypified for each material. Modified hyperbola can be used as a simple means to reduce the small strain

shear modulus (G_0) to secant values of G at working strain levels (G_s), in terms of shear strain γ, or at working load levels, in terms of the mobilized strength (q/q_u). Some of the proposals are available, as the one proposed by Fahey and Carter (1993), in mobilized strength or stress level (q/q_u), with the generalized form is:

$$\frac{E}{E_0} = 1 - f\left(\frac{q}{q_{ult}}\right)^g \tag{22}$$

f and g are fitting parameters (f controls the strain for the peak strength, τ_{max}, and g the shape of the degradation law, being function of the stress level (Figure 37a), the inverse of safety factor. Values of $f = 1$ and $g = 0.3$ appear to be reasonable, for monotonic loading conditions, in unstructured and uncemented geomaterials (Mayne, 2001).

The relationship between G/G0 and normalized shear strain (as suggested by Santos, 1999: $\gamma^* = \gamma/\gamma 0.7$) seems to be very promising as a reference stiffness degradation curve, since, for the range of shear strain tested, it seems scarcely affected by the kind of soils (temperate or tropical soils), plasticity index, confining pressure, overconsolidation ratio an degree of saturation (Figure 37b – for cyclic loading, in resonant column and torsional cyclic stress).

Based on laboratory experimental results from 37 tests, by resonant column, of lateritic and saprolitic soils it was possible to establish the relationship presented in Figure 38, allowing a practical use of these results.

Resonant column tests have been carried out on specimens from Porto experimental sites (Viana da Fonseca, 2006). In these tests, the specimens were reconsolidated under anisotropic conditions for a stress level equal to the best estimate of the in situ stresses. Some curves of the normalized stiffness and damping ratio versus shear strain are shown in Figure 39. The evolution of the curves are similar to sands (Santos, 1999), as it can be observed from the limits plotted together with the experimental data.

Stiffness for serviceability conditions, factoring G_0
Viana da Fonseca (1996, 2001, 2003) presented an analysis of a large scale loading test (circular concrete footing 1.20 m in diameter) and of two other plates of smaller diameter (0.30 m and 0.60 m), performed in the Porto silty sand, lead to the Young's moduli values presented in Table 4, for different loading stages. These results were obtained by back-analysis of the footing loading test (rigid footing), considering a linear elastic layer with constant modulus underlain by a rigid base at 6.0 m depth.

The intermediate stress level (FS = 4) corresponds approximately to the allowable pressure for residual soils (Décourt, 1992).

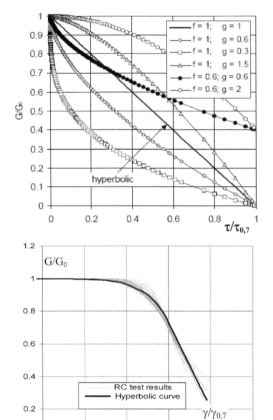

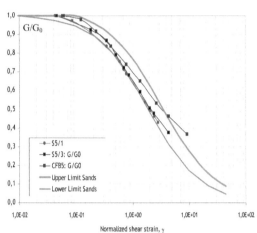

Figure 39. Resonant column results: normalized stiffness and damping ratio versus shear strain.

Table 4. Secant Young's modulus, E_s (MPa) from loading test for different service criteria

Loading Tests	Service Criteria			
	q (s/B = 0.75%)	$q/q_{ult}^{(*)}$ ($F_S = 10$)	q/q_{ult} ($F_S = 4$)	q/q_{ult} ($F_S = 2$)
Footing	17.3	20.7	16.0	11.0
Plate (0,6)	11.9	11.2	12.5	12.7
Plate (0,3)	6.7	6.9	5.9	5.7

(*) Corresponding to allowable pressure for serviceability limit state design.

Figure 37. Hyperbolic fitting in normalized secant shear modulus vs mobilised shear stress (a) ad. Fahey et al. (2003), or normalized shear strain (b) ad. Gomes Correia et al., 2004.

Viana da Fonseca (1996, 2001), for Porto silty sand, using a crossed interpretation of footing and plate loading tests with the SPT values in the settlement influence zone, for service level ($q_s/q_{ult} \cong 10–20\%$), obtained an average ratio between the secant Young's modulus and SPT values of:

$$E(\text{MPa})/ N_{60} \cong 1 \qquad (23)$$

This relationship is similar to the proposal of Stroud (1988) for normally consolidated soils, in identical stress levels.

Sandroni (1991), in the general report on the characteristics of young residual (saprolitic) metamorphic soils, recognizes that "rigidity correlates crudely with SPT". By backcalculating settlement records and correlating them with the average SPT values along the depth of influence measurements, the author analysed 14 cases of superficial loading in residual soils from Brazil and Piedmont, USA, and obtained a relation of:

$$E(\text{MPa}) = 0.4 \text{ to } 0.9 \ (N_{SPT})^{1.4} \qquad (24)$$

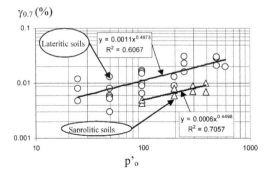

Figure 38. Ratios between $\gamma_{0.7}$ and p'_0 for lateritic and saprolitic soils (Gomes Correia et al. 2004).

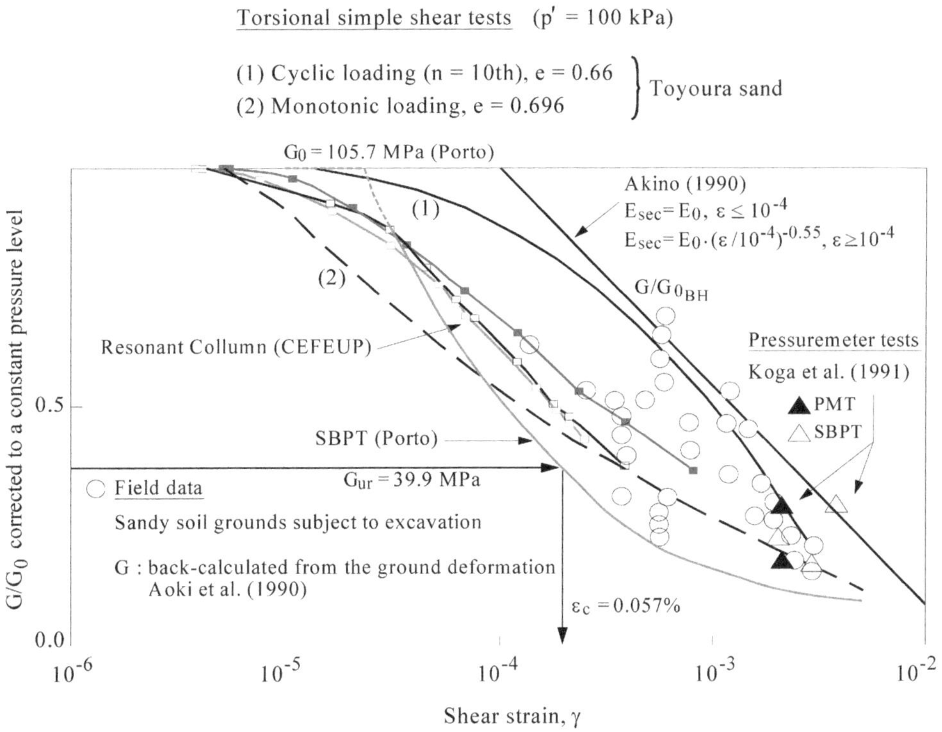

Torsional simple shear tests   (p' = 100 kPa)

(1) Cyclic loading (n = 10th), e = 0.66 ⎫
(2) Monotonic loading, e = 0.696   ⎬ Toyoura sand

G_0 = 105.7 MPa (Porto)

Akino (1990)
$E_{sec}=E_0$, $\varepsilon \leq 10^{-4}$
$E_{sec}=E_0 \cdot (\varepsilon/10^{-4})^{-0.55}$, $\varepsilon \geq 10^{-4}$

G/G_{0BH}

Resonant Collumn (CEFEUP)

Pressuremeter tests
Koga et al. (1991)
▲ PMT
△ SBPT

SBPT (Porto)

○ Field data
Sandy soil grounds subject to excavation
G_{ur} = 39.9 MPa

G : back-calculated from the ground deformation
Aoki et al. (1990)
ε_c =0.057%

G/G₀ corrected to a constant pressure level

Shear strain, γ

Figure 40.   Comparison of secant shear modulus as function of shear stress level: results of resonant column results in residual soils from granite of Porto versus data from other testing conditions in sandy soils (Aoki et al. 1990, apud Tatsuoka & Shibuya, 1992).

being the lower limit mostly associated to micaceous soils, with high fines content, and the average value, of 0.6. The scatter is small, considering the limitations of SPT. The author emphasised that "load intensities" are typically less than 200 kPa, therefore considered "pre-yield".

The possibility of inferring design values for Young's modulus to predict the behavior of load tests, on plates and on a prototype footing, conducted in the vicinity of was these penetration tests was developed and thoroughly discussed elsewhere (summarized in Viana da Fonseca, 2001). The main conclusions drawn for the most common methods are as follows: the Burland and Burbidge (1985) equation based on SPT results led to an overestimation of the observed settlement by a factor of 2 to 3, while the application of the Schmertmann et al.'s (1978) method reproduced accurately the footing settlement for $\alpha = E/q_c$ values in the range of 4.0 to 4.5. Both methods identify this saprolitic soil in the global typology of cemented or overconsolidated granular soils.

Correlations between q_c and Young's modulus, for different stress-strain levels by triaxial tests (CD) with local strain measurements, confirmed the very strong influence of non-linearity on E/q_c ratios as well as

a singular pattern of that variation when compared to proposals for transported soils (Viana da Fonseca et al. 1998a, 2001, 2003).

Experimental in-situ work described by Viana da Fonseca (2003) revealed stiffness from reload-unload cycles of PMT (E_{pmur}) and SBPT tests in saprolitic granite soils apparently very different. In fact, for PMT it were found the following relations: $E_{pmur}/E_{pm} \cong 2$ and $E_0/E_{pm} \cong 18$–20, with E_0 determined on seismic survey (G_0-CH), while for SBPT $G_0/G_{ru} \cong 2,6$ to 3,0. It must be noticed that these last values are substantially lower than the ratio ($\cong 10$), reported by Tatsuoka & Shibuya (1992) on Japanese residual soils from granite. The non-linearity model of Akino – cited by the previous authors – developed for a high range of soil types, including residual soils, is expressed by:

$$E_{sec} = E_0, \quad \varepsilon \leq 10^{-4} \tag{25}$$

$$E_{sec} = E_0 \cdot (\varepsilon/10^{-4})^{-0.55}, \quad \varepsilon \geq 10^{-4} \tag{26}$$

SBPT unload-reload modulus correspond to secant values for shear strain of about 6×10^{-4}, fitting this last very well with the above indicated trends.

227

Figure 40 illustrates the relative positions of test performed in Porto saprolitic soils (Viana da Fonseca et al. 1998a, 2006). The comparison of results of the two types of tests can only be properly discussed if the mean effective stress during the cycle (p') is well estimated and the strain level of the cycle of each test reported.

Using these functions factoring G_0, it is possible to obtain a serviceability shear modulus, or serviceability stiffness, to be used in routine calculations to obtain settlements. Values deduced for G_{ur}/G_0 ratio in the work presented in this conference in residual soils from granite of Porto (Topa Gomes et al. 2008) range from 0.40 to 0.59 with a mean value of 0.50. This ratio between G_0 and G_{ur}, from 1.7 to 2.5, is very much in accordance with the values reported above by Pinto & Abramento (1997) Viana da Fonseca et al. (1998a).

Fahey et al. (2003) suggested that G_{ur} can often be used as a "working stiffness" for design of large spread footings on stiff sandy soils. This leads to a very tentative suggestion that an appropriate "working stiffness" value for these soils could be taken to be about 0.4 to 0.5 times the G_o value. However other signs in more cemented soils, may lead to some care in practical options.

Fahey et al. (2007) compared some very interesting results in two horizons of sands from Perth: the Spearwood Dune sands at Shen-ton Park, showed G_o values somewhat higher than the trend for the Spearwood Dune sands in the CBD, but a significantly greater rate of softening (low g value). This stiffness "brittleness" is most likely to be due to the unsaturated state of the Shenton Park sands, and possibly to slight cementation (perhaps resulting from the partially saturated state). As emphasized, the ratio G_{ur}/G_0 seems to follow the same trend as the softening parameter g. At Shenton Park, low values of G_{ur}/G_0 (~0.2) indicate brittleness in the softening behaviour, and this is also reflected in the observed responses of the footings at this site. This is noticeable finding and relevates the low values of the relation between serviceability stiffness values and G_o value that have been observed from the back-analysis of several geotechnical structures in residual soils.

Self-boring pressuremeter tests were carried in Hong Kong to study the variations in the secant shear modulus with shear strain of decomposed granite (Ng et al. 2000, Ng & Wang 2001). Figure 41a shows the relationship between the normalized secant shear modulus ($G_{sec}/\sqrt{p'}$, being p' the mean effective stress) and the shear strain (ε_s) interpreted from the third cycle of the unload-reload loop of the tests. The variations in $G_{sec}/\sqrt{p\_}$ with ε_s at Yen Chow Street, derived from the last complete unloading loop, are shown in Figure 41b.

It was previously found that the secant shear modulus derived from that unloading loop is consistent with that derived from each unload-reload loop (Ng &

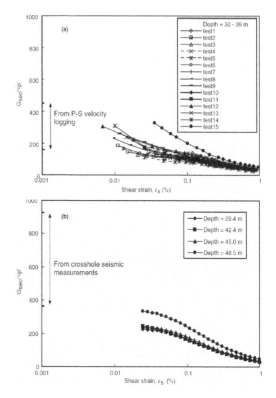

Figure 41. Stiffness-strain relationships of decomposed granite from self-boring pressuremeter tests: (a) Kowloon Bay; (b) Yen Chow Street.

Wang 2001), thus it is believed that the results shown in Figure 42 are comparable, stability of the multicycle analysis that is subjacent to some models above referred (Fahey and Carter, 1993, used by Topa Gomes et al. 2007). The value of $G_{sec}/\sqrt{p'}$ decreases significantly from about 300 to about 50 as εs increases from 0.02% to 1%. The range of the shear modulus at very small strains (G_{vh} or G_{hv}) determined from seismic measurements is also shown in the figures, being clear that the values are higher than the values adjourned from the loops, corroborating the G_0 (G_{vh} or G_{hv})/ G_{ur} trendy are higher than unity.

Saprolitic soils versus lateritic (mature) residual soils

Schnaid & Coutinho (2005) provides an overall picture of the state-of-practice for pressuremeter tests in Brazil. In the past 10 years, the use of pressuremeters has increased considerably and now plays a role in geotechnical characterisation of residual soils in Brazil, particularly in research. Figure 42 illustrates typical data from a set of self-boring pressuremeter tests carried out on a gneissic residual soil in São Paulo

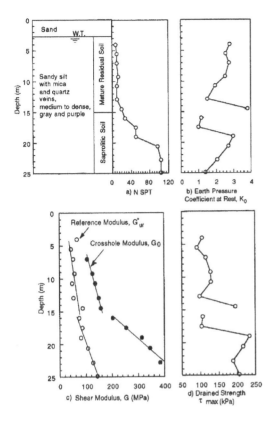

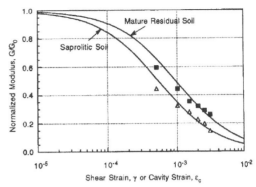

Figure 43. Decrease in shear modulus with shear or cavity strain (Pinto & Abramento, 1997).

Figure 42. Tests results (Pinto & Abramento, 1997).

City, at the site of a subway station (Pinto e Abramento, 1997).

The deposit is typically constituted by a layer of mature residual soil, 12 m thick, followed by an 10 m thick saprolitic soil topping weathered rock. The N_{SPT} blowcounts for the mature residual soil ranged from 8 to 13, whereas for the saprolitic soil it varied from 20 to 100. The water table was at 2.7 m depth, characterising a natural saturated condition for tests performed.

Figure 42 shows that for both mature residual and saprolitic layers, the reference modulus G_{ur}^* increases approximately linearly with depth. However, there is a difference in the linear coefficient for both layers. The same trend can be observed in crosshole test results, G_0.

Figure 43 shows a distinct decrease in shear modulus with shear curves for each typical soil layer mature residual or saprolitic. These curves were obtained supposing that the shear modulus variation with shear strain could be represented by a hyperbolic equation, $G/G_0 = [1 + \gamma(G_0 \tau_{max})]^{-1}$ where G_0 is the shear modulus obtained from crosshole tests, γ is the shear strain and τ_{max} is the drained strength.

On the stiffness derived from flat dilatometer tests (DMT)

The modulus determined by the Marchetti's Flat Dilatometer (DMT), designated M_{DMT}, is the vertical confined (one dimensional) tangent modulus at σ'_{v0} and is said to be the same as E_{oed} ($=1/m_v$) obtained from an oedometer test in the same range of strains. This modulus can be converted in the Young's modulus (E) via the theory of elasticity. For $\nu = 0.25$–0.30 it is possible to write: $E \approx 0.8 M_{DMT}$.

This empirical Marchetti's modulus is applied to predict settlements in sand and clays (Marchetti et al. 2001) and it was validated by different researchers (Schmertmann, 1986 in Marchetti et al. 2001 and Hayes, 1990, in Mayne, 2001).

Viana da Fonseca (1996) and Viana da Fonseca and Ferreira (2002) for characterization of the soil stiffness for shallow foundations settlement assessment, used correlations between the moduli E_{DMT} and Go, $E_{s10\%}$ (secant modulus corresponding to 10% of peak shear strength). The following correlations were obtained:

$$G_0 / E_{DMT} \cong 16.7 - 16.3 \cdot \log_{10}(p_{0N}) \qquad (27)$$

$$E_{s10\%} / E_{DMT} = 2.35 - 2.21 \cdot \log_{10}(p_{0N}) \qquad (28)$$

These formulae are situated between those that are used for NC and OC transported soils (Baldi et al. 1989).

On the stiffness derived from plate loading tests (PLT)

Schnaid et al (2004) have demonstrated the level of uncertainty associated in back-figuring the degradation curve from in situ tests in bonded soils. Typical (simplified) variations of the very small strain Young's modulus, E_0 (inferred from shear wave velocities with a Poisson's ratio of 0.1) of uncemented and cemented Perth sand are presented. The E_{sec} value of the cemented sand is seen to reduce from a high initial value (E_0) of 350 MPa, which prevails until a

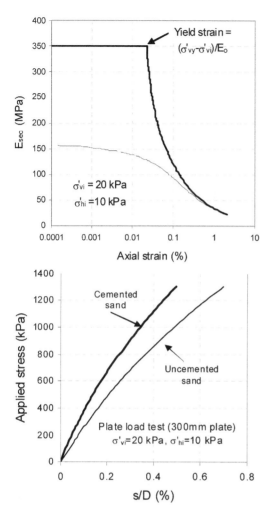

Figure 44. Analysis/back-analysis of plate loading tests for cemented and uncemented sands (from Schnaid et al. 2004).

the stiffness of the cemented and uncemented sands were identical at initial effective stress levels in excess of σ'_{vy} ($=100\,kPa$). The q_{app}-s/D predictions are shown on Figure 44b.

It is apparent that, despite the significant differences in stiffness, the curves are almost linear and not dissimilar. It appears that the increase in E_o with stress level in the uncemented sand almost compensates for its much lower E_o value at the beginning of loading. Significant softening that may be expected on inspection of Figure 44a when the applied stresses exceeded σ'_{vy} is also not apparent for the same reason. This latter observation is consistent with that observed by Viana da Fonseca et al. (1997a) in a footing test on a cemented saprolitic soil, and it would appear that backanalysis of PLTs under drained conditions can lead to the inference of a range of markedly different non-linear stiffness characteristics. Clearly, PLT interpretation would benefit from unload-reload loops.

The intermediate stress level (FS = 4) corresponds approximately to the allowable pressure for residual soils, from Décourt's (1992) criterion. This stipulates that the allowable pressure on a shallow foundation on residual soil should be that which causes a settlement of 6.0 mm for a 0.8 m diameter plate loading test, representing a settlement of 0.75% of the diameter of the loading surface. The possibility of inferring design values for Young's modulus to predict the behavior of load tests, on plates and on a prototype footing, conducted in the vicinity of penetration tests was developed and thoroughly discussed elsewhere (Viana da Fonseca, 2001). The main conclusions drawn for the most common methods are as follows: the Burland and Burbidge (1985) equation based on SPT results led to an overestimation of the observed settlement by a factor of 2 to 3, while the application of the Schmertmann et al.'s (1978) method reproduced accurately the footing settlement for $\alpha = E/q_c$ values in the range of 4.0 to 4.5.

Combining data from various in-situ tests

Fahey et al. (2003) and Lehane & Fahey (2004) – as cited by Schnaid (2005) – describe an approximate means of assessing stiffness non-linearity by combining G_o measurements with trends indicated in SBPTs and DMTs. The approach, developed for Perth sands using a relatively large database of *in situ* test results in Perth, had the following features:

(i) Correlations such as those given in equation (2) and (4) were derived for sands of various ages and converted to equivalent E_o values assuming a Poisson's ratio of 0.1.

(ii) Dilatometer E_D data were found to vary in a similar way with q_c and σ'_v to the G_o data but indicated a relatively low sensitivity to sand age

presumed yield stress (σ'_{vy}) of 100 kPa is exceeded, to a stiffness comparable to that of the uncemented sand at axial strains in excess of 0.4% (Figure 44a). This data and a simplified non-linear settlement prediction method proposed by Lehane & Fahey (2002) were employed to predict the applied stress (q_{app}) – settlement (s) response of a 300 mm diameter (D) plate on the Perth sands. This method incorporates the strain and stress level dependence of stiffness in a computer program, and although it does not model plastic flow and assumes a Boussinesq stress distribution, it has been shown to be a reliable predictive tool at typical working settlements (i.e. s/D <2%). Parameters to match the stiffness characteristics shown on Figure 44a were derived using the procedures described in Lehane & Fahey (2002). The predictions assumed that

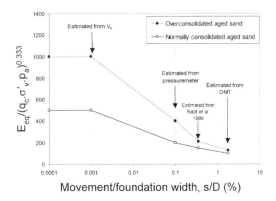

Figure 45. Proposed degradation curve (Schnaid, 2005).

Figure 46. SBP used in residual soils in Porto (Topa Gomes et al. 2008), with details of the tri-cone rock-roller bit and imaging of the type of soils and testing curves.

and stress history. Best-fit correlations indicated E_o/E_D ratios of 11 ± 3 and 7 ± 2 for overconsolidated and normally consolidated sands respectively. Lehane & Fahey (2004) indicate that, as a consequence of a number of compensatory factors, E_D values approximate to about 70% of the in situ vertical operational stiffness at $s/D = 1.8\%$.

(iii) G_{ur} values measured using the procedure suggested by Fahey & Carter (1993) were typically 0.40 ± 0.05 of G_o at cavity strains of $\approx 0.1\%$.

(iv) Correlations proposed by Baldi et al. (1989) for the Young's modulus of sand measured in triaxial compression at an axial strain of 0.1% were converted to a format similar to that of equation (2); these correlations were shown to predict average ratios of E_o to operational stiffness of 5 for both normally and overconsolidated sand.

These features are combined in Figure 45, with due acknowledgement to the strain levels induced in each test, to provide the practitioner with an approximate means of deriving a strain dependent Young's modulus (E_{eq}) operational at a vertical stress level of σ'_v.

Fahey et al. (2003) show that this approach provides a good estimate of stiffnesses backfigured from footings in Perth sand and Lehane & Fahey (2004) show that a refinement of the method describes the non-linear response to load of the Texas experimental footings on sand reported by Briaud & Gibbens (1994), even if only V_s and DMT data are available.

3.4 Overall fitting of SBP pressure-expansion curve as better approach to characterize sensitive cohesive-frictional residual soil

In situ measurement of the variation in the shear modulus with the shear strain can be achieved by performing self-boring pressuremeter (SBP) tests using, for example, the Cambridge-type SBP (Clarke 1996). The self-boring mechanism of the pressuremeter minimizes disturbance to the surrounding soil during installation.

Several unload-reload loops are generally recommended and conducted for the determination of the shear modulus of soil from the stress-strain curve.

From the pressuremeters probes, the self-boring pressuremeter (SBP) is the only one that can allow the measurement of the geostatic total horizontal stress σ_{h0}, offering a better interpretation of test results from small to large strains levels. Nevertheless, the feasible performance of self-boring pressuremeter (SBP) testing is not easy in residual soils, and, unfortunately, not frequent. Test curves resulting from such a difficult self-boring process – somewhat larger in diameter than the body of the instrument – are commonly far from perfect, showing evidence of significant over-drilling.

Topa Gomes et al. (2008) report an interesting case of a set of SBP executed in the vicinity of a recently-constructed metro station, involving residual soils of granite weathered profiles and discuss the quality of results and the efforts done to interpret them, as drilling is often imperfect, due to the fabric, a structural variability and the cohesive-frictional nature of this type of materials. Two distinctly different types of materials were involved in the two boreholes: one is a residual material with a weathering degree W5 (confirmed by the boreholes in the vicinity), while the material in the other borehole is a much more resistant and stiffer material, indicating a weathering degree W4 and, hence, a soft-rock-like material (Figure 46).

Theoretically, the initial slope of a SBP test yields the G_0 value. However, since in practice there is still some disturbance (Fahey & Randolph, 1984), the modulus must be taken from an unload-reload cycle (G_{ur}).

For very overconsolidated soils and cemented geomaterials it could be assumed that $G_{ur} = G_0$ if the strain of one cycle is less than 0.01% (Mayne, 2001). However, that has not been observed in most practical cases in non-textbook materials, such as residual soils (values ranging from 2.6 to 3.0, as reported by Viana da Fonseca, 1998a, and 2.2 to 2.3, by Pinto & Abramento, 1997).

The use of G_{ur} in practice can be done by different approaches:

(i) To link G_{ur} to G_0 using a determined stress-strain relationship (Bellotti et al. 1994; Ghionna et al. 1994);
(ii) To compare G_{ur} values to the degradation modulus curve G/G_0 versus shear strain – γ from laboratory, taking into account the average values of shear strain and mean plane effective stress associated with the soil around the expanded cavity (Bellotti et al. 1994);
(iii) To recourse to traditional empirical techniques and numerical modelling under axysimetric condition, using simple models, such as a numerical hyperbolic Fahey & Carter (1993) model, in order to back analysis the test results and to better understand the soil behaviour. Resistance and stiffness parameters, including the K_0, value may be compared with typical values for these soils, reported from regional experience.

In the cited reported case of SBP tests performed over residual soils from granite of Porto, with different weathering degrees, Topa Gomes (2008) present test curves resulting from such a difficult self-boring process – somewhat larger in diameter than the body of the instrument – were far from perfect, with the test curves showing evidence of significant over-drilling (Figure 47). As a result, the first part of the expansion phase shows no resistance, as the membrane expands to make contact with the over-sized hole. This has severe implications for the determination of K_0 from the tests, as is discussed in the paper.

The interpretation of such tests give *in situ* σ_h (and hence K_0) values problematic, and approaches used in interpreting standard PMT tests were used, assuming, for instance, that the *in situ* horizontal stress corresponds to the intersection of the initial linear elastic stretch of the curve with the roughly horizontal line defined by the stretch in the curve where, after the lift-off, the membrane tends to fit the soil (Marsland and Randolph, 1977).

With this in mind, in the cited work, the authors considered the overall fitting to the pressure-expansion curve, starting from a set of results derived from

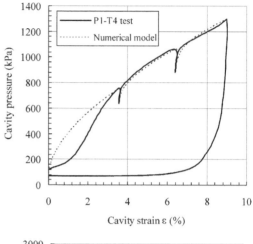

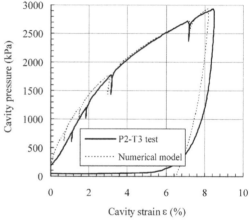

Figure 47. Pressure versus extension curves obtained from difficult self-boring process in residual soils superposed with the numerical back-analysis (Topa Gomes et al. 2008): a) W4 – highly weathered; b) W5 – completely weathered.

laboratory tests and cross-hole seismic tests, and recurring to "distorted" hyperbolic model (Fahey & Carter, 1993) to numerically model the tests, aiming to find the set of parameters that allowed an overall best fit of the observed results. A large number of variables were involved, which turned finding a unique set of parameters to match each curve more complex. However, a pattern was achieved and an overall tendency of the results was found allowing deriving values for the parameters with a certain degree of confidence.

Figure 47 present the fitting curves over the registered data obtained for two of the SBT tests in both profiles, the more resistant and stiffer material – soft-rock W4 (S1T4), for one side, and the more weathered residual soil – a typical W5 saprolitic soil (S2T4), for the other. The results in Table 5 show the clearly different characteristics of the materials in the two

Table 5. Parameters derived from the backanalysis of the pressuremeter tests.

Test	c' (kPa)	$\phi'(°)$	$\psi(°)$	K_0	G_0 (MPa)
S1-T1	10	39	5	0.57	95
S1-T2	10	39	10	0.74	112
S1-T3	0	32	0	0.60	108
S1-T4	50	43	15	0.73	132
S1-T5	50	43	10	0.70	206
S2-T1	1000	45	15	0.80	270
S2-T2	1000	45	15	1.40	268
S2-T3	1000	45	15	1.20	474
S2-T4	1000	45	15	1.10	426
S2-T5	1000	45	15	1.30	346
S2-T6	1000	45	15	1.33	691

boreholes (more details in (Topa Gomes et al. 2008). Material on borehole 2 is much more resistant and stiffer material, indicating a weathering degree W4 and, hence, a rock-like material

The change in the K_0 values allowed a better fit of the tests and the variation on this parameter is quite reduced. For tests on borehole 1, K_0 has an average value of 0.67 ranging from 0.60 to 0.74. This value does not differ very much from the value obtained from the lift-off of the membrane, even bearing in mind the amount of over-drilling involved. However, referring to borehole 2, the values appear to be much higher, with an average value of 1.19. The variation is also higher, ranging from 0.80 to 1.40. If test S2T1 is excluded, due to poor overall fit, the average K_0 value obtained is 1.27, with the range being 1.10 to 1.40. This result suggests a decrease in K_0 value as weathering increases. In addition, the conclusion of a K_0 value greater than 1.0 for the less-weathered granite soils in Porto is novel, although confirming the preliminary indications (Viana da Fonseca & Sousa, 2001).

4 THREE SPECIFIC FEATURES IN RESIDUAL SOILS

4.1 At rest stress state

The evaluation of the at rest coefficient of earth pressure, K_0, is of much importance in several applications in geotechnical analyses, namely numerical, as the mechanical response of geomaterials is dependent of the correct definition of the initial stress state. The value of K_0 in transported soils is often derived from semi-empirical correlations based on physical index factors, but these correlations are misleading in residual soils, or even in specific transported soils, such as those structured, since their cemented microstructure is dominantly responsible for the geomechanical behavior.

The difficulties in directly measuring the value of K_0 – assumed as the ideal procedure for a good determination of this value – on these sensitive soils is very much a consequence of the difficulties in preserving such weak cemented microstructure during an intrusion process (such as in pressuremeters or dilatometers – Clarke, 1996).

Even though, the sensitivity of the cemented structures of natural soils, especially of residual structurally bonded soils, makes this task very unreliable when skilfulness is not present. Simple details, as the bit used for the drilling of the insertion process or the injection pressure of the washing fluid – may result in the inconclusive answer of the results, expressed in non reliable values of lift-off pressure (Figure 47).

Some trends to alternatively evaluate K_0 in laboratory using high quality samples, is always criticisable, since while testing in triaxial cells has the great advantage of ensuring the application of a homogeneous stress field, this is not a fundamental process as it consists of a radial controlled triaxial test over specimens that, having been trimmed from a good quality sample, were previously stress-relieved, even considering that the structural cohesive component present in young residual (saprolitic) soils is very important in maintaining the soil integrity. It should be recall that this process is really – taken as guaranteed that sampling has not damage the essential structure is the soils -, inducing an elastic stress path, where the relations between the principal effective stresses are really ruled by the theory of elasticity. Vaughan & Kwan (1984) have suggested that this structural interpretation of the rock masses weathering process based on the hypothesis that the rock is a quasi-elastic structure. By this, the authors have associated the weathering evolution to a reduction of stiffness and shear strength. The vertical effective stress decrease (associated to loss of mass due to decomposition) leads to a decrease of the horizontal stress, which, considering the stability of structural matrix and ignoring collapsible secondary effects, is governed by the absence of horizontal strains, will be evaluated by elastic formulations, as it follows:

$$\varepsilon_h = \frac{1}{E} \cdot (\sigma'_h - \nu \cdot \sigma'_h - \nu \cdot \sigma'_v) = 0 \Rightarrow K_0 = \frac{\sigma'_h}{\sigma'_v} = \frac{\nu}{1-\nu} \quad (29)$$

being ν the Poisson ratio.

Viana da Fonseca & Almeida e Sousa (2001) have discussed this problem and presented some experience on the subject, giving emphasis to results obtained in experimental sites in Porto's metropolitan area. It should be noted that the relict granitic rock mass in this region has typically stress states with low horizontal stresses, due to the confinement relaxation associated to the steep valleys created by the rivers, which potentate the decrease of σ_h. Experimental studies and FEM back-analyses of several geotechnical works produced very interesting results: low values of K_0 for

233

high weathering degrees (class W_5 of rock classification, ISRM 1981), between 0.35 to 0.50, increasing for moderate weathering levels (W_4-W_3 classes), with K_0 values close to unity. Those values were in close agreement with some of the available results from the most careful self-boring pressuremeter tests, SBPT (Viana da Fonseca et al. 1998a). Other methodologies have revealed inadaptability and some more recent SBP tests have proved how difficult is to assure a good test to have a competent preservation of at rest stress state before cylindrical expansion. The increase of K_0 value with the decreasing of this weathering level, where the same relic cementation is more preserved (values up to 0.9, in W_4-W_3 classes of weathering), was recently corroborated in a new campaign, referred above and included in this conference (Topa Gomes et al. 2008).

Excellent results and no worse impressive interpretative analysis have been carried out in University of São Paulo, in Brazil (Pinto & Abramento, 1997, Abramento & Pinto, 1998). In these papers the authors report very high values of K_0 both for lateritic and saptolitic horizons, tending to decrese from 3.0 at the top the the the former (the mature soil) to 1.0 at the base and ranging between 1.0 to 3.0 in the saprolite. The authors recognize that these values are unexpected, but associate it to the existence of locked-in stress in the residual profiles, which have been inherited from rock masses, therefore, different rock mass in situ stress states, when compared to Porto region. This factor is naturally and important factor to be consider in the analysis.

The use of shear wave velocities determined in Down-Hole (Vsvh) and Cross-Hole (Vshv) tests for the evaluation of K_0

Recently, promising trends in using and interpreting seismic tests, especially those performed in boreholes, have been assumed (Jamiolkowski & Lo Presti, 1994; Sully & Campanella, 1995; Jamiolkowski & Manassero, 1996;). For the estimation of K_0, it has been postulated that when both horizontally and vertically polarized shear waves are measured during Cross-Hole tests, one can attempt to estimate the at rest coefficient of the earth pressure. Several authors, e.g. Hatanaka & Uchida (1996) and Fioravante et al. (1998) have postulated that there is some feasibility in predicting the value of the at rest coefficient of earth pressure by measuring and relating velocities of seismic shear waves under diverse propagation and polarization directions. This may be achieved by measuring shear waves velocities under specific paths, using different techniques (Butcher & Powell, 1996). Figure 48 systematizes some of these conditions.

The evaluation of K_0 from shear wave velocity measurements is based on the dependence of wave velocities propagated in an elastic medium (inherently isotropic) on the principal effective stresses acting

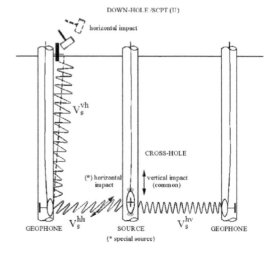

Figure 48. Difference in the directions of propagation and polarization of shear waves.

in the direction of the wave propagation and in that of the particle motion – polarization. The difference in the directions of propagation and polarization of shear waves induced by DH or Seismic Cone (SCPT) and CH methods is reflected in the respective notations: V_s^{vh} (vertical propagation and horizontal particle motion) and V_s^{hv} (the inverse condition). A special Cross-Hole system, as illustrated in Figure 48, will generate a horizontally propagated and polarized shear wave (V_s^{hh}). As expressed by Fioravante et al. (1998), the principle of the dependence of these velocities on stress state (and also on the state indices – here admitted as constant) can be generally explained by the following laws:

$$V_s^{vh} = C_s^{vh} \cdot \sqrt{F(e)} \cdot {\sigma'}_v^{n_a} \cdot {\sigma'}_h^{n_b} \qquad (30)$$

$$V_s^{hv} = C_s^{hv} \cdot \sqrt{F(e)} \cdot {\sigma'}_h^{n_a} \cdot {\sigma'}_v^{n_b} \qquad (31)$$

$$V_s^{hh} = C_s^{hh} \cdot \sqrt{F(e)} \cdot {\sigma'}_h^{n_a} \cdot {\sigma'}_h^{n_b} \qquad (32)$$

with C_s^{vh} and C_s^{hv} as two dimensional material constants, which ratio is a sign of the anisotropy of the soil structure and/or fabric (equal to unity in a continuum and homogeneous medium and for inherently isotropic soils); $F(e)$ is the void ratio function, and n_a and n_b are experimental stress exponents related to the principal stresses acting in the direction of the wave propagation and of the particle motion, respectively (in a isotropic and recent continuous medium $n_a = n_b$ – reconstituted soils). By combining equations (31) and (32), a ratio between these velocities can be generated, resulting in the following expression for K_0 ($=\sigma'_{h0}/\sigma'_{v0}$):

$$\left(\frac{V_s^{hh}}{V_s^{hv}} \cdot \frac{C_s^{hv}}{C_s^{hh}} \right)^{\frac{1}{n}} = K_0 \qquad (33)$$

234

Viana da Fonseca et al. (2004 and 2005) have proven that this hypothesis is not adaptable to the data accumulated in experimental studies over undisturbed samples from residual (saprolitic) soils of granite, which were tested on very careful controlled triaxial conditions, using well calibrated bender elements. The non-consideration of the similarity of n_b and n_a values would imply the evaluation of both V_s^{hh} and V_s^{hv}, as expressed by Sully and Campanella (1995):

$$\left(\frac{V_s^{hh}}{V_s^{hv}} \cdot \frac{C_s^{hv}}{C_s^{hh}}\right)^{\frac{1}{n_a+n_b}} = \frac{2 \cdot K_0}{1+K_0} \tag{34}$$

imposing a non-conventional Cross-Hole test, in order to measure V_s^{hh}. As it was expressed by the mentioned authors (Viana da Fonseca et al. 2004 and 2005) the assessment of K_0 can be made with less degree of confidence by measuring V_s^{hv} and V_s^{hh}, by means of conventional CH tests, for the value of V_s^{hv}, and to DH or SCPT tests, for V_s^{vh}. This implies the assumption of perfect inherent isotropy of the ground, giving similar values for C_s^{hv}, and an unique function of the void ratio, $F(e)$. The value of the coefficient of at rest in situ stress is expressed by:

$$K_0 = \left(\frac{V_s^{vh}}{V_s^{hv}}\right)^{\frac{1}{n_b-n_a}} \tag{35}$$

The values of n_a and n_b may be inferred from laboratory tests over undisturbed samples as described in the above mentioned work. There is an important obstacle to this method, which lays on the low values of the exponents correlating the polarized velocities, requiring a very high reliability and precision in their evaluation. This research is ongoing in the University of Porto.

One important subject should, however, be taken into consideration and this is the dual and independent factors that rule the dependence of V_s^{hh}, V_s^{hv}, or, V_s^{vh} on the level of (micro-) structuring and stress state. This may be expressed, for instance, for V_s^{vh}, by the equation:

$$V_s^{vh} = \overline{C_s^{vh}} + C_s^{vh} \cdot \sqrt{F(e)} \cdot \sigma'^{n_a}_v \cdot \sigma'^{n_b}_h \tag{37}$$

Where $\overline{C_s^{vh}}$ is a factor that is dependent on the cementation-reactive force or tensile force required to break the cement at contacts, independent of the stress-controlled interparticle engagement, as proposed by Santamarina (2001) and described in the introduction of this text. Similar laws should be expressed for V_s^{hh} and V_s^{hv}, with values of the respective constants, being eventually different in value, but to be considered.

This concept, jeopardizes the direct and simplified use of the relations expressed in equations (33) and

(34), but potentates the re-evaluation of the shear wave approach, as far both the constants and exponents due to the granular condition and the one due to bonding forces are associated. These studies are being conducted in the University of Porto, with a significant accumulation of data obtained in true triaxial (cube) cell, with bender elements in the 6 faces and mounted in orthogonal directions (Ferreira, 2008).

Even though the evaluation of coefficient of earth pressure at rest from in situ or laboratory testing is very controversial, due to the level of disturbance induced by penetration/installation of equipments and sampling processes, the fact is that this parameter is often needed for design purposes. A rough experimental estimation is better than only an empirical one!

Baldi et al. (1986) have realized the usefulness of combining CPT(U)+DMT and proposed the following correlation to derive K_0 in granular sedimentary soils:

$$K_0 = C_1 + C_2 \cdot K_D + C_3 \cdot q_c/\sigma'_v \tag{38}$$

where: $C_1 = 0.376$, $C_2 = 0.095$, $C_3 = -0.00172$

q_c representing the CPT tip resistance and σ'_v stands for the effective vertical stress, which can be derived from DMT results.

Taking into consideration the q_c/σ'_v relation equal to 33 K_D, established by Campanella & Robertson (1991) for non-cemented sandy soils, it is clear that this ratio is not representative of the studied soils. Viana da Fonseca (1996), Cruz et al. (1997) and Viana da Fonseca et al. (2001) proposed to correct C_2 constant of equation (38) as follows:

$$C_2 = 0.095 * [(q_c/\sigma'_v) / K_D] / 33 \tag{39}$$

Although available data on K0 is very rare, the analysed data reflects the local experiment (0.35–0.5). It should be noted that direct application of Baldi's correlation would lead to much higher values, usually greater than 1, mostly unreliable to these soils (as reported from regional experience – Viana da Fonseca & Sousa, 2001).

In this conference, Viana da Fonseca et al. (2008) present the analysis of another profile in residual soils, restating that the corrected correlation give more realistic values (Figure 49).

Pinto e Abramento (1997) present very interesting data from a set of self-boring pressuremeter tests carried out on a gneissic residual soil in São Paulo City, at the site of a subway station. The variation of $K_0 = (\sigma'_{ho}/\sigma'_{vo})$ with depth, where σ'_{ho} and σ'_{vo} are the effective horizontal and vertical stresses, tend to decrease from 3.0 at the top of mature residual soil up to 1.0 at its base. One of the tests, at a depth of 14.5 m, yielded a very high value of $K_0 = 3.8$, which is attributed to a local heterogeneity with higher strength as, for example, quartz veins or gravel. For the

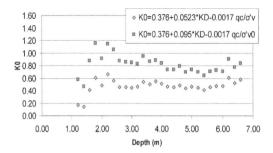

Figure 49. Estimation of the coefficient of earth pressure K0 by specific correlations for residual soils.

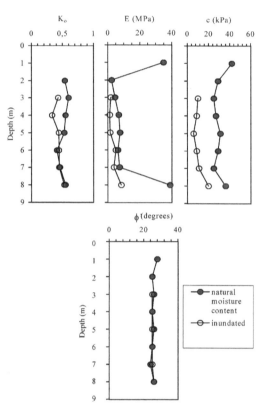

Figure 50. Estimation of K_0 by specific correlations for residual soils (Cunha & Vecchi, 2001).

saprolitic soil, K_o varied between 1.0 and 3.0. These unexpected high K_o values suggest the existence of a strong locked-in stresses in the residual soil layers, which were inherited from the rock mass. It is not unusual to observe high K_o values in bonded materials: Rocha Filho & Queiroz de Carvalho (1990) reported values greater than unity; Árabe (1995) reported values in the range of 1.5–2.7 (with a mean of value of 1.85) in the gneissic residual soil from Rio de Janeiro; Schnaid & Mántaras (2004) calculated values ranging from 0.2 to 1.2 (average of 0.8) for a gneiss residual soil from São Paulo.

K_0 in unstaurated soils
There is not much work in these conditions, as its evaluation is very much dependent on the relative influence of both suction (fundamentaly isotropic) and interpraticle forces (these may be or not isotropic). Cunha & Vecchi (2001) illustrates the applicability of pressuremeters in the unsaturated, tropical and collapsible Brasília clay (the so called Brasília "porous" clay). Figure 50 summarises the results of the main geotechnical parameters obtained in two different soil conditions: at natural moisture contend and after soaking the ground. A considerable decrease in the predicted geotechnical parameters is observed after soaking: K_0 and E can be reduced by around 30%, c (total cohesion) reduces by as much as 60%, whereas the internal friction angle ϕ' remains reasonably constant. The interpretation of the pressuremeter testing curve was done by Cunha et al. (2001) with the use of an elasto-plastic cavity expansion model modified after Yu & Houlsby (1991), using the theoretical approach put forward by Kratz de Oliveira (1999). The predicted K_0 values for the soil at natural moisture content were closer to the upper range of laboratory values than to the lower range.

4.2 Permeability

The importance of a good evaluation of the hydraulic conductivity is related to problems of slope stability and flow through dam foundations as well as abutments, deep excavations and retaining structures. Usually it is the macro-structural variability – mainly in saprolitic soils – that determines the *in situ* behaviour. It is, then, very difficult to evaluate its real patterns by laboratory testing, even with good undisturbed samples. Here the scale effects are of main concern. Being the permeability of residual soils controlled, to a large extent, by the relic structure of the parent rock and for the degree of alteration (the all weathering processes, including the pedogenetic processes). The possibility of the soils present a complex macro and micro structure, with scattered grain size distribution, variations in mineralogy and clay content, containing aggregated particles, can have a dominant effect on its permeability, and hence on in situ behaviour at given loading rates.

Viana da Fonseca (2003) present the synthesis of extensive recent experimental campaigns in the urban area of Porto for the new metro lines, in particular for the evaluation of permeability variation with depth, and its relation to weathering degrees, as this information is important for the calculation of pumping volumes for the design of deep excavations and retaining structures in the future stations. Although it is

Table 6.   Trendy values of permeability by classes of weathering of Porto granite.

Class of rock weathering (W$_i$ from ISRM, 1981)	Permeability (m/s)	
	References (1)	Exp. Results (2)
Decomposed rock – soil with no relic structure (W$_6$)	Low	$\approx 10^{-7}$
Completely weathered rock – saprolitic soil (W$_5$)	Medium	10^{-6}–10^{-5}(3)
Highly weathered (W$_4$) and fractured (F$_4$–F$_5$) rock	High to medium	10^{-5}–10^{-4}(3)
Moderately weathered (W$_3$) and fractured (F$_3$–F$_4$) rock	Medium to high	10^{-5}–10^{-6}
Slightly weathered rock (W$_2$)	Medium	10^{-6}–10^{-7}

(1) Trends as revealed in Deere & Patton (1971), Dearman (1976), Costa Filho & Vargas Jr (1985)
(2) Experimental values obtained in head tests through sections in drillholes or pump tests, with surrounding piezometers.
(3) More kaolinitic matrices present values of permeability that are lower than the indicated ranges by one order of magnitude.

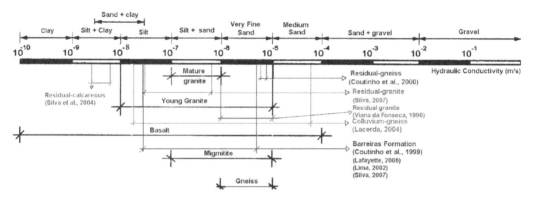

Figure 51.   Typical saturated permeability in natural soils (after Schnaid et al. 2004; Coutinho & Silva, 2006) with inclusion of others Brazilian soils and young residual-Porto granite results.

rather difficult to generalize, Table 6 indicates some of the trends that have been detected in these studies.

Many young residual soils show permeability values in the range of 10–4 to 10–7 m/s, which are geomaterials in the so called intermediate permeability range of 10–5 to 10–7 m/s. (Schnaid et al. 2004; Schnaid, 2005). Coutinho & Silva (2006) included in the Figure 51 results of range of permeability values of other Brazilian unusuals soils, including finer residual soil from calcareous rock, (young or/and mature) residual soils from gneiss and granite and Barreiras formation from Pernambuco, and colluvium material from Rio de Janeiro (see Lacerda, 2004). In this paper is an also included result from young residual soil – Porto granite. The results are in agreement with the general proposal considering the grain size distribution and structure characteristics of the soils investigated. Mature residual soil, in tropical regions; commonly show a granular and porous structure, independent of the texture, which can provide the soil with higher saturated permeability.

An important point in the topic of permeability in residual soils is the possibility of occurrence of partial drainage during field tests. For intermediate soils (K in the range of 10^{-5} to 10^{-8} m/s), the simplest accepted approach of a broad distinction between drained (gravel and sand) and undrained (clay) conditions for the interpretation of in situ tests cannot be applicable since test response can be affected by partial consolidation and consequently existing analytical, numerical or empirical correlations can lead to unrealistic assessment of geotechnical properties (e.g. Schnaid et al. 2004).

Schnaid et al. (2004) and Schnaid (2005) present an important discussion about this topic showing the importance and some proposals to evaluate the partial drainage during in situ tests. For instance, it has been considered that for hydraulic conductivities greater than 10^{-4} m/s cone penetration is fully drained, whereas for hydraulic conductivities less than 10^{-8} m/s (or B$_q$ greater than 0.5) an undrained penetration will take place. It is also presented a dimensionless

237

velocity (Randolph & Hope, 2004) to analyze the rate effects.

$$V = v \cdot d / c_v \qquad (40)$$

Being v the rate of penetration, d the probe diameter and c_v is coefficient of consolidation.

It is recommended to change the penetration ratio for field tests in intermediate permeability soils to avoid tests that yield dimensionless velocities within the range of 0.01 to 10. In this range, partial drainage is expected to occur.

Another important point to note is the variation with suction of the hydraulic conductivity of residual soil when these are in unsaturated condition. The coefficient of permeability is generally assumed to be a constant under all stress states for a saturated soil. However, the coefficient of permeability for an unsaturated soil can vary widely depending on the stress state and therefore takes on the form of a function of mathematical equation. Although any change in the stress states of a soil can affect the coefficient of permeability, it is manly matric suction that influences the amount of water in the soil and therefore has the dominant influence. A further increase in suction after the air entry of the soil causes the coefficient of permeability to drop by several orders of magnitude. The relation between the permeability function and the soil-water characteristic curve can subsequently be used for the estimation of the permeability function (Fredlund, 2006).

Figure 52 shows results from an unsaturated deposit of gneiss residual soils (mature and young soil) from Pernambuco, Brazil (Coutinho et al. 2000). It can be seen that the values of the saturated hydraulic conductivity obtained in the field (Guelph tests) are basically in agreement with the range proposed in Figure 51 and that the mature residual soil, a sandy clay presents a higher saturated hydraulic conductivity ($K_{fs} = 1.1 \pm 0.48 \times 10^{-5}$ m/s) than the young residual soil, a silty sand ($K_{fs} = 0.41 \pm 0.15 \times 10^{-5}$ m/s). This behavior is due to the granular (aggregated particles) and porous structure of the mature residual soil in this site. It is also observed that the variation with suction of the relative hydraulic conductivity to the young residual soil ($K_r = K/K_{fs}$) indicate a strong reduction in the K_r values to small values of suction in this residual soil influencing the flow behavior of the site (hillsides area).

4.3 Collapsibility: identification and classification of collapsible residual soils

The most qualitative criterion for determining the susceptibility to collapse are based on relationships between the porosity, void ratio, water content, or in place dry density (Thornton and Arulanandan, 1975).

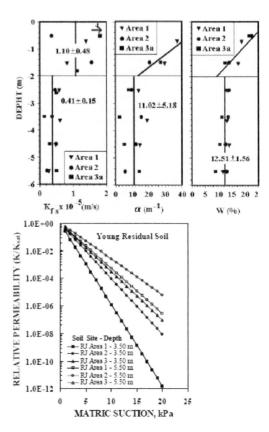

Figure 52. Results of permeability test with Guelph permeameter (Coutinho et al. 2000).

Examples of these criteria can be seen in Gibbs & Bara (1967) to laboratory tests and Décourt and Quaresma Filho, (1994) to in situ tests (SPT-T). Quantitative criterion have been developed from results of double oedometric tests (Reginatto and Ferrero, 1973) and simple oedometric tests (Jennings and Knight, 1975; Lutenegger and Saber, 1988) in order to classify susceptibility to soil collapse.

Jennings and Knight (1975) define collapse potential (CP) as the deformation due wetting obtained through simple oedometric tests according to the expression:

$$CP = \frac{\Delta e}{1 + e_o} 100 \qquad (41)$$

where: Δe and e_o is void ratio variation caused by wetting and e_o the initial void index.

Although SPT is a test broadly utilized world-wide, to the author's knowledge, no identification criterion based on the results from SPT test has been published. However, in Brazil, porous collapsible soils in southern and central-west regions have been characterized for

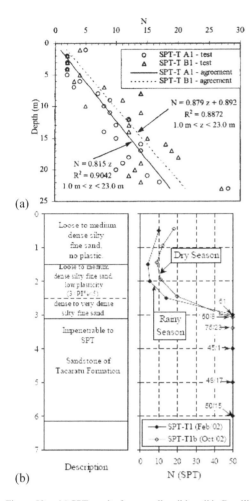

(a)

(b)

Figure 53. (a) SPT results from a collapsible soil in Brasília (Marques et al. 2004); (b) SPT borehole and water content profile (Coutinho et al. 2004b and Souza Neto et al. 2005).

showing low values in N (in general N < 5). A typical example is showed in Figure 53a, by Marques et al. (2004) concerning the porous clay layer in Brasília, Brazil, where the penetration index down to the 7 m in depth fit in this limit. Suctions in these regions are low (<50 kPa) and it may account for the small variations in the results when the test is carried out in the dry or humid season. In Petrolândia City, in the State of Pernambuco, Northeastern region of Brazil, suctions may reach values such as 10 MPa in the dry season, influencing significantly the test results. This site was part of a broader study carried out by Federal University of Pernambuco – UFPE. Typical results are shown in Figure 53b which reflect clearly this aspect, where the results from SPT in the dry season in the collapsible layer (9 < N < 20) was about 100% greater than those obtained in the humid season (4 < N < 13). More information can be seen in Coutinho et al. (2004b), Souza Neto et al. (2005) and Dourado & Coutinho (2007).

Kratz de Oliveira et al. (1999) present a proposal of collapsible soil identification from results of double pressuremeter tests. The methodology consists of a comparison between a test in the natural condition with another in a borehole previously flooded, similar to double oedometer test considered by Jennings and Knight (1957). The pressuremeter collapse is defined by:

$$C_{press} = \frac{r_f^{\,2} - r_i^{\,2}}{r_i^{\,2}} - \frac{r_o^{\,2}{}_{wet} - r_o^{\,2}{}_{nat}}{r_o^{\,2}{}_{nat}} \quad (42)$$

where: r_i and r_f both are the radii of the cavity for the ground under condition of natural water content and for the wetting soil for the level of the creep pressure P_F of the wetting PMT tests; and r_{onat} and r_{owet} are the initial radii of the cavity for the soil conditions of natural water content and wetting, respectively.

Figure 54a presents double PMT tests obtained in a collapsible soil in the northeastern region of Brazil. The wetting results in a great loss of rigidity of the soil, characterized for the difference in the two curves. Other example of reduction in pressuremeter stiffness response, after soaking the area, is showed by Schnaid et al. (2004).

Dourado & Coutinho (2007) utilize this procedure and compares the results of the C_{press} (Eq. 2) with results obtained by Souza Neto et al. (2005) for the same local through simple oedometer tests for a soaking pressure of 200 kPa using values of collapse potential (CP) (Eq. 1) and the classification of the soil collapsibility proposed by Jennings & Knigth (1975). These results are presented in Figure 54b. The values of C_{press} vary between 6.7 and 32.4%, with highest values in the stretch between 1.0 and 1.5 m. Similarity of behavior is observed, when compared with the oedometer tests (CP), characterizing a stretch (between 1.0 and 2.0 m) more susceptible to the collapse. Considering the coherence observed in the results showed in the Figure 53b and the results obtained by Kratz de Oliveira et al. (1999), Dourado & Coutinho (2007) presented a proposal for classification of the soil collapsibility from PMT tests showed in Figure 55.

Ferreira and Lacerda (1995) and Houston et al. (1995) presented two similar – although developed independently – for performing in situ collapse tests. The former authors reported several tests, with the so-called "Expansocollapsometer" (ECT), performed inside an open auger borehole using a rigid circular plate of small diameter (100 mm). Souza Neto et al. (2005) utilized an improved version of this equipment. The in situ tests, basically, consist in the

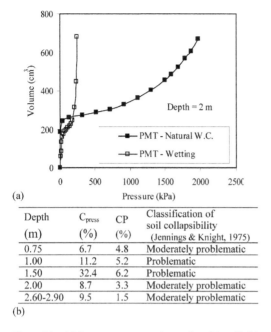

(a)

Depth (m)	C_{press} (%)	CP (%)	Classification of soil collapsibility (Jennings & Knight, 1975)
0.75	6.7	4.8	Moderately problematic
1.00	11.2	5.2	Problematic
1.50	32.4	6.2	Problematic
2.00	8.7	3.3	Moderately problematic
2.60-2.90	9.5	1.5	Moderately problematic

(b)

Figure 54. (a) Pressuremeter tests in a collapsible soil; (b) C_{press}, CP (p/σ_{soak} = 200 kPa), classification of collapsibility (Coutinho et al. 2004b; Dourado & Coutinho, 2007).

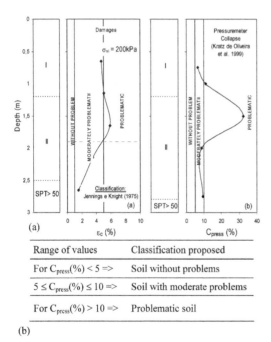

(a)

Range of values	Classification proposed
For C_{press}(%) < 5 =>	Soil without problems
$5 \leq C_{press}$(%) ≤ 10 =>	Soil with moderate problems
For C_{press}(%) > 10 =>	Problematic soil

(b)

Figure 55. (a) Classification of the soil collapsibility according Jennings & Knigth (1975) and variation of the C_{press} versus depth; (b) proposal of classification of the soil collapsibility from PMT results (Dourado & Coutinho, 2007).

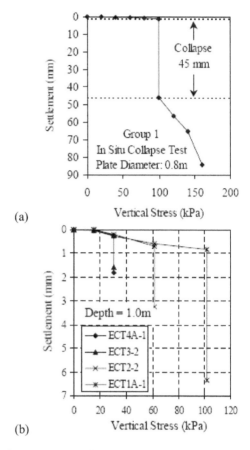

(a)

(b)

Figure 60. (a) In situ collapse test with plate load tests (PLT); (b) Typical result of the in situ collapse test with "expansocollapsometer" (ECT) (Souza Neto et al. 2005).

loading of a plate inside an open auger borehole up to a determined vertical stress. After stabilization of settlements, the soil is soaked, the process being monitored.

Settlements due to soaking (collapse) are measured until their stabilization. Collapse strain is determined by dividing the collapse settlement for the wetting front depth obtained at the end of the test. Ferreira and Lacerda (1995) show that the results from field tests can be correlated to single oedometric tests, thus allowing the criteria of identification and classification (e.g. Jennings and Knight, 1975), based on oedometric tests, to be extended to field tests. Considering the results of UFPE experimental site (Petrolândia-PE) Dourado et al. (2007) present and discuss the most soil collapse criterion (laboratory and in situ tests).

Figure 55 presents results obtained by Souza Neto et al. (2005) at the same area in Petrolândia – PE, Northeaster of Brazil, performing plate load test (Figure 60a) and the ECT test (Figure 60b). After wetting the collapse settlement measured in the plate load test was 45.0 mm. For the ECT test the collapse settlements increase with the soaking vertical stress, as expected, and in the same vertical stress of 100 kPa the settlement predicted (53 mm) by the collapse strain determined was of the order of 18% higher than the collapse settlement measured (plate load test). This result can be considered to be situated in the error range generally obtained by other design methods. Complementary studies involving numerical modeling associated to more results of in situ tests will allow better evaluation of the in situ collapse behavior and, consequently, a more rational interpretation of the in situ collapse tests – ECT.

Dourado & Coutinho (2007) calculated the collapse settlement using the PMT results considering the traditional methodology of Briaud (1992) for footing settlement on sand. The settlements were calculated in both natural water content and wetting conditions and their difference in these two conditions was considered as collapse settlement (r_c) of the soil. It was observed that for any stress that the PMT underestimated the r_c measured in the plate load test. The use of a factor of increase (Fm) of 2.5 in r_c predicted ($F_m \cdot r_c$pred $= r_c$med) by the PMT, would lead to one better approach of r_c measured for the in situ collapse tests. The authors emphasized that it must be seen as a suggestion, having to be applied only to the soil studied.

5 CONCLUSIONS

Residual soils are a product of rock mass weathering and complex diagenesis that generated materials dominated by strong inhomogeneity, fabric and structure (macro and micro). The former may be dealt in diverse ways, but in this paper an emphasis was made in geophysical methods for mapping, zonation and classification. Fabric is a matter of geological perception and insight, and is very much dependent of the visual.

The aspects for classification of these soils may be enumerated as: morphological and mineralogical characteristics, physical identification testing and engineering properties, therefore the following expeditious parametrical evaluations are due: hygrometrical state, colour (or discoloration), strength-consistency (simple methodologies such as penetration of the geological hammer, harrowing between fingers, desegregation in "slake durability tests", or, even, uniaxial compression strength), fabric (taken as the macrostructure particle voids size, arrangement and distribution, as well as fissuring,…), texture (particles sizes and shapes), density, apparent mechanical and hydraulic properties (expeditious field tests), mineralogy (specially on the advent of minerals due to weathering processes). This description has to be involved by a hydrogeological framing.

Structure, specially the effect of interparticle bonding, has a significant influence in the interpretation of test data, especially from in situ tests. These singularities were addressed in theses lines, giving relevance to the consequences in very objective parameters (apart from the starting index properties), useful to geotechnical design, such as: (i) at rest stress state; (ii) stiffness – from the very small strain levels (so important for dynamic analysis, as for competent definition of stress-strain constitutive laws) to serviceability levels (useful for common projects, such as foundations); (iii) non-linearity of stiffness (cyclic and monotonic) or degradation laws (iv) strength, where the triode "cohesive intercept ('effective cohesion') – dilation – and friction ('angle of shearing resistance')" rules the behaviour in failure…; (v) special features, such as permeability and collapsibility and few lines on the very important subject of non-saturated conditions.

The special features that demand for unusual or non-classical approaches in the characterization of residual soils have been recognized to be the key point for consequent analysis and subsequent remarks. This text is limited to authors' knowledge and to what they have managed to recoil. Residual soils are variable in their origins (parent rocks, climate conditions and else) and too much complex in their physical and stress state, to be dealt in such a limited space. Nevertheless, some specific characteristics were identified for such non-textbook approach, such as structured micro organization of their particles – with clear relic bonded matrices -, which is translated in singular non-linear stress-strain behaviours, and unsaturated conditions – which implications with the mechanical behaviour of these soils are strongly interdependent with the former interparticle structure.

ACKNOWLEDGMENTS

The authors want to express their sincere gratitude to António Topa Gomes, Cristiana Ferreira, Jorge Carvalho, Karina Dourado, who helped in compiling some disperse data, and to all the collegues that have shared some of their data and analyses. A special address is due to Martin Fahey, Serge Leroueil and mostly to Fernando Schnaid, for the sharing of their knowledge about the singularities of these soils. Acknowledge is also due to Private Sponsors of the experimental site in FEUP: Mota-Engil, SA, Sopecate, SA; Tecnasol-FGE, SA, and Teixeira Duarte, SA. This work was developed under the research activities of CEC from the FEUP, supported by multi-annual funding from FCT (Portuguese Science and Technology Foundation). The research activities developed in the UFPE – Brazil were supported by PRONEX Program – CNPq / FACEPE (Brazilian Science and Technology Federal and State Foundation).

REFERENCES

Abramento, M. & Pinto, C.S. 1998. Properties of residual soils from gneiss and migmatite determined by the 'Camkometer' self-boring pressuremeter tests. Proc. of the 11th Brazilian Cong. on Soil Mechs and Foundation Eng., Brasília, V. 2, pp. 1037–1046. (In Portuguese)

Airey, D.W. & Fahey, M. 1991. Cyclic Response of Calcareous Soil from the North-West Shelf of Australia, Géotechnique, vol. 41, no. 1, pp. 101–121.

Akino, N. 1990. Non-linear analysis of settlement of building taking into account the non-linear stiffness of soil. (in Japanese) Proc. 25th Japan. Nat. Conf. SMFE, JSSMFE, pp. 565–568

Almeida, F., Hermosilha, H., Carvalho, J.M., Viana da Fonseca, A. & Moura, R. 2004. ISC'2 experimental site investigation and characterization – Part II: From SH waves high resolution shallow reflection to shallower GPR tests. Geotechnical and Geophysical Site Characterization. Ed. A. Viana da Fonseca & P.W. Mayne. Millpress, Rotterdam

Alonso, E.E., Gens, A. & Josa, A. 1990. A Constitutive Model for Partially Saturated Soils. Géotechnique, Vol. 40, No. ° 3, pp. 405–430.

Árabe, L.C.G. 1995. Interpretation of in Situ Test Results in Soft Clay Deposits and Residual Soils. Ph.D. Thesis. PUC/Rio de Janeiro, Brazil. (in Portuguese)

Atkinson, J. 1993. An Introduction to The Mechanics of Soils and Foundations, Mc Graw Hill, New York.

Baldi, G., Bellotti, R., Ghionna, V., Jamiolkowski, M., Marchetti, S. & Pasqualini, E., 1986. Flat dilatometer tests in calibration chambers. Proc. In Situ '86, ASCE, Speciality Conference on Use of In situ Tests on Geotechnical Engineering, Blacksburg, Virginia, pp. 431–446.

Baldi, G., Belloti, R., Ghiona, V.N., Jamiolkowski, M. & Lo Presti, D.C.F. 1989. Modulus of sands from CPT's and DMTs. Proc. XIII ICSMFE, Rio de Janeiro, Vol. 1: pp. 165–170. Balkema, Rotterdam.

Barros, J.M.C. 1997. Dynamic shear modulus in tropical soils, PhD Thesis, São Paulo University, (In Portuguese).

Barros, J.M.C. & Hachich, W. 1996. Foundations subjected to Dynamic Loading. Ed by ABMS – Brazilian Society of Soil Mechanics and Geotechnical Enginering. In Foundations – Theory and Practice, Chapter 10, pp. 409–442, São paulo. Brazil (In Portuguese).

Bellotti, R., Fretti, C., Jamiolkowski, M. & Tanizawa, F. 1994. Flat dilatometer tests in Toyoura sand. Proc. XIII ICSMFE, New Dehli, Vol.4: pp. 1799–1782. Balkema, Rotterdam.

Blight, G.E. 1997. Mechanics of Residual Soils, Balkema, Rotterdam, 237p.

Briaud, J.-L. 1992. The Pressuremeter. Trans Tech Publications. A.A. Balkema, Rotterdam, 322p.

Briaud, J.-L. and Gibbens, R.M. (Eds.) 1994. Predicted and Measured Behaviour of Five Spread Footings on Sand. Proc. ASCE Prediction Symposium, Texas A&M University, June, ASCE Geotechnical Special Publication No. 41.

Burland, J. B. 1989. Ninth Laurits Bjerrum Memorial Lecture: 'Small is beautiful'. The stiffness of soils at small strains. Canadian Geot. J., Vol. 26, pp. 499–516.

Burland, J.B. & Burbidge, M.C. 1985. Settlement of foundations on sand and gravel. Proc. Inst. of Civil Eng. Vol.78: pp. 1325–1381. Thomas Tellford, London.

Burns, S.E. & Mayne, P.W. 1998. Monotonic and dilatory pore pressure decay using piezocone tests. Canadian Geotechnical Journal 35(6), 1063–1073.

Camapum de Carvalho, J. & Leroueil, S. 2004. Transformed soil-water characteristic curve. Solos e Rochas (Soils and Rocks) Journal, São Paulo, 27, (3): 231–242 (in Portuguese).

Campanella, R.G. & Robertson, P.K. 1991. Use and interpretation of a research dilatometer. Canadian Geot. J. No.-28: pp. 113–126. Ottawa.

Carvalho, J.M., Viana da Fonseca, A., Almeida, F. & Hermosilha, H 2004. ISC'2 experimental site invest. and characterization – Part I: Conventional and tomographic P and S waves refraction seismics vs. electrical resistivity. Geotechnical & Geophysical Site Characterizaton. Ed. A. Viana da Fonseca & P.W. Mayne. pp. 433–442. Millpress, Rotterdam.

Carvalho, J.M., Viana da Fonseca, A. & Almeida, F 2007. In-situ geophysical and signal treatment techniques in modeling ISC'2 experimental site. Proc. XIVth European Conf. Soil Mechanics and Geotechnical Engineering. (Ed. Cuéllar, V. et al.) Millpress, Nth. Vol. 3, pp. 1697–1702.

Carvalho, J.M., Viana da Fonseca, A. & Almeida, F. 2008. Seismic cross-hole tomography in modeling ISC'2 experimental site. Accepted to Proceedings of ISC'3.

Cavalcante, E. H.; Danziger, F.A.B. & Danziger, B. R. 2004. Estimating the SPT penetration resistance from rod penetration based on instrumentation. Geotechnical & Geophysical Site Characterizaton. Ed. A. Viana da Fonseca & P.W. Mayne. pp. 293–298. Millpress, Rotterdam.

Chandler R. J. & Gutierrez C. I. 1986. The filter paper method of suction measurement. Géotechnique, 36, 265–268.

Cho, G-C., Dodds, J. & Santamarina, J. C. 2006. Particle Shape Effects on Packing Density, Stiffness, and Strength: Natural and Crushed Sands. Journal of Geotechnical and Geoenvironmental Engineering, ASCE, Vol. 132, No. 5, pp. 591–602.

Clarke, B.G. 1996. Moderator's report on Session 4(b): pressuremeter, permeability and plate tests. Advances in Site Investigation Practice: pp. 623–641. London: Thomas Telford.

Clayton, C.R.I. & Serratrice, J.F. 1997. General report session 2: The mechanical properties of hard soils and soft rocks. Geotechnical Engineering of Hard Soils and Soft Rocks. Vol. 3, pp. 1839–1877. Ed. Anagnastopoulos et al. Balkema, Rotterdam.

Collins, K. 1985. Towards characterisation of tropical soil microstructure. Proc. 1st Int. Conf. on Geomech. in Tropical Lateritic and Saprolitic Soils, Brasilia, Vol. 1, pp. 85–96; Discussion, Vol. 3, pp. 207–209.

Consoli, N.C., Schnaid, F. & Mililitsky, J. 1998. Interpretation of plate load tests on residual soil site. Journal of Geotechnical and Geoenvironmental Engineering, ASCE, Vol. 124, No. 9, pp. 857–867.

Costa, E. 2005. Tests and response analysis of piles in residual soils from granite under vertical loads. MSc Thesis, University of Porto, Portugal (in Portuguese).

Costa Filho, L.M. & Vargas Jr., E.A. 1985. Hydraulic properties. 'Mechanical and Hydraulic Properties of Topical Lateritic and Saprolitic Soils'. Progress Report of the ISSMFE Technical Committee (1985), pp. 67–84. Assoc. Brasileira de Mecânica dos Solos, Brasília.

Costa Filho, L.M., L. Döbereiner, T.M.P. De Campos & E. Vargas Jr. 1989. Fabric and engineering properties of saprolites and laterites. *General Report/Discus. Session 6 – Invited lecture. Proc. 12th ICSMFE.* Rio de Janeiro. Vol. 4, pp. 2463–2476.

Coutinho, R.Q. & Silva, M.M. 2006. Classification and mechanisms of mass movement. IV COBRAE – Barzilian Conference on Solpe Stability. Salvador, Bahia, Brasil. Volume Pos Congress (in Portuguese).

Coutinho, R.Q., Costa, F.Q. & Souza Neto, J. B. 1997. Geotechnical Characterization & Slope in Residual Soil in Pernambuco, Brasil, in *Proc. II PSL – 2nd Pan-American Symposium on Landslides / II COBRAE – 2nd Brazilian Conference on Slope Stability*, ABMS, Rio de Janeiro, Vol. 1, pp. 287–298.

Coutinho, R.Q., Souza Neto, J.B., Barros, M.L.S., Lima, E. S. & Carvalho, H. A. 1998. Geotechnical characterization of a young residual soil/gneissic rock of a slope in Pernambuco, Brazil. 2nd International Symposium on The Geotechnics of Hard Soils and Soft Rocks, Vol. 1, Naples, Italy, 115–126.

Coutinho, R.Q., Souza Neto, J.B., Costa, F.Q. 2000. Design Strength Parameters of a Slope on Unsaturated Gneissic Residual Soil. Advances in Unsaturated Geotechnics, ASCE – Geotechnical Special Publication No. 99, pp. 247–261.

Coutinho, R.Q., Souza Neto, J.B. & Dourado, K.C.A. 2004a. General report: Characterization of non-textbook geomaterials. Proc. ISC-2 on Geotechnical and Geophysical Site Characterization, Porto, Portugal, Vol. 2 pp. 1233–1257.

Coutinho. R.Q., Dourado, K.C.A. & Souza Neto, J.B. 2004b. Evaluation of the collapsibility of a sand by Ménard pressuremeter. ISC'2, Vol. 2, 1267–1273, Porto, Portugal.

Cruz, N. & Viana da Fonseca, A. 2006. Portuguese experience in residual soil characterization by DMT tests. *International Symposium on Flat Plate Dilatometer Conference (DMT 2006)*, Ed. CD-Rom, In Situ Soil Testing, Inc.Washington D.C., USA

Cruz, N., Viana da Fonseca, A., Coelho, P. & Lemos, J. 1997. Evaluation of Geotechnical Parameters by DMT in Portuguese Soils. *XIV Int. Conf. on Soil Mechanics and Foundation Engineering*, Hamburg, Vol. 1, pp. 77–70. Balkema.

Cruz, N., Viana da Fonseca, A. & Neves, E. 2004. Evaluation of effective cohesive intercept on residual soils by DMT data. *Geotechnical & Geophysical Site Characterizaton. Ed. A. Viana da Fonseca & P.W. Mayne*. pp. 1275–1278. Millpress, Rotterdam.

Cuccovillo, T. & Coop, M.R. 1997. Yielding and Pre-Failure Deformation of Structure Sands, Géotechnique, Vol. 47, No. 3, pp. 491–508.

Cui, Y. J. & Delage, P. 1996. Yielding and Behaviour of an Unsaturated Compacted Silt. *Géotechnique*, 46, N. 2, 291–311.

Cunha, R.P. & Camapum de Carvalho, J. 1997. Analysis of the behavior of a drilled pier foundation in a weathered, foliated and folded slate. In: *15th Int. Congress on Soil Mechanics and Foundation Engineering*, Hamburg, 1997. Vol. 2. pp. 785–786.

Cunha, R.P., Pereira, J.F. & Vecchi, P.P.L. 2001. The use of the Menárd pressuremeter test to obtain geotechnical parameters in the unsaturated and tropical Brasília clay. Proc. *Int. Conf. In-Situ Measurement of Soil Properties & Case Histories*, Bali, Indonesia, 599–605. Parayangan Catholic University.

Danzinger, F.A.B., Politano, C.F. & Danzinger, B.R. 1998. CPT-SPT correlations for some Brazilian residual soils. *First International Conference on Site Characterization – ISC'98*. Atlanta, Vol. 2, pp. 907–912. Balkema, Rotterdam.

De Mello, V.F.B. 1972. Thoughts on soil engineering applicable to residual soils. Proc. Third Southeast Asian Conference on Soil Engineering, Hong Kong.

Dearman, W.R. 1976. Weathering classification in the characterisation of rock: a revision. Bull. Int. Assoc. Eng. Geol., No. 13, pp. 123–127.

Décourt, L. 1989. The standard penetration test. State of the Art Report. *Proc. XII ICSMFE*, Rio de Janeiro. Balkema, Rotterdam. Vol. 4: 2405–2416.

Décourt, L. 1992. SPT in non classical material. *Proc. US/Brazil Geot. Workshop on Applicab. of Classical Soil Mechanics Principles in Structured Soil'*, *Belo Horizonte*, pp. 67–100

Décourt. L. & Quaresma Filho, A.R. 1994. Practical applications of the standard penetration complemented by torque measurements, SPT-T; present stage and future trends. Proc. XIII ICSMFE, New Delhi, India, Vol. 1, pp. 143–146.

Deere, D.U. & Patton, F.D. 1971. Slope stability in residual soils. *Proc. Pan-American Conf. on SMFE*. Vol. 4, pp. 87–170. Puerto Rico, ISSMFE.

Dourado, K.C.A. & Coutinho, R.Q. 2007. Identification, classification and evaluation of soil collapsibility by Ménard pressuremeter. *XIII Panamerican Conference on Soil Mechanic and Foundation Engineering*. Isla de Margarita, Venezuela, pp. 724–730.

Dourado, K.C.A., Souza Neto, J.B. & Coutinho, R.Q. 2007. Identification and classification of a sandy soil located at Petrolândia City – PE, by laboratory and in situ tests. *VI Brazilian Symposium on Unsaturated Soils*, Bahia, Brazil. Accepted to Proceedings of NSat'2007. (in Portuguese).

Eslaamizaad, S. & Robertson, P.K. 1996. A Framework for *Insitu* determination of Sand Compressibility 49th Canadian Geotechnical Conference; St John's Newfoundland

Eslami, A. & Fellenius, B.H. 1997. *Pile Capacity by Direct CPT and CPTU Methods Applied to 102 Case Histories.* Can. Geotech. J. 34: 886–904.

Fahey, M. 1998. Deformation and *in situ* stress measurement. Invited Theme Lecture, *Geotechnical Site Characterisation: Proc. 1st International Conference on Site Characterisation (ISC '98)*, Atlanta, Georgia, Vol. 1, 49–68, Balkema, Rotterdam.

Fahey, M. 2001. Soil stiffness values for foundation settlement analysis. Proc. 2nd Int. Conf. on Pre-failure Deformation Characteristics of Geomaterials, Torino, Italy, Vol. 2: 1325–1332, Balkema, Lisse.

Fahey, M. & Randolph, M.F. 1984. Effect of disturbance on parameters derived from self-boring pressuremeter tests in sand. *Géotechnique* 34 (1): 81–97.

Fahey, M, & Carter, J.P. 1993. A finite element study of the pressuremeter test in sand using a non-linear elastic plastic model. *Canadian Geotechnical Journal*, 30, 348–362.

Fahey, M., Lehane, B. & Stewart, D.P. 2003. Soil stiffness for shallow foundation design in the Perth CBD. *Australian Geomechanics*, 38(3): 61–89.

243

Fahey, M., Schneider, J.M. & Lehane, B. 2007. Self boring pressuremeter testing in Spearwood dune sand. *Australian Geomechanics*, in press – December.

Ferreira, C. 2003. Implementation and Application of Piezo-electric Transducers for the Determination of Seismic Wave Velocities in Soil Specimens. Assessment of Sampling Quality in Residual Soil FEUP MSc. Thesis. University of Porto (in Portuguese).

Ferreira, C. 2008. Seismic wave velocities applied to the definition of state parameters and dynamic properties of residual soils. *PhD Thesis*, University of Porto. (in press).

Ferreira, R.S. & Robertson, P.K. 1992. Interpretation of undrained selfboring pressuremeter test results incorporating unloading. *Can. Geotech. J.*, 29: 918–928

Ferreira, S.R.M. & Lacerda, W.A. 1995. Volume Change Measurements in Collapsible Soil by Laboratory and Field Tests. Unsaturated Soils, Alonso & Delage eds., Vol. 2, pp. 847–854.

Ferreira, C., Viana da Fonseca, A. & Santos, J.A. 2007. Comparison of simultaneous bender elements and resonant-column tests on Porto residual soil and Toyoura sand. *Geomechanics: Laboratory Testing, Modeling and Applications – A Collection of Papers of the Geotechnical Symposium in Rome*, March 16–17, 2006. Ling, Callisto, Leshchinsky & Koseki (Eds.). Springer, ISBN 978-1-4020-6145-5. pp. 523–535.

Feuerharmel, C., Gehling, W.Y.Y. & Bica, A.V.D. (2007). The use of filter-paper and suction-plate methods for determining de soil-water characteristic curve of undisturbed colluvium soils. Geotechnical Testing Journal, Vol. 29, No. 5, 419–425.

Finke, K.A. & Mayne, P.W. 1999. Piezocone tests in US Atlantic Piedmont residual soils. Proc.s, XI Pan American Conf. on Soil Mech. & Geot. Eng., Vol.1, Foz do Iguassu, Brazil, 329–334.

Fioravante, V., Jamiolkowski, M., Lo Presti, D.C.F., Manfredini, G. and Pedroni, S. 1998. Assessment of the coefficient of earth pressure at rest from shear wave velocity measurements, *Géotechnique*, Vol. 48, No. 5, pp. 657–666.

Fonseca, A.P. 2006. Analysis of slope mechanism associated to erosion in Bacia do Bananal (SP/RJ). PhD thesis. COPPE/ UFRJ, Rio de Janeiro (in Portuguese).

Fonseca, A.P. & Lacerda, W.A. 2004. Thoughts on residual strength of lateritic soils. In: *Landslides. Advances in Evaluation and Stabilization*. Proc. IX Intenational Symposium on Landslides, Rio de Janeiro. Balkema, Netherlands. Vol. 1. pp. 669–673.

Fredlund, D.G. 1979. Appropriate Concepts and Technology for Unsaturated Soil. Canadian Geotechnical Journal, vol. 15, pp. 313–321.

Fredlund, D. G. 2006. Unsaturated Soil Mechanics in Engineering Practice. J. Geotechnical and Geoenvironmental Engr., ASCE, Vol. 132, No. 3, pp. 286–321.

Futai, M.M., Almeida, M.S.S. & Lacerda, W.A. 2007. The laboratory behaviour of a residual tropical soil. *Characterisation and Engineering Properties of Natural Soils – Tan, Phoon, Hight & Leroueil (eds) Taylor & Francis, London, Vol. 4, pp. 2477–2505.

Futai, M.M., Almeida, M.S.S., Silva Filho, F.C. & Conciani, W. 1999. XI PCSMGE. Foz do Iguassu, Brazil, Vol. 2, pp. 267–274.

Futai, M.M., Almeida, M.S.S., Lacerda, W.A. 2004. Yield, strength and critical state conditions of a tropical saturated soil. J. Geotech. Geoenviron. Engng 130, No. 11, pp. 1169–1179.

GCO 1987. Guide to Site Investigation. Reprint. *Geotechnical Engineering Office.* Hong Kong.

Giacheti, H. L. & De Mio, G. (2008): Seismic cone tests in tropical soils and the Go/qc ratio. Accepted to Proceedings of ISC'3.

Gibbs, H.J. & Bara, J.P. 1967. Stabililty Problems of Collapsing Soil. Journal of the Mech. Soil and Fround. ASCE, Vol. 93, No. 4, pp. 577–594.

Greening, P.D., Nash, P.F.T, Benahmed, N., Ferreira, C. & Viana da Fonseca, A. 2003. "Comparison of shear wave velocity measurements in different materials using time and frequency domain techniques". *Pre-failure Deformation Characteristics of Geomaterials, IS Torino 99.* Eds Jamiolkowski, Lancellota & Lo Presti, Vol. 1, pp. 381–386. Balkema, Lisse.

GSEGWP 1990. Report on tropical residual soils – Geological Society Engineering Group Working Party. The Quarterly Journal of Engineering Geology, Vol. 23, No. 1, pp. 1–101.

Hardin, B.O. 1978. The nature of stress-strain behavior of soils. Geotechnical Div. *Specialty Conf. on Earthquake Engineering and Soil Dynamics Pasadena*, California. ASCE, Vol. 1, pp. 3–90.

Hardin, B.O. & Richart, F.E., Jr. 1963. Elastic wave velocities in granular soils, *Journal of Soil Mechanics and Foundation Division*, ASCE, 89 (SM1), 33–65.

Harris, D.E. & Mayne, P.W. 1994. Axial compression behavior of two drilled shafts in Piedmont residual soils, Proc.s Int. Conference on *Design and Construction of Deep Foundations*, Vol. 2, Federal Highway Administration, Washington, D.C., 352–367.

Hatanaka, M., & Uchida, A. 1996. Empirical Correlation Between Penetration Resistance and Internal Friction Angle of Sandy Soils. Soils and Foundations, Vol. 36, No. 4, pp. 1–9.

Hight, D.W. 2000. Sampling methods: evaluation of disturbance and new practical techniqeus for high quality sampling in soils. Keynote lecture. 7th National Congress of the Portuguese Society of Geotechnics, Porto, 10–13th April. SPG, Lisbon.

Houston, S.L., Mahmoud, H.H. & Houston, W.N. 1995. Down-hole collapse test system. Journal of Geotechnical Engineering, ASCE, Vol. 121, No. 4. pp. 341–349.

Howie, J.A. 2004. Discussion report: Characterization of non-textbook geomaterials. Proc. ISC-2 on Geotechnical and Geophysical Site Characterization, Porto, Portugal, Vol. 2, pp. 1259–1261.

Hughes, J.M.O, Wroth, C.P. & Windle, D.W. 1977. Pressuremeter tests in sands. *Géotechnique*, 27(4), 455–477.

Ishihara, K. 1986. Evaluation of soil properties for use in earthquake response analysis. *Geot. Mod. Earthq. Eng. Pr.*: pp. 241–275. Balkema, Rotterdam.

ISRM 1981. Rock Characterization Testing Monitoring. ISRM Suggested methods. Edition ET Brown.

Jamiolkowski, M. & Lo Presti, D. 1994. Validity of in situ tests related to real behaviour. Proc. XIII Int. Conf. on Soil Mechanics and Foundation Engineering, New Dehli, Vol. 5, pp. 51–55.

Jamiolkowski, M. & Manassero, N. 1996. The role of in situ testing in geotechnical – thoughts about the future. Closing address. *Advances in Site Investigation Practice*, pp. 929–951. London: Thomas Telford.

Jamiolkowski, M., Ladd, C.C., Germaine, J.T. & Lancellotta, R. 1985. New Developments in Field and Laboratory Testing of Soils, Proc. 11th ICSMFE, San Francisco, Vol. 1, pp. 67–153.

Jamiolkowski, M., Lancellotta, R. & Lo Presti, D.C.F. 1995. Remarks on the stiffness at small strains of six Italian clays. Keynote Lecture 3, *Proc. Int. Symp. on Pre-Failure Deformation Charact. of Geomaterials*, Sapporo, Vol. 2: 817–836.

Jardine R.J., Kuwano, R., Zdravkovic, L. & Thornton, C. 2001. Some fundamental aspects of the pre-failure behaviour of granular soils. *Proc. 2nd Int. Symp. Prefailure Deformation Characteristics of Geomaterials, IS Torino 99*. Ed Jamiolkowski, Lancellota & Lo Presti, Vol. 2, pp. 1077–1111. Balkema, Lisse, Nth.

Jefferies, M.G. 1988. Determination of horizontal geostatic stress in clay with self-bored pressuremeter. *Can. Geotech. J.*, 25, pp. 559–573.

Jennings, J.E. & Knight, K. 1975. A Guide to construction on or with materials exhibiting additional settlement due to a collapse of grain structure. *4th Confer. for Africa on Soil Mech. and Found. Engineer.* Durban, pp. 99–105.

Kratz de Oliveira, L.A., Schnaid, F. & Gehling, W.Y.Y. 1999. Use of Pressuremeter Tests in the Prediction of Collapse Potential Soils. Soils and Rocks Magazine, São Paulo – Brazil, Vol. 22, No. 3, pp. 143–165.

Kulhawy, F.H. & Mayne, P.W. 1990. Manual on Estimating Soil Properties for Foundation Design. Report EL-6800, Electric Power Research Institute, Palo Alta, CA, August, 306 p.

Kwong, J.S.M. 1998. *Pilot study of the use of suspension PS logging*. Technical Note No. TN2/98, Geotechnical Engineering Office, Civil Engineering Department, The Government of the Hong Kong Special Administrative Region.

Lacerda, W.A. 2004. The behavior of colluvial slopes in a tropical environment. In IX International Symposium on Landslides. Rio de Janeiro, Brasil. Vol. 2, pp. 1315–1342.

Lacerda, W.A. & Fonseca, A.P. 2003. Residual Strength of Colluvial & Residual Soils. III Pan American Conf. Soil Mechanics and Geotechnical Engineering, Boston, Vol. 1, pp. 485–488.

Lacerda, W., Sandroni, S.S., Collins, K., Dias, R.D. & Prusza, Z.V. 1985. Compressibility properties of lateritic and saprolitic soils. *Mechanical and Hydraulic Properties of Tropical Lateritic and Saprolitic Soils'*. Progress Report (1982–1985), pp. 85–113. *ICSMFE Tech. Com.* A.B.M.S., Brasília.

Lacerda, W. A. & Almeida, M. S. S. 1995. State-of-the-Art-Lecture: Engineering Properties of Regional Soils: Residual Soils and Soft Clays. IX PCSFE. Guadalajara/Mexico.

Ladd, C.C. & Lambe, T.W. 1963. The strength of Undisturbed Clay Determined from Undrained Tests, NRC-ASTM Symposium on Laboratory Shear Testing of Soils, Ottawa, ASTM STP 361, pp. 342–371.

Lade, P.V. & Overton, D. D. 1989. Cementation effects in Frictional Materials, ASCE Journal of Geotechnical Engineering. 115, pp. 1373–1387.

Lafayette, K.P.V. 2006. Geologic and Geotechnical Study of Erosives Processes in Slopes at the Metropolitan Park Armando de Holanda Cavalcanti – Cabo de Santo Agostinho/PE. PhD Thesis. Federal University of Pernambuco, 358p. (in Portuguese).

Lambe, T.W. & Whitman, R.V. 1969. Soil Mechanics, J. Wiley, New York.

Lehane, B. & Fahey, M. 2002. A simplified non-linear settlement prediction model for foundations on sand. *Canadian Geotechnical Journal*, Vol. 39, No. 2, 293–303.

Leroueil, S & Vaughan, P.R. 1990. The general and congruent effects of structure in natural clays and weak rocks. Géotechnique, Vol. 40, pp. 467–488.

Leroueil, S. & Barbosa, P.S.A. 2000. Combined effect of fabric, bonding and partial saturation on yielding of soils. Proc. *Asian Conf. on Unsaturated Soils*, Singapore, pp. 527–532.

Leroueil, S. & Hight, D.W. 2003. Behaviour and properties of natural and soft rocks. *Characterization and Engineering Properties of Natural Soils*. Eds. Tan et al. Vol. 1, pp. 29–254. Swets & Zeitlinger, Lisse.

Lopes, I., Strobbia, C., Almeida, I. Teves-Costa, P., Deidda, G.P., Mendes, M. & Santos, J.A. 2004. *Joint acquisition of SWM and other seismic techniques in the ISC'2 experimental site*. Proceedings ISC'2 on Geotechnical and Geophysical Characterization, Viana da Fonseca & Mayne (eds), Porto, Milpress, Vol. 1, pp. 521–530.

Lumb, P. 1962. The properties of decomposed granite. *Géotechnique*. Vol.12, No. 3, pp. 226–243. London.

Lunne, T., Robertson, P.K. & Powell, J.J.M. 1997. *Cone Penetration Testing in Geotechnical Practice*. Blackie, London.

Lutenegger, A.J. & Saber, R.T. 1988. Determination of Collapse Potential of Soils. Geotechnical Testing Journal, ASTM, Vol. 11, no. 3, September, pp. 173–178.

Maâtouk, A., Leroueil, S. & La Rochelle, P. 1995. Yielding and critical state of a collapsible unsaturated silty soil. *Géotechnique*, 45, no. 3, 465–477.

Machado, S. L. & Vilar, O. M. 2003. Geotechnical characteristics of an unsaturated soil deposit at São Carlos, Brazil. Characterisation and Engineering Properties of Natural Soils – Tan et al. (eds.), Swets & Zeitlinger, Lisse, ISBN 90 5809 537 1.

Machado, S. L., Carneiro, B. J. I., Vilar, O. M. & Cintra, J. C. A. 2002. Soil plasticity aspects applied in the prediction of the behaviour of field structures. Proc. *3rd Int. Conf. on Unsaturated Soils*. Recife, Vol. 2, pp. 697–702.

Mancuso, C., Vassallo, R. & d'Onofrio, A. 2000. Soil behaviour in suction controlled cyclic and dynamic torsional shear tests. *Proc. 1st Asian Reg. Conf.Unsaturated Soils, Singapore*: 539–544.

Mántaras, F.M. & Schnaid, F. 2002. Cavity expansion in dilatant cohesive-frictional soils. *Géotechnique*, 52(5), 337–348.

Marchetti, S. 1980. *In situ tests by flat dilatometer*. J. Geotech. Eng. Div., ASCE. Vol. 106, GT3, 299–321.

Marchetti, S. 1997. The Flat Dilatometer Design Applications. *III Geotechnical Engineering Conference*, Cairo University.

Marchetti, S. 2001. The Flat Dilatometer Test (DMT) in Soil Investigation. *ISSMGE TC 16 Report*. Proc. Int. Conf. In-Situ Measurement of Soil Properties & Case Histories, Bali, Indonesia, 1–26. Parayangan Catholic University.

Marques, F.E.R., Almeida e Souza, J.A.; Santos, C.B., Assis, A.P. & Cunha, R.P. 2004. In-situ geotechnical characterisation of the Brasília porous clay. *Geotechnical & Geophysical Site Characterizaton. Ed. A. Viana da Fonseca & P.W. Mayne*. Vol. 2, pp. 1311–1309. Millpress, Rotterdam.

Marsland, A. & Randolph, M.F.R. 1977. Comparisons of the results from pressuremeter tests and large in situ plate tests in London clay. *Géotechnique*, 27(2), 217–243.

Martin, R.E. 1987. Estimating foundation settlement in residual soils. *J. Geot. Eng. Div.*, 103, GT3, pp. 197–212. ASCE, New York.

Mayne, P.W. 1995. CPT determination of OCR and K_0 in clean quartz sands. Procs, Symposium on *Cone Penetration Testing*, Vol. 2, Swedish Geotechnical Society, Linköping, 215–220.

Mayne, P.W. 2001. Stress-strain-strength-flow parameters from enhanced in-situ tests. Proc. Intl. Conf. on In-Situ Measurement of Soil Properties & Case Histories, Bali, Indonesia, 27–48. Parayangan Catholic University.

Mayne, P.W. & Brown, D.A. 2003. Site characterization of Piedmont residuum of North America. *Characterization and Engineering Properties of Natural Soils*, Vol. 2, pp.1323–1339. Swets & Zeitlinger, Lisse.

Mayne, P.W. & Frost, D.D. 1988. Dilatometer experience in Washington, D.C. *Transp. Research Record 1169*, National Academy Press, Washington, D.C., 16–23. Eng. 122(10), 813–821.

Mayne, P.W. & Schneider, J.A. 2001. Evaluating axial drilled shaft response by seismic cone. Foundations & Ground Improvement. *Geot. Sp. Pub. GSP 113*, ASCE, Reston/VA, 655–669.

Mayne, P.W., Kulhawy, F.H. & Kay, J.N. 1990. Observations on the development of porewater pressures during piezocone penetration in clays. *Canadian Geotechnical Journal* 27 (4), 418–428.

Mello, V.F.B. de 1972. Thoughts on soil engineering applicable to residual soils. *Proc. of 3rd Southeast Asian Conference on Soil Engineering*, 5–34.

Mitchell, J.K. 1993. Fundamentals of Soil Behavior, Second Edition, Wiley, New York.

Mitchell, J.K. & Coutinho, R.Q. 1991. Special Lecture: Occurrence, geotechnical properties, and special problems of some soils of America. IX Panamerican Conference on soil Mechanics and Foundation Engineering. Chile, Vol. IV, pp. 1651–1741.

Mota, N.M.B., Guimarães, R.C., Costa, C.M.C., Cunha, R.P. & Camapum de Carvalho, J.P. 2007. Behaviour of bored and floating deep foundations in typical unsaturated soil of Brasília as function of the local seasonal variations. *Brazilian Symposium on Unsaturated Soils*, Bahia, Brazil. Accepted to Proceedings of NSat'2007. (in Portuguese).

Ng, C.W.W. & Wang, Y. 2001. Field and laboratory measurements of small strain stiffness of decomposed granites. *Soils and Foundations* 41(3): 57–71.

Ng, C.W.W., Pun, W.K. & Pang, R. P. L. 2000. Small strain stiffness of natural granitic saprolite in Hong Kong." *J. Geotech.Geoenviron. Eng.*, 126(9), 819–833.

Ng, C.W.W. & Leung, E.H.Y. 2007a. Small-strain stiffness of granitic and volcanic saprolites in Hong Kong. *Characterization and Engineering Properties of Natural Soils*. Tan, Phoon,Hight & Leroueil (eds.) Vol. 4, Taylor & Francis Group, London, pp. 2507–2538.

Ng, C.W.W. & Leung, E.H.Y. 2007b. Determination of Shear-Wave Velocities and Shear Moduli of Completely Decomposed Tuff. Journal of Geotechnical and Geoenvironmental Engineering, Vol. 133, No. 6, pp. 630–640. ASCE.

Novais Ferreira, H. 1985. Characterisation, identification and classification of tropical lateritic and saprolitic soils for geotechnical purposes. Proc. First Int. Conf. Trop. Saprolitic and Lateritic Soils – Brasília. Report N. 76/85, LNEC, Lisbon.

Odebrecht, E., Schnaid, E., Rocha, M.M. & Bernardes, G.P. 2004. Energy measurements for Standard Penetration Tests and the effects of the length of rods. *Geotechnical & Geophysical Site Characterizaton*. Ed. Viana da Fonseca & P.W. Mayne. pp. 351–358. Millpress, Rotterdam.

Ohsaki, Y. & Iwasaki, R. 1973. On dynamic shear moduli and Poisson's ratio of soil deposits. *Soils and Foundations, JSSMFE*, Vol. 14, No. 4, Dec. pp. 59–73.

Pinyol, N., Vaunat, J. & Alonso, E.E. 2007. A constitutive model for soft clayey rocks that includes weathering effects. *Géotechnique* 57(2), 137–151.

Pinto, C.S. & Abramento, M. 1997. Pressuremeter tests on gneissic residual soil in São Pualo, Brazil. *14th Int. Conf. Soil Mech. Found. Engng*, Hamburg, 1: 175–176.

Randolph, M.F. & Hope, S. 2004. Effect of cone velocity on cone resistance and excess pore pressures. *Proc. Int. Symp. on Engng. Practice and Performance of Soft Deposits*. Osaka.

Reddy, K.R. & Saxena, S.K. 1993. Effects of Cementation on Stress-Strain and Strength Characteristics of Sands, Soils and Foundations, Vol. 33, No. 4, pp. 121–134.

Reginatto, A.R. & Ferrero 1973. Collapse Potential of Soils and Water Chemistry. Proceedings of the VIII International Conference on Soil Mechanics and Foundation Engineering, Moscow, Vol. 2, pp. 177–183.

Rios Silva, S. 2007. Modelling of a supported excavation in an access trench to the Casa da Música station in "Metro do Porto". MSc Thesis, University of Porto (In Portuguese).

Rix, G.J. & Stokoe, K.H. 1992. Correlations of initial tangent modulus and cone resistance. *Proc. Int. Symp. Calibration Chamber Testing*. Potsdam, New York: 351–362. Elsevier.

Robertson, P.K. 1990. Soil classification using the cone penetration test. *Canadian Geot. J.* Vol. 27: 151–158.

Robertson, P.K. 1991. Estimation of foundation settlements in sand from CPT. Proc. Geot. Special Pub. 27, Vol. II: pp. 764–778. ASCE, New York.

Robertson, P.K. & Campanella, R.G. 1983. Interpretation of cone penetration tests. Part I: Sand; Part II: Clay. *Canadian Geot. J.*, Vol. 20, No. 4: 718–745.

Robertson, P.K., Sasitharan, S., Cunning, J.C. & Segs, D.C. 1995. Shear wave velocity to evaluate flow liquefaction. *J. Geotechnical Engineering*, ASCE, 121(3), pp. 262–73.

Rocha Filho, P., Antunes, F.S. & Falcão, M.F.G. 1985. Qualitative influence of the weathering degree upon the mechanical properties of a young gneisseic residual soil. *Tropical'85, First Int. Conf. on Geomechanics in Tropical Lateritic and Saprolitic Soils*, Brasília, Vol. 1, pp. 281–294.

Roesler, S.K. 1979. Anisotropic shear modulus due to stress anisotropy. *Journal of the Geotechnical Engineering Division* 105(7): 871–880.

Sabatini, P.J., Bachus, R.C., Mayne, P.W., Schneider, J.A. & Zettler, T.E. 2002. Evaluation of Soil and Rock Properties. *Technical Manual. FHWA-IF-02-034.* Federal Highway Admin., Washington.

Sandroni, S.S. 1977. Micromorphology and the interpretation of a landslide in gneissic residual soil. Int. Symp. on the Geotechnics of Structurally Complex formations. Ass. Geotecnica Italiana. Capri, Vol. 1, 423–431.

Sandroni, S.S. 1981. Residual soils: research developed at PUC-RJ. *Proc. Simpósio Brasileiro de Solos Tropicais em Engenharia,* COPPE, Rio de Janeiro, Vol. 2, pp. 30–65 (In Portuguese).

Sandroni, S.S. 1991. General Report: Young Metamorphic Residual Soils, *Proc. IX Pan-American Conf. on SMFE*, Vol. 4, pp. 1771–1788, Viña Del Mar, Chile.

Santamarina, J.C. 1997. Cohesive Soil: A Dangerous Oxymoron EJGE/Magazine iGEM Article (http://www.ejge. com/iGEM/oxymoron/Dangeoxi.htm)

Santamarina, J.C. 2001. Soil Behavior at the Microscale: Particle Forces. Proc. Symp. Soil Behavior and Soft Ground Construction, in honor of Charles C. Ladd – October 2001, MIT.

Santamarina, J.C. 2005. The role of geophysics in geotechnical engineering. Presentation in at the Microscale: Particle Forces. *SAGEEP, 2005 – Symposium on the Application of Geophysics to Engineering and Environmental Problems* (http://pmrl.ce.gatech.edu/presentations. html)

Santos, J.A. 1999. Soil characterisation by dynamic and cyclic torsional shear tests. Application to the study of piles under lateral static and dynamic loadings. *PhD Thesis*, Technical University of Lisbon, Portugal (in Portuguese).

Schnaid, F. 2005. Geo-characterisation and properties of natural soils by in situ tests. Keynote *Lecture.* 16th ICSMGE, Osaka, (1), 3–45. Millpress, Rotterdam.

Schnaid, F. & Mántaras, F.M. 2003 Cavity expansion in cemented materials: structure degradation effects. *Géotechnique*, 53(9): 797–807.

Schnaid, F. & Coutinho, R.Q. 2005. Pressuremeter Tests in Brazil (National Report). In: International Symposium 50 Years of Pressuremeters, (2) 305–318.

Schnaid, F. & Mántaras, F.M. 2004. Interpretation of pressuremeter tests in a gneiss residual soil from São Paulo, Brazil. *Geotechnical & Geophysical Site Characterization.* Ed. A. Viana da Fonseca & P.W. Mayne. Vol. 2, pp. 1353–1359. Millpress, Rotterdam.

Schnaid, F., Ortigão, J.R., Mántaras, F.M., Cunha, R.P. & McGregor, I. (2000) Analysis of self-boring pressuremeter (SBPM) and Marchetti dilatometer (DMT) tests in granite saprolites. *Can Geotech. J.* 37, 1–15.

Schnaid, F., Fahey, M. & Lehane, B. 2004. In situ test characterisation of unusual geomaterial. *Keynote Lecture. Geotechnical & Geophysical Site Characterization*. Ed. A. Viana da Fonseca & P.W. Mayne. Vol. 1, pp. 49–74. Millpress, Rotterdam.

Silva, M.M. (2007). Geological and geotechnical characterization of a landslide in a slope from Camaragibe City, Pernambuco. PhD Thesis, Federal University of Pernambuco (in Portuguese).

Silva, M.M., Coutinho, R.Q. & Lacerda, W. A. 2004. Residual Shear Strength of a Calcareous Soil from a Landslide in Pernambuco, Brazil. In IX International Symposium on Landslides. Rio de Janeiro, Brasil. Vol. 1, pp. 675–680.

Souza Neto, J.B., Coutinho, R.Q. & Lacerda, W.A 2005. Evaluation of the Collapsibility of a Sandy by In Situ Collapse Test. In: 16th ICSMGE, Osaka, (2) 735–738.

Stokoe, K.H. & J.C. Santamarina 2000. Seismic-Wave-Based Testing in Geotechnical Engineering, GeoEng 2000, Melbourne, Australia, November, pp. 1490–1536.

Strobbia, C. & Foti, S. 2006. *Multi-Offset Phase Analysis of Surface Wave Data.* Journal of Applied Geophysics 59, pp. 300–313.

Stroud, M.A. 1988. The standard penetration test – its application and interpretation. *Proc. Geot. Conf. Penetration Testing in U.K.*, Birmingham. Thomas Telford, London., pp. 24–49.

Tatsuoka, F., Jardine, R.J., Lo Presti, D., Di Benedetto, H. & Kodaka, T. 1997. Theme Lecture: Characterising the prefailure deformation properties of geomaterials. *14th Int. Conf. Soil Mech. Found. Engng*, Hamburg, 4: 2129–216

Tatsuoka, F. & Shibuya, S. 1992. Deformation Characteristics of Soils and Rocks from Field and laboratory Tests, Keynote lecture, 9th Asian Reg. Conf. SMFE., Bangkok, Vol. 2, pp. 101–170. A.A. Balkema, Rotterdam; Report Inst. Ind. Science, Univ. of Tokyo, Vol. 37, No. 1, Serie 235.

Thornton, S.I. & Arulanandan, K. 1975. Collapsible Soils State-of-the-Art. Highway Geology Symposium, Vol. 26, pp. 205–219.

Topa Gomes, A., Viana da Fonseca, A. & Fahey, M. 2008. Self-boring pressuremeter tests in Porto residual soil: results and numerical modelling. Accepted to Proceedings of ISC'3.

Vargas, M. 1985. The Concept of Tropical Soils. First Int. Conf. on *Geomachanics in Tropical Lateritic and Saprolitic Soils*, Brasilia, Brazil, Vol. 3, pp. 101–134.

Vaughan, P.R. 1985. "Characterising the mechanical properties of in-situ residual soils". *Geomechanics in Tropical Soil.* Keynote paper, Proc. 2nd Int. Conf., Singapore, Vol. 2, pp. 469–486. A.A. Balkema, Rotterdam.

Vaughan, P.R. 1988. Characterising the mechanical properties of in-situ residual soils. *Geomechanics in Tropical Soil.* Proc. Sec. Int. Conf., Singapore, Keynote paper, Vol. 2, pp. 469–486. A.A. Balkema, Rotterdam.

Vaughan, P.R. & Kwan, C.W. 1984. Weathering, structure and in situ stress in residual soils. Géotechnique 34(1): 43–59

Viana da Fonseca, A. 1996. Geomechanics in Residual Soils from Porto Granite. Criteria for the Design of Shallow Foundations. *PhD Thesis*, University of Porto. (in Portuguese)

Viana da Fonseca, A. 1998. Identifying the reserve of strength and stiffness characteristics due to cemented structure of a saprolitic soil from granite. *Proc. 2nd International Symposium on Hard Soils – Soft Rocks*. Naples. Vol. 1: pp. 361–372. Balkema, Rotterdam.

Viana da Fonseca, A. 2001. Load Tests on residual soil and settlement prediction on shallow foundation. *J. Geotechnical and Geoenvironmental Eng., The Geo-Inst.* ASCE. Vol. 127, No. 10, pp. 869–883. New York.

Viana da Fonseca, A. 2003. Characterizing and deriving engineering properties of a saprolitic soil from granite, in Porto. *Characterization and Engineering Properties of*

Natural Soils. Eds. Tan et al. Vol. 2, pp. 1341–1378. Swets & Zeitlinger, Lisse.

Viana da Fonseca, A. & Almeida e Sousa, J. 2001. At rest coefficient of earth pressure in saprolitic soils from granite. *Proc. XIV ICSMFE*, Istambul, Vol. 1: 397–400.

Viana da Fonseca, A. & Ferreira, C. 2002. Management of sampling quality on residual soils and soft clayey soils. Comparative analysis of in situ and laboratory seismic waves velocities. (in Portuguese) *Proc. Workshop Sampling Techniques for Soils and Soft Rocks & Quality Control*. FEUP, Porto.

Viana da Fonseca, A. & Almeida e Sousa, J. 2002. Hyperbolic model parameters for FEM analysis of a footing load test on a residual soil from granite. PARAM 2002: Int. Symposium on Identification and determination of soil and rock parameters for geotechnical design. Vol. 1, pp. 429–443, Ed. J-P Magnan, Presses L'ENPC, Paris.

Viana da Fonseca, A.M. Matos Fernandes, A.S. Cardoso & J.B. Martins 1994. Portuguese experience on geotechnical characterization of residual soils from granite. *Proc. 13th ICSMFE, New Delhi*. Vol. 1, pp. 377–380. Balkema, Rotterdam.

Viana da Fonseca, A., M. Matos Fernandes, A.S. Cardoso 1997a. Interpretation of a footing load test on a saprolitic soil from granite. *Géotechnique*, Vol. 47, No. 3: pp. 633–651. London.

Viana da Fonseca, A., Matos Fernandes, M., Cardoso, A.S. 1997b. Correlations between SPT, CPT and Cross-Hole testing results over the granite residual soil of Porto. *Proc. 14th ICSMFE.*, Hamburg, Vol. 1, pp. 619–622. Balkema, Rotterdam.

Viana da Fonseca, A., Matos Fernandes, M. & Cardoso, A.S. 1998a. Characterization of a saprolitic soil from Porto granite by in situ testing, *First Int. Conf. on Site Characterization – ISC'98*. Atlanta, Vol. 2, pp. 1381–1388. Balkema, Rotterdam.

Viana da Fonseca, A., Almeida e Sousa, J., Cardoso, A.S. & Matos Fernandes, M. 1998b. Finite element analyses of a shallow foundation on a residual soil from granite using Lade's model. 'Applications of Computational Mechanics in Geotechnical Engineering', Fernandes et al. (Ed), pp. 69–77. 2001. A. A. Balkema, Rotterdam.

Viana da Fonseca, A., Vieira de Sousa, J. & Cruz, N. 2001 Correlations between SPT, CPT, DPL, PMT, DMT, CH, SP and PLT Tests Results on Typical Profiles of Saprolitic Soils from Granite. Proc. Intl. Conf. on In-Situ Measurement of Soil Properties & Case Histories, Bali, Indonesia, pp. 577–584, Parayangan Catholic University.

Viana da Fonseca, A., Vieira de Sousa, J. & Ferreira, C. 2003. Deriving stiffness parameters from "simple" in situ tests and relating tehm with reference values on saprolitic soils from granite. Proc. 12th Panamerican Conf. Soil Mechan-

ics and Geotechnical Eng., MIT, Cambridge, USA, Vol. 1, pp. 321–328, Verlag Guckuf GmbH, Essen.

Viana da Fonseca, A., Carvalho, J., Ferreira, C., Tuna, C., Costa, E. & Santos, J. 2004. Geotechnical characterization of a residual soil profile: the ISC'2 experimental site, Porto. *Geotechnical & Geophysical Site Characterizaton. Ed. A. Viana da Fonseca & P.W. Mayne*. Vol. 2, pp. 1361–1370. Millpress, Rotterdam.

Viana da Fonseca, A., Ferreira, C. & Carvalho, J. 2004b. Tentative evaluation of K_0 from shear waves velocities determined in Down-Hole (V_s^{vh}) and Cross-Hole (V_s^{hv}) tests on a residual soil. *Geotechnical and Geophysical Site Characterization*. Vol. 2, pp. 1755–1764. Eds. A. Viana da Fonseca & P.W. Mayne. Millpress, Rotterdam.

Viana da Fonseca, A., Ferreira, C. & Carvalho, J. 2005. The use of shear wave velocities determined in Down-Hole (Vsvh) and Cross-Hole (Vshv) tests for the evaluation of K_0. *Soils and Rocks*, Latin-American Journal of Geotechnics, Vol. 28, No. 3, pp 271–281, São Paulo.

Viana da Fonseca, A., Carvalho, J., Ferreira, C., Santos, J. A., Almeida, F., Pereira, E., Feliciano, J., Grade, J. & Oliveira, A. & 2006. Characterization of a profile of residual soil from granite combining geological, geophysical, and mechanical testing techniques. *Geotechnical and Geological Engineering*, Vol. 14, No. 5, pp. 1307–1348. Springer, Netherlands.

Viana da Fonseca, A., Ferreira, C. & Consoli, N. 2008. Stiffness vs stress state parameters for residual soil. *Soils and Rocks, Int. Geotechnical and Geoenvironmental Journal* (submitted).

Vilar, O.M. 2007. An expedite method to predict the shear strength of unsaturated soils. Soils and Rocks, Int. Geotechnical and Geoenvironmental Journal, Vol. 30, N. 1, pp. 51–61.

Wesley, L.D. 1990. Influence of structure and composition on residual soils. J. Geot. Eng. Div., Vol. 116, GT4, pp. 589–603. ASCE, New York

Wheeler, S.J. & Sivakumar, V. 1995. An Elasto-plastic Critical State Framework for Unsaturated Soil. Géotechnique, Vol. 45, No. 1, pp. 35–53.

Wong, Y.L., Lam, E.S.S., Zhao, J.X. & Chau, K.T. 1998. Assessing Seismic Response of Soft Soil Sites in Hong Kong Using Microtremor Records, *The HKIE Transaction* 5(3): 70–78.

Wood, D.M. 1990. Soil Behavior and Critical State Soil Mechanics, Cambridge U. Press, Cambridge, UK.

Yu, H.S. & Houlsby, G.T. 1991. Finite cavity expansion in dilatant soils: loading analysis. *Géotechnique* 41(2): 173–183.

Zhang Z. & Tumay M.T. 1999. Statistical to fuzzy approach toward CPT soil classification. *ASCE Journal of Geotech. & Geoenvir. Engineering*. Vol. 125, No. 3.

Author Index

*For Product Safety Concerns and Information please contact
our EU representative GPSR@taylorandfrancis.com Taylor & Francis
Verlag GmbH, Kaufingerstraße 24, 80331 München, Germany*

T - #0065 - 160425 - C0 - 246/174/14 [16] - CB - 9780415469364 - Gloss Lamination